FUNDAMENTALS
OF SOIL SCIENCE

FUNDAMENTALS OF SOIL SCIENCE

SEVENTH EDITION

HENRY D. FOTH

Professor of Soil Science
Michigan State University

JOHN WILEY & SONS

New York   Chichester   Brisbane   Toronto   Singapore

Library of Congress Cataloging in Publication Data

Foth, H. D.
    Fundamentals of soil science.

    Includes bibliographies and index.
    1. Soil science.   I. Title.
S591.F67 1984   631.4   83-23383
ISBN 0-471-88926-1

Printed in the United States of America

10 9 8 7 6 5 4 3 2 1

This book is dedicated to the late

Dr. Guy D. Smith

who devoted 30 years of his life
to the development of Soil Taxonomy.

PREFACE

Over the past several decades there has been a rapid increase in the knowledge about soil science. This Seventh Edition was extensively revised to illustrate that increase. The soil orders of *Soil Taxonomy* are introduced in Chapter 1 and are integrated into the chapters that follow. The concept of cation exchange is introduced in Chapter 2 and is used throughout the text. The energy relationships of soil water are discussed in terms of water potentials, and the material on irrigation and salinity is expanded. Soil mineralogy is covered in a separate, updated chapter (Chapter 7); it includes a more unified treatment of clay structures. This is followed by Chapter 8 on soil chemistry, which includes a unified discussion of soil pH in terms of a continuum. The roles of exchangeable aluminum and oxidic clays

are examined in detail. Chapters 9 to 14 have been thoroughly updated and have additional material on fertilizers. The last chapter is entirely new and deals with the world population-food-land problem.

Throughout the book, there is a greater emphasis on soils from a global perspective. The numerous nonagricultural illustrations on the role of soils have been retained.

Many of my colleagues have engaged me in discussions that have helped to clarify difficult concepts. Special thanks go to Nate Rufe and Lynn Foth, who printed the new photographs used in this edition, and to Mary Foth for the cover photograph.

East Lansing *Henry D. Foth*
Michigan

CONTENTS

CHAPTER 1

CONCEPTS OF SOIL

SOIL: Can you think of a substance that has had more meaning for humanity? The close bond that ancient civilizations had with the soil was expressed by the writer of Genesis in these words:

the Lord God formed Man from the dust of the earth—and Man became a living being.

There has been, and is, a reverence for soil or the earth. Someone has said that the fabric of human life is woven on earthen looms everywhere it smells of clay. Even today, most of the world's population are tillers of the soil and live close to the soil that they depend on for food and fiber (*see* Fig. 1-1).

Since the development of agriculture, the most important concept of soil has been the concept of soil as a natural medium for plant growth. When cities developed, soil became important as an engineering material to support roads and buildings. Now, soil serves many engineering uses, including landfills for waste disposal. The concept of soil as an engineering material is related to soil as a mantle of weathered rock or regolith—a concept developed by geologists in the late eighteenth century. Since the late nineteenth century, soil scientists developed the concept of soil as an *organized natural body.*

Figure 1-1
One half of the world's population are farmers who are closely tied to the land and make their living producing crops with simple tools.

Soil as an Organized Natural Body

The rapid accumulation of knowledge about soils during the nineteenth century created a need for a concept of soil that would accommodate the new facts. A revolutionary way of looking at soil was developed about 1870 in Russia by Dokuchaev. As he traveled about, he observed many different kinds of soils and noted that a given soil was found repeatedly in a given situation. Dokuchaev saw that each kind of soil had a unique morphology resulting from a unique combination of climate, living matter (plants and animals), earthy parent material, topography, and age of the land. The soil was the product of evolution and changed over time. This dynamic and evolutionary nature is embodied in a definition of soil as:

> *unconsolidated mineral matter on the surface of the earth that has been subjected to and influenced by genetic and environmental factors of:* parent material, climate *(including moisture and temperature effects),* macro- *and* micro-organisms, *and* topography, *all acting over a period of time and producing a product-soil-that differs from the material from which it is derived in many physical, chemical, and biological properties, and characteristics.*[1]

Soil Genesis Processes

Soils are products of evolution and have a unique organization consisting of genetically developed layers or horizons. Soil genesis or horizon development processes can be viewed as *additions, losses, transformations,* or *translocations*. Plants and animals find a habitat in all soils and become a part of the organic matter. Carbon in organic matter is lost from soil as carbon dioxide that results from microbial decomposition. Nitrogen is transformed from the organic to inorganic forms. Furthermore, organic matter is subject to translocation from place to place

[1]From *Glossary of Soil Science Terms.* Soil Science Society of America, Madison, Wis., October, 1979.

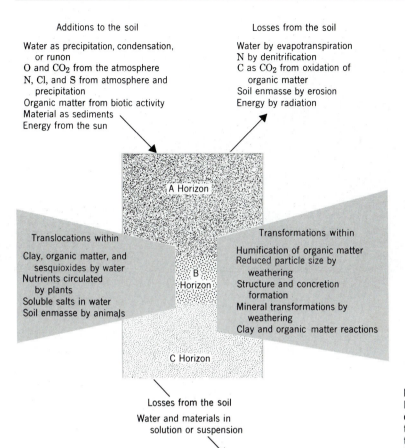

Additions to the soil

Water as precipitation, condensation,
    or runon
O and CO_2 from the atmosphere
N, Cl, and S from atmosphere and
    precipitation
Organic matter from biotic activity
Material as sediments
Energy from the sun

Losses from the soil

Water by evapotranspiration
N by denitrification
C as CO_2 from oxidation of
    organic matter
Soil enmasse by erosion
Energy by radiation

A Horizon

Translocations within

Clay, organic matter, and
    sesquioxides by water
Nutrients circulated
    by plants
Soluble salts in water
Soil enmasse by animals

B Horizon

Transformations within

Humification of organic matter
Reduced particle size by
    weathering
Structure and concretion
    formation
Mineral transformations by
    weathering
Clay and organic matter reactions

C Horizon

Losses from the soil

Water and materials in
    solution or suspension

Figure 1-2
Diagrammatic presentation
of additions, losses,
translocations, and
transformations involved in
horizon differentiation.

in the soil by means of water and animal activity.

Mineral constituents undergo changes that can be similarly considered. In all soils, minerals weather with the simultaneous formation of secondary minerals and other compounds of varying solubility that may be moved from one horizon to another. In humid regions, water migrates down and through the soil and removes soluble material. Many soils receive additions of dust, volcanic ash, or sediments eroded from higher land. A summary of these processes is presented in Fig. 1-2.

Soil Horizon Evolution

Weathering of bedrock produces unconsolidated debris that serves as the *parent material* for the evolution of soils that eventually reflect the integrated effect of climate, living matter, relief, and time. Exposure of parent material to the weather, under favorable conditions, will result in the establishment of plants. Plant growth results in the accumulation of organic residues. Animals, bacteria, and fungi join the biological community and feed on these organic remains. Breakdown of organic matter sets free the nutrients con-

A Horizon

_____

R Horizon (Bedrock)

Figure 1-3
An AR soil. The A horizon is about 30 centimeters (or 1 foot, scale is in feet) thick and developed in parent material formed by the direct weathering of sandstone. Soils such as this may develop in as little as 100 years or 100,000 years or longer, depending on the hardness of the rock and the environmental conditions.

tained therein for another plant growth cycle. The microorganisms and animals feeding on the organic debris become a part of the total organic matter complex. When the surface layer attains a reasonable thickness and assumes a darkened color because of the accumulation of organic matter, an A *horizon* comes into existence. A soil horizon is a layer approximately parallel to the earth's surface that is the product of evolution; it has properties differing from adjacent horizons.

Soils developing under grass typically have thick, dark-colored A horizons that result from the profuse growth of roots to a considerable depth. In the forest, the addition of organic matter results largely from leaves and wood. The addition of leaves and wood on top of the soil promotes the development of a thin, dark-colored A horizon enriched with organic

matter. A soil with two horizons, an A horizon overlying an R horizon (bedrock), is shown in Fig. 1-3.

The soil in Fig. 1-3 may eventually become over 100 centimeters thick[2] if the rate of soil removal by erosion is less than the rate of weathering that converts bedrock into soil-parent material. It appears, however, that most soils have formed in sediments produced by rock weathering and were transported to their present sites before the current cycle of soil formation began. Where soil evolution occurs in sediments, horizon evolution may proceed rapidly by comparison to evolu-

[2]There are 2.54 centimeters per inch and 100 centimeters equals 1 meter, or 39.37 inches. For easy conversion, remember that 30 centimeters equals about 1 foot, and 1 meter is about equal to 1 yard. (*See* the table of conversion factors on the front inside cover of the book.)

Figure 1-4
Soil developed under forest with a thin A horizon that overlies E, B, and C horizons, respectively. The B horizon has been enriched with clay (Bt).

designated the *B horizon*. The most common colloidal particles that accumulate in B horizons are clay, organic matter, and oxides of iron and aluminum (sesquioxides).

The translocation of colloids from the A horizon results in a concentration of sand and silt-sized particles of quartz and other resistant minerals in the upper part of many soils. In soils with thin A horizons, a lighter-colored layer, low in organic matter, may develop below the A horizon and above the B horizon. This horizon, commonly grayish in color, is the *E horizon*. The symbol E is derived from *eluvial*, meaning "washed out." Both A and E horizons are eluvial in a given soil, but the main feature of A is organic matter and a dark color, while that of E is a lighter color and concentration of silt and sand-sized particles of quartz and other resistant minerals.

The *C horizon* is a layer that is hardly affected by the soil-forming processes and lacks the properties of the other horizons. The C horizon commonly consists of sediments or material weathered directly from underlying bedrock. The most weathered upper soil horizons, above the C horizon, comprise the *solum*. A soil with A, E, B, and C horizons is shown in Fig. 1-4.

Master Soil Horizons

The master horizons are indicated by capital letters; the A, E, B, C, and R horizons have been described and their genesis discussed. In addition, there is the *O horizon*, which is dominated by organic matter. The mineral fraction is only a small percentage of the volume and generally much less than half the weight.

Some O horizons or layers such as muck and peat develop where the envi-

tion directly from hard bedrock. Pore spaces in sediments permit deep rooting by plants and facilitate removal of soluble compounds by percolating water. Suspended colloidal-sized particles are translocated by percolating water; however, the suspended colloidal particles tend to move only a short distance, commonly 15 to 50 centimeters, before the particles become lodged or precipitated. The process of deposition of material in a horizon that has been moved from some other horizon is *illuviation*. Illuviation, in this case, produces a zone under the A horizon where colloidal particles accumulate. This zone is ronment is water saturated for long periods of time. Much of the organic matter produced fails to decompose due to a lack of oxygen for organic matter decomposers. In forests of cold and humid regions, O horizons develop on top of the mineral soil horizons where conditions such as acidity and low temperature greatly inhibit organic matter decomposition.

Sometimes a soil horizon is dominated by the properties of one master horizon, but has the subordinate properties of another. Two capital letters are used, as in the case of AB. The first letter of AB indicates that the properties are more like the A than the B horizon. A hypothetical soil profile, with all master horizons and some transitional horizons, is shown in Fig. 1-5.

Subordinate Distinctions within Master Horizons. Lower case letters are used as suffixes to designate specific kinds of master horizons. The symbols and their meanings are as follows:

a—Highly decomposed organic material (contrast with e and i).

b—Buried genetic horizon.

c—Concretions or hard nonconcretionary nodules (iron, aluminum, manganese or titanium).

e—Organic material of intermediate decomposition.

f—Frozen soil (permanent ice).

g—Strong gleying (reduction of iron and other compounds and development of gray colors due to poor drainage).

h—Illuvial accumulation of organic matter.

i—Slightly decomposed organic material.

k—Accumulation of carbonates.

m—Cementation or induration.

n—Accumulation of sodium.

o—Residual accumulation of sesquioxides (mainly oxides of iron and aluminum).

p—Plowing or other disturbance.

q—Accumulation of silica.

r—Weathered or soft bedrock.

s—Illuvial accumulation of sesquioxides and organic matter.

t—Accumulation of silicate clay.

v—Plinthite (subsoil material enriched with iron becoming hard or bricklike due to repeated drying and wetting).

w—Development of color or structure.

x—Fragipan character (brittle with high bulk density).

y—Accumulation of gypsum.

z—Accumulation of salts more soluble than gypsum.

An example of the use of suffixes is the case of a Ckm horizon, which indicates a C horizon with an accumulation of carbonates that is cemented or indurated. The soil in Fig. 1-4 has a Bt horizon due to an illuvial accumulation of clay particles that have been moved downward from the A and E horizons. When more than one suffix is used, the following letters, if used, are written first: a, e, i, h, r, s, t, and w.

Soil Orders

There is great diversity among the soil-forming factors; the result is that hundreds of thousands of different soils have been recognized in the world. These soils are classified into orders. The order is the most general category of the soil

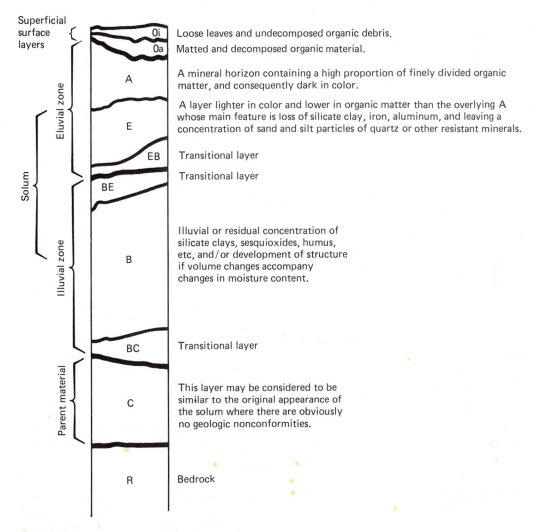

Superficial surface layers

Oi — Loose leaves and undecomposed organic debris.

Oa — Matted and decomposed organic material.

A — A mineral horizon containing a high proportion of finely divided organic matter, and consequently dark in color.

E — A layer lighter in color and lower in organic matter than the overlying A whose main feature is loss of silicate clay, iron, aluminum, and leaving a concentration of sand and silt particles of quartz or other resistant minerals.

Eluvial zone

EB — Transitional layer

Solum

BE — Transitional layer

B — Illuvial or residual concentration of silicate clays, sesquioxides, humus, etc, and/or development of structure if volume changes accompany changes in moisture content.

Illuvial zone

BC — Transitional layer

Parent material

C — This layer may be considered to be similar to the original appearance of the solum where there are obviously no geologic nonconformities.

R — Bedrock

Figure 1-5
A hypothetical soil profile with all master horizons and some transitional horizons. The thickness of the horizons varies as indicated.

classification system (*Soil Taxonomy*, 1975). The 10 orders have been developed mainly on the basis of the kinds of horizons found in soils and the properties of these horizons. The soil order name consists of a prefix and ends with *sol*. The order names, their derivation, and meaning are given in Table 1-1.

Our knowledge and theories of soil genesis make it possible to develop a schematic diagram that relates the orders to the soil-forming factors as shown in Fig. 1-6. Histosols develop from organic parent materials that frequently form where the environment is water saturated, as in ponds and lakes. Histosols are character-

Table 1-1

Derivation and Meaning of Soil Order Names

Order	Derivation	Meaning
Histosol	Gr. *histos,* tissue	Tissue or organic soil.
Vertisol	L. *verto,* turn	Inverted soil
Entisol	Coined syllable	Recent soil
Spodosol	Gr. *spodos,* wood ash	Ashy soil
Inceptisol	L. *inceptum,* beginning	Inception or young soil
Alfisol	Coined syllable	Pedalfer[a] soil
Ultisol	L. *ultimus,* last	Ultimately leached soil
Oxisol	F. *oxide,* oxide	Oxide soil
Mollisol	L. *mollis,* soft	Soft soil
Aridisol	L. *aridus,* arid	Arid soil

[a]A term used by C.F. Marbut for soils in which there was a downward movement of aluminum (Al) and iron (Fe) and no lime accumulation.

ized by thick O horizons (*see* 1 of Fig. 1-6). Included in Histosols are soils commonly called peat and muck.

The other nine orders evolve from mineral parent materials that are not subjected to water saturation or submergence for long periods of time. These materials consist of weathered bedrock, volcanic ash, and sediments produced by the activities of water, wind, ice, and gravity. Parent material high in content of expanding clay, and with alternating wet and dry seasons, produces Vertisols (*see* 2 in Fig. 1-6 and Color Plate 1). Large cracks form in the dry season and soil material falls in the cracks. During the wet season, the "extra" material at the bottom of the cracks expands causing an outward and upward pressure that slowly inverts the soil. This constant inversion of the soil prevents the development of B horizons. AC and AR soils develop from the other parent materials. These young or recent soils are Entisols (*see* 3 in Fig. 1-6). Entisols that are formed from quartzitic sand in humid regions, where forest is the common vegetation, experience intense leaching as a result of both high precipitation and very permeable soil. Oxides of iron and aluminum, along with colloidal humus, commonly accumulate in the subsoil to form Bs and/or Bhs horizons. Development of a B horizon coincides with the development of a nearly white- or ashy-colored E horizon. These soils are Spodosols (*see* 4 in Fig. 1-6 and Color Plate 1).

Entisols that develop from parent materials finer than sand may develop B horizons and become Inceptisols. Inceptisols are slightly more developed than Entisols and have weakly developed B horizons (*see* 5 in Fig. 1-6). Entisols and Inceptisols occur in all climatic zones ranging from tundra to tropics.

If conditions are favorable for their continued development, Inceptisols may develop into one of the other five orders. Alfisols develop in forested humid regions where clay migration produces a Bt horizon that has 20 percent or more clay than the A horizon, and the soil is only mod-

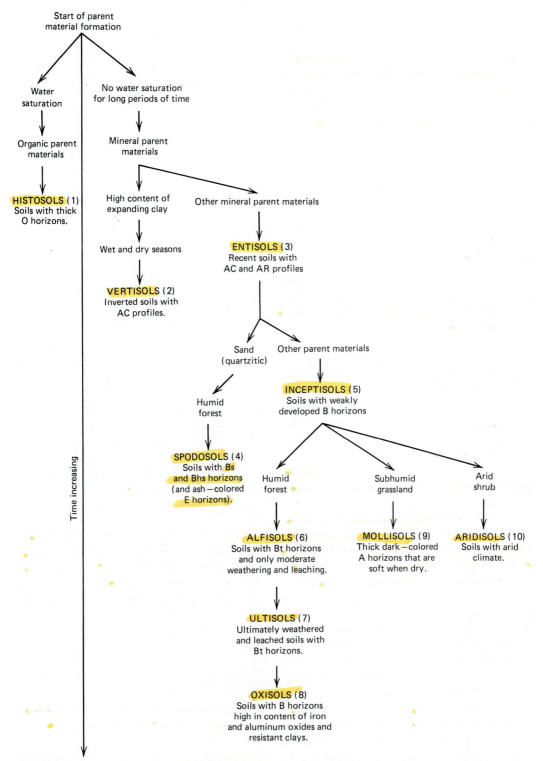

Figure 1-6
Major factors affecting the development of soil orders and some routes for their genesis.

erately leached and weathered. Over time, with continued weathering and leaching, Alfisols become ultimately weathered and leached, thus forming Ultisols (*see* 6 and 7 in Fig. 1-6). Ultisols are very acidic and have a low fertility for agricultural crops. If time and other conditions are conducive for even more intensive weathering, the Bt horizon may be destroyed; the B horizon is composed mainly of iron and aluminum oxides plus some clay that has resisted weathering. These soils are the oxide soils or the Oxisols (*see* 8 in Fig. 1-6). The Oxisols and Ultisols are commonly found in humid tropics; both are very acidic and infertile for agricultural use. Oxisols represent the most weathered or oldest soils. Examples of an Alfisol, Ultisol, and Oxisol are shown in Color Plate 1 opposite p. 20, and Color Plate 2 opposite p. 21.

In subhumid and arid climates, there is less water for weathering and leaching. Soluble materials tend to remain in the solum, and soils tend to remain neutral or alkaline. Grass vegetation in subhumid regions promotes development of thick, dark-colored A horizons that are soft even when dry due to the profuse growth of grass roots. These soft soils are Mollisols (*see* 9 in Fig. 1-6). Mollisols are generally quite fertile for production of cereal grain crops. Aridisols develop in arid regions; the soils are characterized by dryness. The soils are commonly fertile, but require irrigation for agricultural crops (*see* 10 in Fig. 1-6). Examples of two Mollisols and one Aridisol are shown in Color Plate 2.

This scheme, showing the genesis of the orders, is very general. It does not intend to suggest that these are the only routes for their genesis; there are many exceptions. For example, not all Oxisols have evolved from Ultisols, and not all Ultisols are evolved from Alfisols. If Mollisols in a subhumid climate are encroached on by trees after the climate becomes more humid, the Mollisols may be converted over time into Alfisols. Furthermore, not all soils recognized as Mollisols have developed beneath grass vegetation. As we begin our study of soil science, this scheme is very useful in ordering our knowledge about soils and developing a concept of soils as being natural, organized bodies.

Soil Bodies as Parts of Landscapes

At any given location where there is a particular kind of soil, the properties of that soil may remain fairly constant for some distance in all directions. The area in which soil properties remain similar or constant constitutes a soil body. Eventually, however, a significant change in one or more of the soil-forming factors will occur, thus causing a significant change in soil properties. This causes the soil to be a continuum with properties that gradually change in all directions. Discrete, individual soils comparable to individual plants and animals do not exist. Lateral changes from one soil to another are commonly associated with changes in slope and parent material. Typically, a transition zone exists between adjacent soils in which properties of both soils can be observed. A sharper than usual boundary exists between the two soils shown in Fig. 1-7, because the change in slope is abrupt. The landscape as a whole can be viewed as being composed of many different soil bodies; with each contribution to the whole as a piece of the overall pattern.

The Pedon and Polypedon

Soil bodies are large; there is a need for a smaller unit of soil that can be the object of scientific study. The *pedon* is this unit. A soil pedon is the smallest volume that

Figure 1-7
A sharp boundary exists between two soils (polypedons) in this field. An organic soil (Histosol) is in the foreground and a mineral soil (Alfisol) is on the slope.

can be called a soil, and it is roughly polygonal in shape. The lower limit is the somewhat vague boundary between soil and nonsoil or the approximate depth of root penetration. Lateral dimensions are large enough to represent any horizon. The area of a pedon ranges from 1 to 10 square meters, depending on the variability of the soil. The pedon is to a soil body what an oak tree is to an oak forest. A soil body is composed of many pedons (Fig. 1-8); therefore, a soil body is called a *polypedon*.

Naming Polypedons

Each polypedon is given a *series* name; there are about 12,000 series recognized in the United States. The series name is an abstract name usually taken from the name of a town or landscape feature near the place where the series was first recognized and described. All soils of the same series have the same horizon sequence and nearly identical properties of the horizons.

Examples of series include Walla Walla in Washington, Molokai in Hawaii, and Okeechobee in Florida. Walla Walla soils are deep-brown silty soils formed from wind-deposited silt (loess) and some volcanic ash. Molokai soils are red in color and intensely weathered, as are the Nipe soils of Puerto Rico. Okeechobee soils are composed mainly of organic matter and are found in the Everglades of Florida in the United States.

Soil as a Medium for Plant Growth

Soil is the interface between the living and the dead—where plants combine solar energy and carbon dioxide of the atmosphere with nutrients and water from the soil to form living tissue. Even though a significant amount of photosynthesis on earth occurs in the sea, 99 percent of our food is produced from the land.

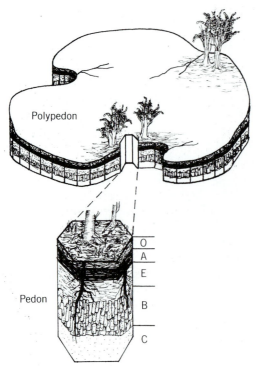

Figure 1-8
Diagram showing the concept of the pedon and its relationship to the polypedon.

Factors of Plant Growth

Basically, plants growing on land depend on soil for water and nutrients. Beyond this, the soil must provide an environment in which roots can function. This requires pore space for root extension. Oxygen must be available for root respiration, and the carbon dioxide that is produced must diffuse out of the soil instead of accumulating in it. An absence of inhibitory factors (such as a toxic concentration of soluble salts), or a toxicity of elements (such as aluminum), or extreme temperature changes and pathogens, is essential. Roots anchored in the soil also hold the plant erect.

Support. One of the most obvious functions of soil is to support the plant. Roots anchored in soil enable growing plants to remain upright. Plants grown by hydroponics are commonly supported by a wire network. There are soils in which the impermeability of the subsoil or B horizon, or the presence of a water table close to the surface of the soil, often induces shallow rooting. Shallow-rooted trees are easily blown over by wind in a phenomenon called *windthrow.* Windthrow causes disruption of soil horizons near the base of trees.

Essential Nutrient Elements. At least 16 elements are currently considered necessary for the growth of vascular plants[3]. Carbon, hydrogen, and oxygen combined in photosynthetic reactions, are obtained from air and water. They compose 90 percent or more of dry matter. The remaining 13 elements are obtained largely from soil. Nitrogen, phosphorus, potassium, calcium, magnesium, and sulfur are required in large quantities, and are referred to as the *macroelements* or *macronutrients.* Elements required in considerably smaller quantitites are called the *microelements* or *micronutrients.* They include manganese, iron, boron, zinc, copper, molybdenum, and chlorine.

More than 40 additional elements have been found in plants. Some plants accumulate elements that are not essential but increase plant growth or quality. The absorption of sodium by celery is an example, in this case resulting in an improvement in flavor. Sodium can also be a substitute for potassium in some plants, if

[3]Based on the definition of Arnon that an essential element for higher plants is needed to complete the life cycle, and a deficiency can not be corrected by any other element.

potassium is in low supply. Silicon fertilization of rice frequently increases stem strength, disease resistance, and growth. In some cases, cobalt and vanadium increase plant growth.

Most nutrients in soils exist in mineral and organic matter and, as such, are insoluble and unavailable to plants. Nutrients become available through weathering and organic matter decay or decomposition. It is a rare soil indeed that is capable of supplying all of the essential elements for long periods of time in the quantities needed to produce high crop yields.

Nutrients are absorbed mostly from soil solution or from colloid surfaces as cations and anions. Cations are positively charged; anions are negatively charged. Table 1-2 shows 13 essential elements, their chemical symbols, and the forms in which they are commonly absorbed by plant roots.

Water Requirements of Plants. About 500 grams of water are required to produce 1 gram of dry plant material. About 5 grams, or 1 percent, of this water becomes an integral part of the plant. The remainder is lost through the stomates of the leaves during absorption of carbon dioxide. Atmospheric conditions, such as relative humidity and temperature, play a major role in determining how quickly water is lost and the amount of water plants require.

Since the growth of virtually all economic crop plants will be curtailed when a shortage of water occurs, even though it may be temporary and the plants are in no danger of dying, the ability of the soil to hold water against gravity becomes very important unless rainfall or irrigation is frequent. The need for removal of excess water from soils is related to the need for oxygen and is discussed in the following paragraph.

Oxygen Requirements of Plants. Roots have openings called *lenticels* that permit gas exchange. Oxygen diffuses into the roots cells and is used for respiration, whereas the carbon dioxide diffuses into the soil. Respiration releases energy that the plant needs for synthesis and translocation of organic compounds and for the active accumulation of nutrient ions against a concentration gradient.

Some plants (water lilies and rice, for example) can grow in standing water because they have morphological structures that permit the internal diffusion of atmospheric oxygen down into the root tissues. Successful production of most plants

Table 1-2

Chemical Symbols and Common Forms of the Essential Elements Absorbed by Plant Roots from Soils

Nutrient	Chemical Symbol	Forms Commonly Absorbed by Plants
Macronutrients		
Nitrogen	N	NO_3^-, NH_4^+
Phosphorus	P	$H_2PO_4^-$, HPO_4^{2-}
Potassium	K	K^+
Calcium	Ca	Ca^{2+}
Magnesium	Mg	Mg^{2+}
Sulfur	S	SO_4^{2-}
Micronutrients		
Manganese	Mn	Mn^{2+}
Iron	Fe	Fe^{2+}
Boron	B	H_3BO_3
Zinc	Zn	Zn^{2+}
Copper	Cu	Cu^{2+}
Molybdenum	Mo	MoO_4^{2-}
Chlorine	Cl	Cl^-

in water culture requires aeration of the solution. Great differences exist between plants in their ability to tolerate low oxygen levels. Sensitive plants may be wilted or killed by saturating the soil with water for a day, as shown in Fig. 1-9. The wilting is believed to result from a decrease in the permeability of the root cells to water, which is a result of a disturbance of metabolic processes due to an oxygen deficiency.

Aerobic microorganisms, bacteria, actinomycetes, and fungi utilize oxygen from the soil atmosphere and are primarily responsible for the conversion of nutrients in organic matter into soluble forms that plants can reuse.

Freedom From Inhibitory Factors. A soil should provide an environment free of inhibiting factors such as extreme acidity or basicity, disease organisms, toxic substances, excess salts, or impenetrable layers (Fig. 1-10).

Usage of the Soil by Plants

The density and distribution of roots affect the plants' efficiency in using a soil. Perennials such as oak or alfalfa do not reestablish a completely new root system each year; this gives them a distinct advantage over annuals such as corn or cotton. This partly explains the difference in the water and nutrient needs of plants.

Figure 1-9
The soil in which these tomato plants were growing was saturated with water. The stopper at the bottom of the right crock was immediately removed, and excess water quickly drained away. The plant on the left became severely wilted within 24 hours by the saturation treatment.

Figure 1-10
Soil salinity (soluble salt) has seriously affected the growth of sugar beets in the foreground of this irrigated field.

Let us look at the extensiveness of root systems and the extent to which the soil is in direct contact with root surfaces.

Extensiveness of Root Systems. It is only reasonable to assume that there is as much difference in root systems as in the tops of plants. Root growth is influenced by environment; therefore, root distribution and density are a function of both the kind of plant and nature of the root environment.

Root extension occurs by cell division and cell elongation directly behind the root cap. Actively growing roots have been known to increase several centimeters in length within 24 hours. It is not surprising that the lateral and vertical extension of the roots of some plants is great. By the time corn (maize) is "knee high," roots may have ramified soil midway between rows spaced 100 centimeters apart. By late summer, they may extend more than 2 meters deep in well-drained, permeable soil. Alfalfa taproots commonly penetrate 2 to 3 meters deep and have been known to reach a depth of 7 meters. Cereal crops such as oats have a moderately deep root system; they effectively use the soil to a depth of about 1 meter. Tree roots commonly extend 15 meters from the base of the tree at shallow depths (*see* Fig. 1-11).

The soybean root system shown in Fig. 1-12 was collected by using a large metal frame to obtain a 10-centimeter-thick slab of soil. The soil slab was cut into small blocks and running water was used to sep-

Figure 1-11
Abundant root growth occurs in the upper layer of many forest soils because of an abundant supply of water, air, and nutrients. The tree is a 10-year old loblolly pine. (Photo courtesy Ed Kerr of USDA Forest Service.)

arate roots from soil. The roots appear to have been effective to a depth of about 1 meter. The roots extended 35 centimeters laterally to each side of the row, which indicates that the soil between the rows was used effectively.

Considerable uniformity in the lateral distribution of roots through much of the soil (*see* Fig. 1-12) is explained on the basis of two factors. First, there is a random, lateral distribution of large pore spaces, resulting from cracks or channels formed by roots of previous crops or earthworm activity. Second, as roots permeate soil, they remove water and nutrients, which makes such soil a less favorable location for additional root growth. Roots then

preferentially grow in areas of the soil where roots have not yet grown. In this way, the soil is permeated uniformly by plant roots.

Extent of Root and Soil Contact. A rye plant was grown in 1 cubic foot of soil for 4 months at the University of Iowa by Dittmer (*see* Bibliography). The root system was carefully removed from the soil by using a stream of running water, and the roots were counted and measured for size and length. It was determined that the plant had hundreds of kilometers or miles of roots. Based on an assumed value for the surface area of the soil, it was calculated that about 1 percent or less of the soil surface was in direct contact with roots. Through much of the soil, the distance between roots is of the order of 1 centimeter. The movement of water and nutrients dissolved in the water, even over very small distances, enables plants to use the soil water and nutrients to a remarkable extent.

Pattern of Soil Usage By Plants. A seed is a dormant plant. Placed in moist soil, where the temperature is favorable, it may absorb water by osmosis and swell. Enzymes then become activated, and food reserves (carbohydrate, etc.) in the endosperm move to the embryo and are used for growth. With the development of green leaves and the initiation of photosynthesis, the plant becomes independent of the seed for its nutrient elements. The plant becomes totally dependent on the atmosphere and soil for its sustenance.

In a sense, this is a critical period in the life of the plant, because the root system is small. In addition, the nutrients and the water retained in the soil are quite immobile. As a consequence, there must be a continual ramification of the soil by the

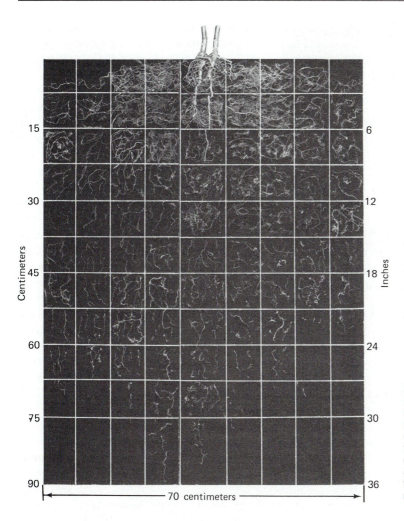

Figure 1-12
Root distribution of mature soybean plants grown in rows spaced 70 centimeters apart. Note the quite uniform, lateral distribution between 15 and 60 centimeters and the penetration to about 1 meter.

roots in order to sustain the growth of an annual plant.

As the plant continues to grow, root extension into the subsoil will probably occur. The subsoil environment will be different in terms of the supply of water, nutrients, and oxygen, and in other growth factors. This causes the roots in different horizons to perform different functions or the same functions to varying degrees. For example, most of the nitrogen will probably be absorbed by roots from the A horizon or the plowed layer (Ap horizon), because most of the organic matter is concentrated there and nitrogen is made available by organic matter decomposition. In contrast, deeply penetrating roots might penetrate less weathered and leached horizons where little available nitrogen is found, but where there is an abundance of calcium. Roots here will tend to absorb a disproportionately large amount of calcium.

The upper soil layer frequently becomes depleted of moisture in dry periods, while an abundance of moisture still exists in underlying horizons. This results in a relatively greater dependence on nutrient and water absorption from the subsoil. Subsequent rains that remoisten the upper layer of soil cause a shift to greater dependence again on the surface soil. This is probably due, in part, to better aeration nearer the surface of the soil. Thus, we see that the manner in which an annual plant uses the soil is complex and changes continually through the season. In this regard, the plant may be defined as *an integrator of a complex and ever-changing set of environmental conditions.*

The Concept of Soil Productivity

From the foregoing section, it is apparent that the use of the soil by plants is complex. Add to this the fact that plant requirements are diverse; it is readily seen that it is impossible for a given soil to be productive for the growth of all plant species. A brief discussion of some of the differences in plant requirements will aid our understanding of the soil productivity concept.

Diversity of Plant Requirements. The requirements of many economic plants will be satisfactorily met if the soil is well aerated, near neutral or slightly acidic in reaction, without layers that inhibit root penetration, without excess salt, and if it has sufficient water and an ample nutrient supply. Corn and sugar cane are profitably grown under a very wide range of soil conditions, but blueberries and azaleas are capable of survival only under a narrow range of conditions that include very acidic soil. Alfalfa, red clover, and table beets are only slightly tolerant of acidity and require nearly neutral soils for highest yields.

Trees such as willow, black spruce, white cedar, and tamarack can tolerate wet soil conditions for long periods of time, whereas maple, red pine, and fruit trees require well-drained and well-aerated soil environments. In addition, red pine cannot tolerate alkaline soil, whereas white cedar thrives very well (*see* Fig. 1-13).

Soil Productivity Defined. Soil *productivity* is defined as "the capability of a soil

Figure 1-13
Red pine was planted along the edges of this nursery to serve as a windbreak. The red pine in the foreground (and rear) failed to grow, because a limestone gravel road went diagonally through the nursery many years ago. White cedar was then planted; it grows well on the alkaline soil of the "old" road.

to produce a specified plant (or sequence of plants) under a specified system of management." For example, the productivity of soil for cotton is commonly expressed as kilos of cotton per hectare when using a particular management system which specifies things such as planting date, fertilization, irrigation schedule, tillage, and pest control. Soil scientists determine soil productivity ratings of soils for various crops by measuring yields (including tree growth or timber production) over a period of time in a "reasonable" number of management systems that are currently relevant. Included in the measurement of productivity are the influence of climate and the nature and aspect of slope. Thus, soil productivity is an expression of all the factors, soil and nonsoil, that influence crop yields.

Soil productivity is basically an economic concept and not a soil property. Three things are involved: (1) inputs (a specified management system), (2) outputs (yields of particular crops), and (3) soil type. By assigning costs and prices, net profit can be calculated and used as a basis for determining land value, which is important in loan appraising and tax as-

sessment. For planning management programs, two important aspects of soil productivity are presented in Fig. 1-14. First, different soils have different capacities to absorb inputs for profit maximization. Second, different crops have different capacities to absorb management inputs for profit maximization on a given soil type.

Soil Fertility Versus Soil Productivity. Soil *fertility* is defined as "*the quality* that enables a soil to provide the proper nutrients, in the proper amounts, and in the proper balance, for the growth of specified plants when temperature and other factors are favorable." Soil *productivity*, on the other hand, is defined as *the capability of a soil for producing a specified plant or sequence of plants under a specified system of management.* For a soil to be productive, it must be fertile. It does not follow, however, that a fertile soil is productive. Many fertile soils exist in arid regions but, under systems of management that do not include irrigation, they cannot be productive for corn (maize) or rice.

Bibliography

Arnon, D. I., "Mineral Nutrition of Plants," *Annual Rev. Biochem.*, 12:493–528, 1943.

Cline, Marlin G., "The Changing Model of Soil," *Soil Sci. Soc. Am. Proc.*, 25:441–451, 1961.

Dittmer, H. J., "A Quantitative Study of the Roots and Root Hairs of a Winter Rye Plant," *Am. Jour. Bot.*, 24:417–420, 1937.

Foth, Henry D., "Root and Top Growth of Corn," *Agron. Jour.*, 54:49–52, 1962.

Harvard University, "The Harvard Forest, 1968–69," *Harvard Black Rock Forest Annual Report,* 1968–1969.

Kellogg, C. E., "Modern Soil Science," *Am. Scientist*, 36:517–536, 1948.

Simonson, Roy W., "Outline of a Generalized Theory of Soil Genesis," *Soil Sci. Soc. Am. Proc.*, 23:161–164, 1959.

Figure 1-14
On the left, as inputs are increased, the yield from soil A increases more rapidly than from soil B. Soil A is a more productive soil than soil B. On the right, the soil represented is more productive for crop A than crop B; crop A has a greater profit potential than crop B.

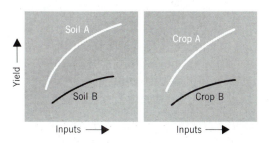

Simonson, Roy W., "Concept of Soil," in *Advances in Agronomy*, Vol. 20, Academic, New York, 1968, pp. 1–47.

Soil Science Society of America, *Glossary of Soil Science Terms*, Madison, Wis., 1979.

Soil Survey Staff, *Soil Taxonomy*, Agr. Handbook 436, USDA, Washington, D.C., 1975.

Wadleigh, C. H., "Growth of Plants," in *Soil*, USDA Yearbook, Washington, D.C., 1957, pp. 38–49.

COLOR PLATE 1

Vertisol

Developed from clay parent material where climate has distinct wet and dry seasons. Scale in centimeters.

Spodosol

Developed from sand parent material under forest in humid climate. Scale in feet.

Alfisol

Developed from loamy parent material under forest in humid climate. Scale in feet.

Ultisol

Developed from loamy parent material under forest in humid climate for very long time. Scale in centimeters.

COLOR PLATE 2

Oxisol

Developed over a long period of time from parent material weathered from basalt in Hawaii. Scale in feet.

Mollisol

Developed from fine-loamy parent material under short grass in subhumid climate. Scale in centimeters.

Mollisol

Developed from loamy parent material under short grass in semi-arid climate. Scale in centimeters.

Aridisol

Developed from loamy parent material under short grass and shrubs in arid climate. Scale in feet.

CHAPTER 2

PHYSICAL PROPERTIES OF SOILS

The physical properties of a soil have much to do with its suitability for the many uses to which it is put. The rigidity and supporting power, drainage and moisture-storage capacity, plasticity, ease of penetration by roots, aeration, and retention of plant nutrients are all intimately connected with the physical condition of the soil. It is pertinent, therefore, that persons dealing with soils know to what extent and by what means those properties can be altered. This is true whether the soil is to be used as a medium for plant growth or as a structural material in the making of highways, dams, and foundations for buildings, in building golf courses and athletic fields, or for waste disposal systems. Texture is perhaps the most permanent and important characteristic of soil and will be discussed first.

Soil Texture

Soil texture refers to the coarseness or fineness of the soil. Specifically, texture is *the relative proportions of sand, silt and clay* or the particle-sized groups smaller than gravel (less than 2 millimeters in diameter). In many soils, gravel, stones, and bedrock outcrops also affect texture and influence land use as shown in Fig. 2-1.

Figure 2-1
A coarse-textured soil containing all particle size groups from clay (the smallest) to large stones, which make the land unsuited for cultivated crops.

The Soil Separates

Soil separates are usually considered to be the size groups of mineral particles less than 2 millimeters in diameter or the size groups that are smaller than gravel. The diameter and the number and surface area per gram are given in Table 2-1. Sand is the 2.0 to 0.05 millimeter-sized fraction and, according to the USDA system, is subdivided into very fine, fine, medium, coarse, and very coarse sand separates. Silt is the 0.05- to 0.002 millimeter-sized fraction. At the 0.05 millimeter particle-size separation between sand and silt, it is difficult to distinguish by hand individual particles. Very fine sand feels slightly abrasive, while silt feels smooth, like powder. The data in Table 2-1 show a significant increase in surface area per gram for silt compared to sand. Soils high in silt content have the greatest capacity for retaining available water for plant growth due to a unique combination of surface area and pore sizes.

There is another important difference between sand and silt in many soils that is related to a given soil's ability to supply essential plant elements (soil fertility). Soil-parent materials of continental land masses tend to be granitic, and they contain a significant amount of quartz (SiO_2). The quartz particles are very resistant to weathering and become the major component of the sand of many parent materials and soils. Minerals such as feldspar and mica contain essential plant elements and may also be present as sand. The generally greater nutrient content of silt particles, their greater surface area per gram, and more rapid weathering rate compared to sand causes silty soils to be more fertile than sandy soils.

The weathering of silt may reduce silt to clay size. Mica and feldspar are not uncommon in both silt and clay separates of many soils. Although some minerals appear in both silt and clay, the great difference in surface area per gram between silt and clay (shown in Table 2-1) suggests that the clay fraction is dominated by a unique suite of minerals. Chemical weathering of minerals in sand and silt sepa-

Table 2-1
Some Characteristics of Soil Separates

Separate	Diameter, mm[a]	Diameter, mm[b]	Number of Particles per Gram	Surface Area in 1 Gram, cm^2
Very coarse sand	2.00–1.00	—	90	11
Coarse sand	1.00–0.50	2.00–0.20	720	23
Medium sand	0.50–0.25	—	5,700	45
Fine sand	0.25–0.10	0.20–0.02	46,000	91
Very fine sand	0.10–0.05	—	722,000	227
Silt	0.05–0.002	0.02–0.002	5,776,000	454
Clay	Below 0.002	Below 0.002	90,260,853,000	8,000,000[c]

[a]United States Department of Agriculture System.
[b]International Soil Science Society System.
[c]The surface area of platy-shaped montmorillonite clay particles determined by the glycol retention method by Sor and Kemper. (See Soil Science Society of America Proceedings, Vol. 23, p. 106, 1959.) The number of particles per gram and surface area of silt and the other separates are based on the assumption that particles are spheres and the largest particle size permissible for the separate.

rates produces ions that recombine to form new crystals of secondary minerals, which have very small particle sizes and occur in the clay separate. These minerals tend to be plate shaped; some of the very fine-sized particles expand and contract with wetting and drying. These features are associated with a very high surface area per gram and an ability to retain a large amount of water. Most clay particles also have a net negative charge that is satisfied by the adsorption of cations (positively charged ions), which neutralize the negative charge of the clay. Many of these cations (Ca^{++}, Mg^{++}, and K^+, for example) are essential elements. The adsorbed cations are in motion near the clay surface and exchange places with each other, and with cations, in the soil solution. The capacity of soil to adsorb exchangeable cations is the *cation exchange capacity*. Adsorbed, exchangeable cations resist being

leached from the soil and are quite available for plant use. Thus, soils with a high clay content tend to have a large capacity for retaining both water and available nutrients.

The clay fraction of Oxisols is dominated by minerals that have a low surface area and cation exchange capacity. Consequently, an Oxisol may have a high clay content, be very infertile, and have a low capacity for retaining water for plant growth. The kind of minerals found in sand, silt, and clay separates determines how these separates affect soil properties. The discussion in Chapter 7 will provide more clarification.

Particle Size Analysis

Bouyoucos devised the hydrometer method for determining the content of sand, silt, and clay without separating them. The soil sample is first soaked overnight in a sodium pyrophosphate solution in order to facilitate dispersion. Then, it is placed in a metal cup with baffles on the inside, and it is dispersed for several minutes by the soil mixer running at a speed of 16,000 revolutions per minute. The soil mixture is poured into the cylinder and distilled water is added to bring the contents up to volume. With the help of a stirrer, the soil suspension is thoroughly resuspended and the time is immediately noted. The rate of fall of suspended particles is related to size—sand settling faster than silt and silt settling faster than clay.[1]

Two hydrometer readings are taken of the soil suspension using a special soil hydrometer. A reading taken at 40 seconds determines the grams of silt and clay remaining in suspension, since the sand (2.0 to 0.05 millimeters) has settled to the bottom. Subtraction of the 40-second reading from the sample weight gives the grams of sand. The 2-hour reading determines the grams of clay (below 0.002 millimeters) in the sample (see Fig. 2-2). The silt (0.05 to 0.002 millimeters) is calculated by difference—add the percent of sand and the percent of clay and subtract from 100.

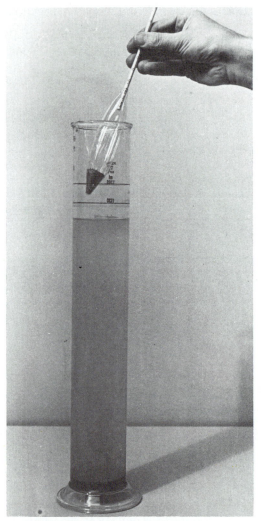

Figure 2-2
Inserting hydrometer for the 2-hour reading. The sand and silt have settled. The 2-hour reading is measured in grams of clay in suspension and used to calculate the percentage of clay.

[1]Stokes' law relates the terminal settling velocity of a smooth, rigid sphere in a viscous fluid of known density and viscosity to the diameter of the sphere when subjected to a known force field. Used in particle-size analysis of soils by the pipette, hydrometer, or centrifuge methods. The equation is

$$V = \frac{2gr^2(d_1 - d_2)}{9\eta}$$

where
V = velocity of fall (cm sec^{-1}),
g = acceleration of gravity (cm sec^{-2}),
r = "equivalent" radius of particle (cm),
d_1 = density of particle (gram cm^{-3}),
d_2 = density of medium (gram cm^{-3}), and
η = viscosity of medium (dyne sec cm^{-2}).

After the hydrometer readings have been obtained, the soil suspension can be poured over a screen to recover the entire sand fraction. After drying, the sands are sieved to obtain the various sand separates listed in Table 2-1.

Soil Classes Used to Designate Texture

Suppose the results of a particle-size analysis showed that a soil contained 15 percent clay, 65 percent sand, and 20 percent silt. The logical question is, "What is the textural class of the soil?"

The proportions of the separates in *classes* commonly used in describing soils are given in the textural triangle shown in Fig. 2-3. The sum of the percentages of sand, silt, and clay at any point in the triangle is 100. Point A represents 15 percent clay, 65 percent sand, and 20 percent silt; the textural class name for this sample is *sandy loam*. A soil containing equal amounts of the three separates is a *clay loam* (point B in Fig. 2-3). The various soil classes are separated from one another by definite lines of division in Fig. 2-3. Their properties do not change abruptly at these boundary lines, however, but one class grades into the adjoining classes of coarser or finer texture.

A loam, according to the textural triangle, is a soil with 7 to 27 percent clay, 28 to 50 percent silt, and less than 52 percent sand. Loams are soils in which all three separates have an important influence on soil properties. The following classification has been devised for grouping soils texturally:

Determining Soil Class by the Field Method

Estimates or determinations of soil texture are often necessary when examining soils in the field. When soil scientists map soils, they use the field method to determine the texture of the various horizons of pedons to identify soils, and to distinguish between different soils on the landscape.

A small quantity of soil is moistened with water and kneaded to the consistency of putty to determine how well the soil forms a ribbon. The kind of ribbon formed is related to the clay content, and it is used to categorize soils as loams, clay loams, and clays (*see* Fig. 2-4). A loam that feels very gritty or sandy is a sandy loam. Smooth-feeling loams are high in silt content and are silt loams. The same applies to clay loams and clays. Sands are loose and incoherent and do not form ribbons.

Influence of Coarse Fragments on the Class Name

Some soils contain significant amounts of gravel, stones, or other coarse fragments that are larger than the size of sand grains. An appropriate adjective is added to the class name in these cases. For ex-

Sandy soils.—*Coarse-textured soils* .... Sands.
Loamy sands.

Loamy soils.—
  Moderately coarse-textured soils .... Sandy loam.
  Fine sandy loam.

  Medium-textured soils .... Very fine sandy loam.
  Loam.
  Silt loam.
  Silt.

  Moderately fine-textured soils .... Clay loam.
  Sandy clay loam.
  Silty clay loam.

Clayey soils.—*Fine-textured soils* .... Sandy clay.
  Silty clay.
  Clay.

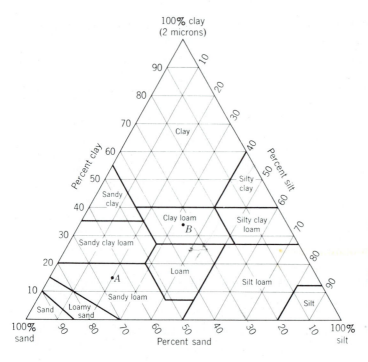

Figure 2-3
The textural triangle
shows the limits of sand,
silt, and clay contents of
the various texture
classes. Point A
represents a soil that
contains 65 percent sand,
15 percent clay, and 20
percent silt, and is a
sandy loam.

ample, a sandy loam in which 20 to 50 percent of the volume is made up of gravel is a gravelly sandy loam. If 50 to 90 percent of the volume was gravel, it would be a very gravelly sandy loam. Cobbly and stony are used for fragments 7.5 to 25 centimeters and over 25 centimeters in di-ameter, respectively. Rockiness is used to express the amount of land surface composed of exposed bedrock. The soil in Fig. 2-1 is a stony sandy loam.

Texture and Use of Soils

Strictly speaking, the class name describes only the particle-size distribution. Plasticity, rigidity, permeability, ease of tillage, droughtiness, fertility, and productivity may be closely related to the textural classes in a given geographical region, but because of the great variation that exists in the mineralogical composition of the separates no broad generalizations of the world's soils can be made. However, many useful crop yield and soil texture relationships have been established for certain geographic areas. The relative yield or growth of red pine *(Pinus resinosa)* and corn (maize) for a region are shown in

Figure 2-4
Determination of texture by feel method. Loam soil on left forms a good cast when moist. Clay loam in center forms ribbon that breaks easily. Clay on right forms a long flexible ribbon.

Fig. 2-5. The highest production of red pine occurred on the sandy loam soil. This reflects the fact that the combined integrated effect of the nutrients, water, and aeration, were most desirable on the sandy loam. Production of corn without irrigation was greatest on the loam-textured soils. With irrigation and fertilization, corn grew best in sandy soil. The world's record corn yield, set in 1977, occurred on an irrigated sandy soil in southern Michigan. Data such as these are useful in appraising land for loan or taxation purposes, and in planning cropping systems or reforestation programs.

The ability of the soil to support machinery and the hooves of grazing cattle is related to both texture and water content. Soils with high contents of organic matter and clay have a low load-bearing capacity when wet.

Soils affect trees importantly through air and water relations. The ability of the soil to store water between rains determines the seasonal supply of soil moisture and commonly determines what species grow in a forest and the growth rate. On the glaciated uplands of the northern central region of the United States, oaks are commonly found on sands, hickory and oaks on sandy loams, and maple and beech on loams and clay loams. Ash and elm grow on soils that are wet and poorly aerated because of topography.

The soil affects tree growth and the presence of the forest, in turn, influences the growth of trees. The presence of trees modifies sun and wind conditions, which modify the effects of soils on tree growth. Where reforestation occurs after a fire, conditions for establishment of seedlings are severe. Seedling survival is highly dependent on water supply and is closely related to soil texture. Some recommendations made for reforestation in

Figure 2-5
Estimates of relative growth of red pine and corn (maize) on soils of varying texture in Michigan.

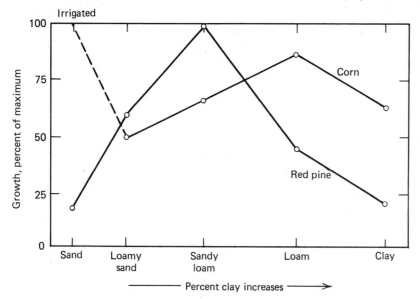

Wisconsin in relation to texture are given in Table 2-2.

The recommendations in Table 2-2 assume a uniform texture exists with increasing soil depth, which is not the common case in the field. Species that are more demanding of water and nutrients can be planted if the underlying soil horizons have a finer texture. In some cases, an underlying soil layer inhibits root penetration, thus creating a "shallow" soil with limited water and nutrient supplies.

Nature and Evolution of Argillic Horizons

Clay particles are moved by percolating water from A horizons and are deposited in B horizons. The result is pedons with horizons that have different texture—a common feature throughout the world. Clay accumulation by movement is indicated by the subscript t, as in Bt. The symbol t comes from the German "ton,"

meaning clay. When the Bt horizon of loamy soils has at least 1.2 times more clay than the A horizon above it, the horizon qualifies as an *argillic* horizon. For our purposes, we can use Bt and argillic horizons interchangeably. Soils classified as Alfisols and Ultisols have argillic horizons. Many Mollisols and Aridisols also have argillic horizons, but these soils are not required to have an argillic horizon.

The formation of argillic horizons requires parent material that contains clay or that weathers to form clay. Alternating periods of wetting and drying seem necessary. Some clay particles are believed to disperse when dry soil is wetted; the clay particles migrate downward in the water. When the percolating water encounters dry soil, the water is withdrawn into dry soil and the clay is deposited on the walls of pore spaces. Repeated cycles of wetting and drying build up layers of oriented clay particles, which are termed clay films

Table 2-2
Recommendations for Reforestation on Upland Soils in Wisconsin in Relation to Soil Texture[a]

Percent Silt Plus Clay		Species Recommended
Less than 5		(Only for wind erosion control)
5–10		Jack pine, Red cedar
10–15		Red pine, Scotch pine, Jack pine
15–25	Increasing demand for water and nutrients	All pines
25–35		White pine, European larch, Yellow birch, White elm, Red oak, Shagbark hickory, Black locust
Over 35		White spruce, Norway spruce, White cedar, White ash, Basswood, Hard maple, White oak, Black walnut

[a]Wilde, S. A., "The Significance of Soil Texture in Forestry, and Its Determination by a Rapid Field Method," *Journal of Forestry, 33*:503–508, 1935. By permission *Journal of Forestry.*

or clay skins. Clay skins are commonly found in argillic horizons, and they can be observed in the field with a 10-power hand lens. Clay skins in the argillic horizons of Cecil soils in North Carolina were found to act as a barrier for both root penetration and nutrient diffusion from pore spaces into the soil matrix.

Thousands of years and alternating wet and dry periods are required to develop argillic horizons. These conditions are well satisfied in the eastern side of the Central Valley of California. Summers are dry and winters are rainy for clay formation and downward movement of clay. Great differences exist in the age of land surfaces, including both alluvial floodplains less than 1,000 years old and alluvial fans of varying age—some over 650,000 years old. Argillic horizon development as a function of time or soil age is presented in Fig. 2-6.

Soils 1,000 years or younger (Hanford) show little or no evidence of clay translo-cation. Argillic horizons, however, had formed in 10,000 years (Greenfield). Maximum clay content in the argillic horizon increased with time, and the depth to maximum clay content decreased with time. The ratio of the maximum clay content of the argillic horizon and the clay content of the surface horizon increased over time to a maximum of 3.8 in the Redding soil. Furthermore, the San Joaquin and Redding soils had developed an iron- and silica-cemented layer impermeable to roots and water below the argillic horizon.

Forest soils 10,000 to 15,000 years old in the glaciated regions of Western Europe and the northern portions of the United States commonly have argillic horizons. The clay content of argillic horizon is about two times greater than the clay content in the A horizon. This clay distribution is similar to the Snelling and typical of many mature soils. Soils with different textures in the various horizons, as in

Figure 2-6
Development of argillic horizons (and cemented horizons) as a function of time in some California soils developed from granitic parent materials. (Data from Arkley, 1964.)

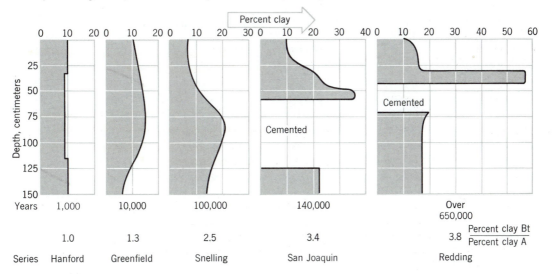

the case of soils with argillic horizons, have a *texture profile.*

Influence of Argillic Horizons on Plant Growth

The presence of an argillic horizon can be beneficial or detrimental, depending on the degree to which it has developed. Up to a certain point, an increase in the amount of clay in subsoil is desirable. It can increase the amount of water and nutrients stored in that zone. By slightly reducing the rate of water movement through the soil, it will reduce the rate of nutrient loss through leaching. If, however, the accumulation of clays is great, as in the case of a clay-pan soil, it will severely restrict the movement of air and water and the penetration of roots in the Bt horizon. It will also tend to increase the amount of water from rainfall or irrigation that will occur as runoff on sloping land. By contrast, clay-pan soils found on the nearly level marine terraces in Louisiana have argillic horizons that restrict the downward movement of irrigation water and are widely used for rice production.

The development of argillic horizons does not restrict root penetration in the Greenfield and Snelling soils of California. Roots penetrate over 2 meters deep and are able to use 20 to 25 centimeters of water stored in the root zone. Rooting depth in the San Joaquin and Redding soils is 50 centimeters or less; only 8 to 10 centimeters of water is in the root zone available to plants. Root zones become saturated and oxygen deficient in wet years, and soils are droughty in dry years. San Joaquin and Redding soils have an agricultural rating (Storie Index) of 30 and 25, respectively, compared to 95 or 100 (maximum rating) for the Hanford, Greenfield, and Snelling soils.

Alteration of Soil Texture

Fabrication of soils to meet certain textural specifications is common for golf greens, tree nurseries, and greenhouse use. Alteration of texture in the field is only occasionally attempted because of the enormous weight of such large quantities of soil. Deep plowing is used in some cases to break up root-inhibiting layers and to control wind erosion.

An interesting case of texture alteration occurs in the Netherlands. The low load-bearing capacity of peat lands limits the number of cattle that can be grazed in pastures. Cattle hooves make deep holes in the peat soil, destroying the sod. Where these peat lands have sand within 1 meter of the surface, giant plows are used to bring up the sand to form a surface with increased load-bearing potential, thus increasing the number of cattle that can be pastured per area. The high value of this land for pasture also makes it profitable to use a machine that augers sand to the surface from depths as great as 3 meters.

Soil Structure

The term *texture* is used in reference to the size of soil particles. However, when the arrangement of the particles is being considered, the term *structure* is used. Structure refers to the combination or arrangement of primary soil particles (sand, silt, clay) into secondary particles or peds (also called aggregates). These units, or peds, are separated from the adjoining units, or peds, by surfaces of weakness. The structure of the different horizons of a soil profile is an essential characteristic of the soil, as are color, texture, or chemical composition.

Importance of Structure

Structure modifies the influence of texture with regard to moisture and air relationships. The macroscopic size of most peds results in the existence of interped spaces much larger than those that can exist between adjacent sand, silt, and clay particles within peds. It is this structural effect on the pore space relationships that makes structure so important. Movement of air and water is facilitated. A good example of this occurs in the Vertisols of the Blackland prairies of Texas where the content of expanding, highly plastic clay is as much as 60 percent. These soils would be of limited value for crop production if they did not have a well-developed granular structure, which facilitates aeration and water movement. The interped spaces also serve as corridors for root extension and as pathways for small animals. The effect of structure on the gross morphology of oat roots is presented in Fig. 2-7.

Structure Types, Classes, and Grades

Field descriptions of soil structures include: (1) the type, which notes the shape and arrangement of peds, (2) the class, which indicates ped size, and (3) the grade, which indicates the distinctiveness of the peds. Soil peds are classified on the basis of shape as spheroidal, plate-like, block-like, or prism-like. These four basic shapes give rise to seven commonly recognized types, as listed in Table 2-3. A brief description and the horizon in which they are commonly found is also given in the table. Types of structure found in the horizons of a forest soil are given in Fig. 2-8.

Grade of structure is the degree of ag-

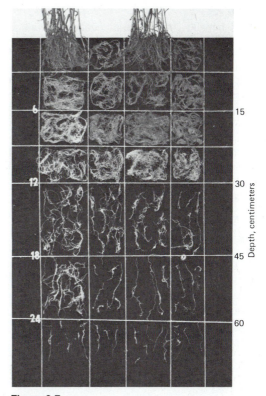

Figure 2-7
Oat roots recovered from a slab of soil 10 centimeters thick. Note the marked change in amount and kind of roots at a depth of 30 centimeters. This is the depth where the granular structure of the plow layer changes to the blocky structure of the Bt horizon. Roots in the B horizon were fewer in number, larger in diameter, and were growing primarily in the vertical spaces between structural units, old root channels, and in earthworm channels.

gregation or structure development; it expresses the differential between cohesion within peds and adhesion between peds. The grade is determined in the field mainly by noting the durability of the peds and the proportion between aggregated and unaggregated material, which results when the peds are displaced or gently crushed. Grade of structure varies with soil moisture content and tends to be

Table 2-3
Diagrammatic Definition and Location of Various Types of Soil Structure

Structure Type	Ped Description	Diagrammatic	Common Horizon Location
Granular	Relatively nonporous, small, and spheroidal peds; not fitted to adjoining peds.		A horizon
Crumb	Relatively porous, small and spheroidal peds; not fitted to adjoining peds.		A horizon
Platy	Peds are plate-like. Plates often overlap and impair permeability.		E horizon in forest and clay-pan soils
Blocky	Block-like peds bounded by other peds whose sharp angular faces form the cast for the ped. The peds often break into smaller blocky peds.		Bt horizon
Subangular blocky	Block-like peds bounded by other peds whose rounded subangular faces form the cast for the ped.		Bt horizon
Prismatic	Column-like peds without rounded caps. Other prismatic peds form the cast for the ped. Some prismatic peds break into smaller blocky peds.		Bt horizon
Columnar	Column-like peds with rounded caps bounded laterally by other columnar peds that form the cast for the peds.		Bt horizon

Adapted from *Soils Laboratory Exercise Source Book,* Am. Soc. of Agron., 1964.

stronger as the soil dries. Terms for grade of structure are as follows:

0. *Structureless*—no observable aggregation or no definite and orderly arrangement of natural lines of weakness. *Massive,* if coherent; *single-grain,* if noncoherent.

1. *Weak*—poorly formed indistinct peds, barely observable in place.

2. *Moderate*—well-formed distinct peds,

Figure 2-8
Structure of the horizons of a forest soil with an argillic horizon (Miami loam, an Alfisol). Scale in centimeters.

moderately durable and evident, but not distinct in undisturbed soil.

3. *Strong*—durable peds that are quite evident in undisturbed soil, adhere weakly to one another, withstand displacement, and become separated when the soil is disturbed.

The sequence followed in combining the three terms to form compound names is: (1) grade, (2) class, and (3) type. Many soil horizons have compound structures consisting of smaller peds held together into larger peds. An example is compound moderate, very coarse prismatic, and moderate medium granular structure.

Ped Formation

Structure develops from either a single-grained or massive condition. To produce peds, there must be some mechanism that groups particles into clusters, and some means by which the clusters are firmly bound so that the peds persist. Plant roots are the primary agent for moving soil particles into close contact with each other, resulting from the invasion of roots into a soil region and their subsequent enlargement. The removal of water by roots causes soil shrinkage and soil cracking, which also aids in ped formation. Other agents active in ped formation include animal activity, wetting and drying, and freezing and thawing. The freezing must be slow enough so that ice-lenses form, pushing particles about rather than fast freezing, which just enmeshes particles in a mass of ice.

The permanence of peds depends on two conditions: (1) The soil along the ped faces must not disperse during rewetting or rehydration, and (2) the colloids must be able to hold together the particles within the peds when the soil becomes wet. Wet-sieving is commonly used to measure ped stability. Dry soil is placed on a sieve that is lowered and raised in water. When the dry soil is immersed in water, the water moves into the aggregates or peds from all directions and compresses the air in the pore spaces. Peds unable to withstand the pressure exerted by the entrapped air are disrupted and fall through the sieve, along with the unaggregated soil. The peds remaining on the sieve after a standard period of time are considered water stable.

Researchers at the University of Wisconsin measured the development of peds in four soils in which they also determined the content of microbial gum, iron oxide, organic carbon (organic matter), and clay. The results are presented in Table 2-4. Microbial gum was the most important agent for producing aggregation in three of the four soils. Iron oxide was most important in one soil and second most important in three soils. Many red tropical soils (Oxisols) are composed mainly of sand-sized aggregates that are very stable because of the high iron oxide content. Subsoils have less organic matter than surface soils, and oxides of iron and manganese tend to be the important binding agents.

Of the four surface soils grouped together, the order of importance was microbial gum, iron oxide, organic carbon (organic matter), and clay. The great importance of microbial gum warrants consideration of the importance of microbial activity in ped formation.

As shown in many experiments, including the one from Wisconsin, ped formation and stability are not closely related to the total soil organic matter or carbon content. Rather, a relatively small number of a particular group of organic com-

Table 2-4

Order of Importance of Soil Constituents in Formation of Peds over 0.5 mm in Diameter in the A Horizons of Four Wisconsin Soils

Soil Type	Order of Importance of Soil Constituents
Parr silt loam	Microbial gum > clay > iron oxide > organic carbon
Almena silt loam	Microbial gum > iron oxide
Miami silt loam	Microbial gum > iron oxide > organic carbon
Kewaunee silt loam	Iron oxide > clay > microbial gum
All soils	Microbial gum > iron oxide > organic carbon > clay

Adapted from Chesters, G., O. J., Attoe, and O. N., Allen, "Soil Aggregation in Relation to Various Soil Constituents," *Soil Science Society of America Proceedings* Vol. 21, 1957, p. 276, by permission of the Soil Science Society of America.

pounds, the polysaccharides, appear to be effective stabilizers of soil peds. Polysaccharides include exocellular microbial gums and microbial cell-wall materials. These organic polymers have high molecular weight, are present as ropes and nets, and are widely distributed through the soil fabric. Clay particles in most soils are flocculated and exist in a somewhat orderly fashion as cores or domains, with a considerable overlapping of their flat surfaces. On drying, the rope or net-like organic molecules are brought into very close contact with the surfaces of the clay particles. The hydroxyls of the polymers and the exposed oxygen atoms of the clay surfaces form hydrogen bonds that bind clay domains both to each other and to sand and silt particles. Polymers also bind silt and sand particles to each other if they also expose oxygens. This is the case with quartz particles, as shown in Fig. 2-9.

Polysaccharides largely appear to be of bacterial origin, and to have low resistance to microbial decomposition. However, on drying, the binding material may become physically inaccessible to the microbial population and/or denatured so that it is not readily decomposed. Fungi and actinomycetes appear to be effective in stabi-

lizing peds due to their filamentous mycelia, which stick or are adsorbed to mineral soil particles.

Soil Consistence

Consistence is the resistance of the soil to deformation or rupture. It is determined by the cohesive and adhesive properties of

Figure 2-9

Some arrangements of clay domains, organic matter, and quartz particles in a soil ped (modified from Emerson, 1959).
A. quartz-organic matter-quartz
B. quartz-organic matter-clay domain
C. clay domain-organic matter-clay domain
D. clay domain-clay domain

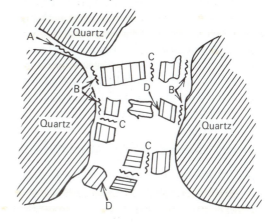

the entire soil mass. Whereas structure deals with the shape, size, and distinctness of natural soil aggregates, consistence deals with the strength and nature of the forces between particles. Consistence is important for tillage and traffic considerations. Dune sand exhibits minimal cohesive and adhesive properties, and it is so easily deformed that automobiles easily get stuck in it. Clay soils can become so sticky when wet as to make hoeing or plowing difficult.

Soil Consistence Terms

Consistence is described for three moisture levels: wet, moist, and dry. A given soil may be sticky when wet, firm when moist, and hard when dry. The terms used to describe consistence include:

1. *Wet soil*—nonsticky, sticky, nonplastic, plastic.
2. *Moist soil*—loose, friable, firm.
3. *Dry soil*—loose, soft, hard.

Cementation is also a type of consistence and is caused by cementing agents, such as calcium carbonate, silica, or oxides of iron and aluminum. Cementation is little affected by moisture content. Cemented and indurated are terms used to describe cementation.

Indurated soil is so hard that a sharp blow of a hammer is required to break the soil apart; generally, the hammer will ring as a result of the blow. *Cemented* horizons exist in San Joaquin and Redding soils (*see* Fig. 2-6). Rippers are used to break up the layers to improve rooting.

Density and Weight Relationships

Two terms are used to express soil density. *Particle* density is a measure of the density of the soil particles and *bulk* density is the density of the soil in its natural state, including the pore space.

Particle Density

In determining the particle density of soil, consideration is given only to solid particles. Thus, the particle density of any soil is a constant and does not vary with the amount of space between the particles. It is defined as the mass (weight) per unit volume of soil particles (soil solids), and it is frequently expressed as grams per cubic centimeter. For many mineral soils, the particle density will average about 2.6 grams per cubic centimeter. It does not vary a great deal for different soils unless there is a considerable variation in content of organic matter or mineralogical composition.

Bulk Density

The bulk density is the weight per unit volume of *oven dry* soil, which is commonly expressed as grams per cubic centimeter. Core samples used to determine bulk density are obtained with the equipment shown in Fig. 2-10. Care is exercised in the collection of cores, so that the natural structure of the soil is preserved. Any change in the structure of the soil is likely to alter the amount of pore space and, likewise, the weight per unit volume. Four or more cores are usually obtained from each soil horizon to obtain a reliable average value.

Bulk density cores obtained in the field are brought to the laboratory for oven drying and weighing. The bulk density is calculated as follows:

$$\text{Bulk density} = \frac{\text{weight of oven dry soil}}{\text{volume of oven dry soil}} = \frac{\text{grams}}{\text{cm}^3}$$

Figure 2-10
Technique for obtaining bulk density cores. The light-colored metal core fits into the core sampler just to the left. This unit is then driven into the soil in pile-driver fashion, using the handle and weight unit.

show that the C horizon (parent material) is the densest layer. It has a bulk density of 1.7 grams per cubic centimeter. Formation of structure during soil development caused the overlying horizons to have lower bulk densities than the original parent material.

The Bt horizon in Miami loam has a greater clay content than the A horizon. Its bulk density is greater than the A horizon and, thus, has a lower percentage of pore space. Clay deposition in the Bt horizon filled some of the pore space and made the horizon more dense as the clay content increased. The general rule that fine-textured soils have more pore space and lower bulk densities than coarse-textured soils may hold when comparable structural conditions exist, as is the case when samples from plow layers are compared. This point should provide a basis for understanding that the greatest amount of available water may not necessarily be in the horizon with the highest clay content, since water is stored in the pore space.

Organic soils or Histosols have a very low bulk density compared to mineral soils. Considerable variations exist depending on the nature of the organic matter and the moisture content at the time of sampling to determine bulk density. Values ranging from 0.1 to 0.6 grams per centimeter are common.

For 600 grams of oven dry soil that fills a 400 cubic centimeter core, the bulk density is 1.5 grams per cubic centimeter.

When expressed in grams per cubic centimeter, the bulk density of granulated clay *surface* soils will commonly be in the range 1.0 to 1.3. Coarse-textured surface soils will usually be in the range 1.3 to 1.8. The greater development of structure in the fine-textured surface soils accounts for their lower bulk density, as compared to more sandy soils.

The bulk densities of the various horizons of Miami loam given in Fig. 2-11

Weight per Acre Furrow Slice

Weight per acre furrow slice is the oven dry weight of soil over 1 acre to a depth of 6 to 7 inches. A soil with a bulk density of 1.5 grams per cubic centimeter would have a density 1.5 times greater than water. The weight of 1 cubic foot of water is the well-known value of 62.4 pounds. Thus, a soil with a bulk density of 1.5

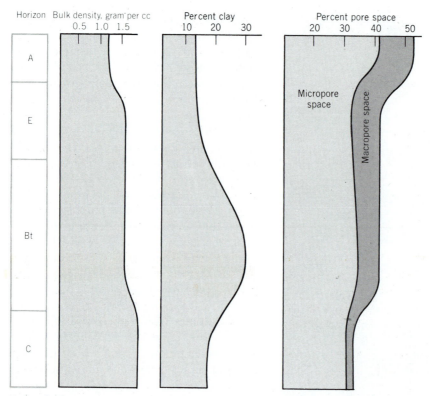

Figure 2-11
Bulk density, clay content, and pore space of the horizons of a soil with an argillic horizon (Miami loam, an Alfisol). (Data from Wascher, 1960.)

grams per cubic centimeter weigh 93.6 pounds (1.5 × 62.4 pounds) per cubic foot. A 7-inch-plowed layer for 1 acre or acre furrow slice weighs:

$$(1.5)(62.4) \times (43560)(7/12)$$
(pounds per   (cubic feet in
 cubic foot)   acre furrow slice)
$$= 2,378,376 \text{ pounds}$$

It is customary to consider that an average acre furrow slice has a bulk density of 1.3 grams per cubic centimeter and weighs about 2,000,000 pounds or 1,000 tons. On this basis, an acre furrow slice that contains 1 percent of organic matter on a weight basis would contain 20,000 pounds of organic matter. Soil losses are usually expressed as tons per acre. An average annual soil loss of 10 tons per acre would result in the removal of soil equivalent to the furrow slice about every 100 years. From these examples, it should be clear that the weight or mass per acre furrow slice, or some other suitable volume, is a useful soil characteristic.

Weight of Soil for a Hectare

A hectare is an area 100 meters square; thus, it has 10,000 square meters of area. The volume of soil 20 centimeter thick over 1 hectare is equal to:

$$10,000 \text{ m}^2 \times 0.2 \text{ m} = 2,000 \text{ m}^3$$

A cubic meter of water weighs 1,000 kilograms. Therefore, if a soil had the same density as water (1 gram per cubic centimeter) the weight of the 20-centimeter-thick layer for 1 hectare would weigh:

$$2,000 \text{ m}^3 \times 1,000 \text{ kg} = 2,000,000 \text{ kg}$$

A soil with a bulk density of 1.3 grams per cubic centimeter would weigh 2,600,000 kilograms per hectare 20 centimeters.

Pore Space and Porosity

Total pore space is the volume of soil occupied by air and water. The volume percentage of total pore space is the *porosity.* To determine porosity, soil cores are placed in a pan of water until completely saturated, and then the cores are weighed. The difference in weight between saturated and oven dry cores represents a volume of the pore space in the soil. For a 400 cubic centimeter core that contained 200 grams (200 cubic centimeters) of water at saturation, the porosity of the soil would be 50 percent, calculated as followed:

$$\frac{\text{cm}^3 \text{ pore space}}{\text{cm}^3 \text{ soil volume}} \times 100$$

$$= \frac{200 \text{ cm}^3}{400 \text{ cm}^3} \times 100 = 50\%$$

Calculation of Porosity Based on Bulk Density and Particle Density

Conceive of the impossible situation where the bulk density (BD) and particle density (PD) are the same. The BD/PD ratio would be 1:0. If the entire volume was occupied by solids, the pore space volume would be zero. As the bulk density de-

creases, pore space increases because a given volume of soil weighs less. As the bulk density decreases (PD remaining the same), the BD/PD ratio also decreases in proportion to the decrease in volume occupied by solids. Therefore, the BD/PD ratio is a measure of the soil volume occupied by solids. For a soil with a bulk density of 1.56 grams per cubic centimeter and a particle density of 2.6, the volume of solids is:

$$\frac{1.56 \text{ gram/cm}^3}{2.6 \text{ gram/cm}^3} \times 100$$

$$= 60\% \text{ solid matter}$$

The pore space is:

$$100\% - 60\% = 40\% \text{ pore space}$$

The following formula, therefore, is used to calculate soil porosity:

$$100\% - \left(\frac{\text{BD}}{\text{PD}} \times 100 \right) = \% \text{ pore space}$$

Substitution in the formula of 1.3 for BD and 2.6 for PD gives 50 percent total pore space, which is considered rather typical for medium-texture plow layers. Using the same formula, the C horizon of the Miami loam (*see* Fig. 2-11) has only 35 percent total pore space.

Effect of Texture and Structure on Pore Space

Spherical particles in closest packing result in a 26 percent total pore space; open packing results in a 48 percent pore space. This is true regardless of sphere size. Single-grained sands have a total pore space of about 40 percent. This suggests that sand particles are not perfect spheres, and that packing is not perfectly closed. The porosity of single-grained sands is low and closely related to texture.

Fine-textured soils have a wide range of particle sizes and shapes. The particles are not in closest packing, and the soil usually has peds. Soils with structural peds have pore space because of the spaces between textured particles and between peds. A clay-textured A horizon with a granular structure may have 60 percent porosity, which results from the interactive effects of texture and structure. This is consistent with the fact that A horizons of clays commonly have low bulk density and sands have high bulk density. Movement and deposition of clay in argillic or Bt horizons decreases pore space and increases soil density.

It has been pointed out that sandy surface soils have a lower porosity than clayey soils. This means that sandy soils have less volume occupied by pore space. Yet our everyday experiences tell us that water usually moves much faster through a sandy soil than a clayey soil. The explanation for this apparent paradox lies in the size of the pores that are found in each soil.

The total pore space in a sandy soil may be low, but a great proportion of it is composed of large pores that are very efficient in the movement of water and air. The percentage of volume occupied by small pores in sandy soils is low, which accounts for their low water-holding capacity. In contrast, the fine-textured surface soils have more total pore space, and a relatively large proportion of it is composed of small pores. The result is a soil with a higher water-holding capacity. Water and air move through the soil with difficulty, because there are few large pores. Thus, we see that the size of the pore spaces in the soil is as important as the total amount of pore space.

In a moist, well-drained soil, the large pore spaces are usually filled with air; consequently, they have been called aeration pores or *macropores*. The small pores usually tend to be filled with water and are commonly called capillary or *micropores*.

Distribution of Pore Space in the Soil

The pore space distribution in the profile of a mature soil is shown in Fig. 2-11. Development of structure in the A horizon results in high total porosity as well as in favorable amounts of both micropore and macropore space. As pointed out earlier, clay deposition in the pores of the Bt horizon has been responsible for decreasing porosity. The Bt horizon of this soil has less of both types of pore spaces than does the A horizon.

Soil Permeability and Hydraulic Conductivity

Water in a capillary tube does not move or drain out, because the attraction between water and glass offers greater resistance than can be overcome by gravity. As a result, a substance can be very porous, and yet be slowly permeable to water. In large (noncapillary size) tubes or pipes, water movement varies as the fourth power of the radius. Thus, as the diameter of a pipe is doubled, the rate of water flow increases 16 times (2^4). Since water molecules are strongly adsorbed onto soil surfaces, as is the case with glass, pore size is of great importance with regard to the flow or movement of water into (*infiltration*) and through (*percolation*) soil. By contrast, the insignificant attraction between soil particles and air results in the air movement being primarily related to the volume of the vacant soil pores (not to the size of the pores) and to the continuity of the pore spaces.

Permeability is the ease with which liquids, gases, and roots pass through the soil. The permeability of the soil for water is the hydraulic conductivity. The various hydraulic conductivity classes for vertical water movement in water-saturated soils are given in Table 2-5.

Important decisions that are based on a knowledge of soil hydraulic conductivity include: (1) determination of the distance between lines of drainage tile, (2) size of area of seepage beds for septic tank systems, (3) size of terrace ridges and the slope of terrace channels for erosion control, and (4) length and gradient of irrigation furrows. For example, note the characteristics of a clay-pan soil, as presented in Fig. 2-12. The high clay content of the Bt horizon is associated with little aeration pore space and very little water permeability. The clay-pan soil is unsuited for drainage by tile. If the soil existed on a slope, the claypan soil would be susceptible to erosion, resulting from a high water runoff rate. During periods of rainy weather, the lower part of the A horizon would become saturated with water. The lack of oxygen in the saturated soil would inhibit root growth. Let us consider next the relationship of soil aeration and plant growth.

Soil Aeration and Plant Growth

The atmosphere contains by volume, about 79 percent nitrogen, 21 percent oxygen, and 0.03 percent carbon dioxide. Respiration of roots and other organisms consumes oxygen and produces carbon dioxide; this causes the soil air to contain, commonly, 10 to 100 times greater concentration of carbon dioxide and slightly less oxygen, than the atmosphere (nitrogen remains nearly constant at 79 percent in both soil air and the atmosphere). Differences in the pressures of the two gases are created, causing oxygen to diffuse from the atmosphere into the soil and carbon dioxide to diffuse from the soil into the atmosphere. Normally, this diffusion is sufficient to prevent oxygen deficiencies or carbon dioxide excesses to the point of toxicity.

We have noted that gas diffusion is re-

Table 2-5
Hydraulic Conductivity Classes

Classes	Saturated Conductivity	
	Micrometers per second (μm/s)	Centimeters per Hour
Very high	over 100	36
High	10–100	3.6–36
Moderate	1–10	0.36–3.6
Moderately low	0.1–1	0.036–.36
Low	0.01–0.1	0.0036–0.036
Very low	Less than 0.01	Less than 0.0036

[a]A micrometer is 1 millionth of a meter or 10^{-6} meter. Micrometers per second multiplied by 0.36 equals centimeters per hour.

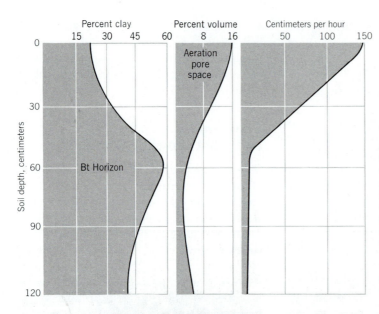

Figure 2-12
Distribution with depth of clay, aeration pore space, and hydraulic conductivity in a clay-pan soil (Edina silt loam). During a prolonged rain, the infiltration of water into the surface of the soil will be limited by the conductivity fo the least-permeable horizon, the Bt horizon. (Data from Ulrich, 1950.)

lated to the volume of gas-filled pore space and not to pore size. Poor soil aeration is caused more by the presence of water than by the amount and size of pores. Clay soils are particularly susceptible to poor aeration when wet, because most of the pore space is water filled and the spaces or avenues for gas diffusion become discontinuous.

The oxygen diffusion rate through water is about 10,000 times slower than through air-filled space. As the water content of the soil increases, the diffusion path of oxygen to the root surfaces increases in length, causing a decrease in availability of oxygen for root respiration. Experiments have shown that roots of many plants fail to penetrate soil when the oxygen diffusion rate is less than 20×10^{-8} grams per square centimeter per minute. Oxygen deficiencies are created when soils become saturated with water; then, plants commonly die (*see* Fig. 2-13). That plants should respond to oxygen this

way is not surprising, because other forms of life are killed most quickly by suffocation, then by thirst, and slowest of all, by starvation.

The data plotted in Fig. 2-14 show that the growth of peas and tomatoes was reduced by an oxygen deficiency that lasted only 24 hours. It is interesting to note that a deficiency early in the season was the most detrimental for tomatoes; for peas, the greatest reduction in growth resulted from a deficiency occurring later in the season (near blossom time). Thus, two important aspects of oxygen deficiency are the length of the oxygen-deficient period and the stage of plant growth when the oxygen deficiency occurs.

Tomato and pea plants are known for their susceptibility to oxygen deficiency. The reduced ability of tomato roots to absorb water from water-saturated soil, which causes wilting (wet wilting), is shown in Fig. 1-9. Oxygen deficiency also impairs the ability of roots to absorb nu-

Figure 2-13
Both of these yews were planted along the west side of a new house. The failure to install a pipe on the right, to carry away the down-spout water, caused saturated soil and killed the plant.

Figure 2-14
The effect of a 24-hour, oxygen-deficient period on the growth of peas and tomatoes. Note that tomatoes were most sensitive in the early part of the season, whereas peas were injured the most near flowering time. (From data of Erickson and Van Doren, 1960.)

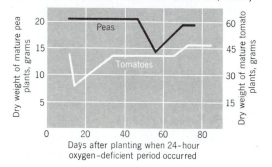

trients. Paddy rice, by contrast, is grown most of the season on water-saturated soil. Rice roots obtain atmospheric oxygen by the downward diffusion of oxygen through air spaces in the stems and roots. Many plants are intermediate and can make some adjustment to poor soil aeration. Even so, gas diffusion or aeration is one of the most important soil properties for plant growth.

Effects of Tillage on Soils and Plant Growth

The beginning of agriculture marks the beginning of soil tillage. Crude sticks were probably the first tillage tools used to establish crops. Paintings on the walls of ancient Egyptian tombs, dating back 5,000 years, depict oxen yoked together by the horns, drawing a plow made from a forked tree. The Greeks improved the Egyptian plow by adding a metal point. By Roman times, tillage tools and techniques had advanced to the point where thorough tillage was a recommended practice for crop production.

Development and improvement of tillage tools and methods contributed to greater food production and an increase in the human population. Childe considered the discovery and development of the plow one of the 19 most important discoveries or applications of science in the development of civilization.

The use of tillage tools requires power in the form of man, beast, or machine and produces traffic that has potential for compacting soils. As machines become larger, the potential for soil compaction increases. Recreational use of land has been marked by increases in foot and vehicular traffic. This section will discuss the

Figure 2-15
A chisel planter designed to plant corn in the residues of the previous crop without plowing the land. In one operation, the land is prepared, planted, fertilized, and herbicide is applied. (Photo courtesy of USDA.)

need for tillage and the effects of tillage and traffic on soil properties and plant growth.

Definition of, and Purposes of Tillage

Tillage is the mechanical manipulation of soil for any purpose; but, in agriculture and forestry it is usually restricted to the modifying of soil conditions for crop production.

There are three commonly accepted purposes of tillage: (1) to kill weeds, (2) to manage crop residues, and (3) to alter soil structure, especially preparation for planting seeds or seedlings. Let us consider next the reasons for tillage and some modern tillage techniques and problems.

Tillage and Weed Control

Weeds compete with crop plants for nutrients, water, and light. If weeds are eliminated without tillage, can cultivation of row crops be eliminated? Data from many experiments support the conclusion that the major benefit of cultivating corn is weed control. Cultivation late in the growing season, however, may prune roots and reduce yields. On many soils, herbicides have been used for weed control with good success. There are, however, some instances where cultivation during the growing season may be justified to improve soil aeration or to increase the infiltration of water by breaking surface crusts.

Tillage and Management of Crop Residues

Crops are generally grown on land that contains the plant residues of a previous crop. The moldboard plow is widely used for burying such crop residues in the hu-

mid region. Fields free of trash permit precision placement of seed and fertilizer at planting, and an easy cultivation of the crop during the growing season. In the subhumid and semiarid regions, by contrast, the need for wind erosion control and conservation of moisture have led to the development of machines that can successfully establish crops without plowing (Fig. 2-15). The plant tops remain on the surface and provide some protection from water and wind erosion. The plant residues left on the land over the winter may also cause snow to become lodged; this later melts and increases the water content of the soil.

Effect of Tillage on Soil Structure

All tillage operations change the structure of soil. The lifting, twisting, and turning action of the moldboard plow leaves the soil in an aggregated and loose condition. Ped stability, however, remains unchanged. Cultivators, discs, and packers crush some of the soil aggregates. Cultivation of a field to kill weeds may have the immediate effect of loosening the soil, increasing soil aeration, and infiltrating water. The long-time (few weeks or months) effect of cultivation resulting from crushing of soil peds is a less well-structured and more compact soil. Exposed cultivated land also suffers from disruption of peds by raindrop impact in the absence of a vegetative cover. Cropping systems with the least frequent tillage are associated with highest percentage of aggregated soil.

The Minimum Tillage Concept

It is obvious that plants grow without tillage of the soil. Sooner or later, it was inevitable that the extent to which tillage was necessary would be questioned in the search for ways to maintain the soil in good physical condition, and to produce high yields at minimum cost. Even though plants may grow very well in experiments without tillage, production of most crops will generally require at least some tillage. For the production of sugar beets, it is interesting to observe in Table 2-6 that on an experimental field, the yields were lowest when the land was not worked at all between plowing and planting and when the land was worked four or more times. The highest yields occurred where the land was worked only one or two times between plowing and planting. This experiment shows that excess tillage is detrimental. Some tillage is necessary for the practical production of crops, but the concept of "thorough tillage" is declining. The result has been considerable progress in the development of minimum tillage systems.

Minimum tillage systems employ fewer operations to produce crops. Machinery as shown in Fig. 2-15 prepares the land, plants the seed, and applies fertilizer and herbicide in one trip over the field. Under

favorable conditions, no further tillage will be required during the growing season. Press wheels behind the planter shoes pack soil only where seed is placed, leaving most of the soil surface in a state very conducive to water infiltration. Experimental evidence shows that minimum tillage reduces water erosion on sloping land. It also makes weed control easier, because planting immediately follows plowing. Weed seeds left in the loose soil are at a maximum disadvantage. Crop yields are similar to those where more tillage is used, but the use of minimum tillage in crop production reduces costs.

Within the past 20 years, much research has been directed toward the development of *no-till* systems of row crop production. In no-till, a narrow slot is made in the untilled soil so that seed can be planted where moisture is adequate for germination. Weed control is entirely by herbicides, and this cost partially offsets the savings of fewer tillage operations.

From this discussion of tillage, it should be obvious that tillage is not a requirement of plant growth. If you have a garden, you do not have to plow or engage in tillage operations that require a tractor or plow.

Table 2-6
Beet Yields as Affected by the Number of Times a Field Was Worked Prior to Planting on Lake Plain Soils

Times Worked	Tons per Acre	Metric Tons per Hectare
None	14.0	31.4
One	16.8	37.6
Two	16.7	37.4
Three	15.2	34.0
Four	14.8	33.2
Five	14.2	31.8
Six	14.3	32.0

Adapted from Cook, et al., 1959.

Effect of Long-Time Cultivation on Pore Space

When forest or grassland soils are converted to use for crop production, there is a decline in soil aggregation; soils become more compact. Compaction pushes aggregates and soil particles closer together. Obviously, the total volume of pore spaces decreases, and the bulk density increases. Pushing particles together results in a decrease in the average pore size. Some of the macropore spaces are reduced to micropores. The result is an increase in the volume of micropore space. For a sand, this could be desirable because the soil could retain more water. By contrast, the increase in micropore space or water-filled space in fine-textured soils is generally detrimental because of reduced aeration and water movement. The decrease in total pore space with compaction results from a greater decrease in macropore space than the increase in micropore space.

An interesting study was made of the effects of growing cotton for 90 years on clay-textured Houston soils in the Texas blacklands. Soil samples from cultivated fields were compared with samples obtained from an adjacent area that had remained in grass. Long-time cultivation caused a significant decrease in soil aggregation and total pore space and an increase in bulk density. These changes are shown in Fig. 2-16. The most striking and important change was the decrease in macropore space, which was reduced to about half of that of the uncultivated soil. Note that these changes occurred to a

depth of 75 centimeters. Changes in total pore space parallel the changes in bulk density.

Surface Soil Crusts

During crop production, the bare surface soil is exposed for varying periods of time without the protection of vegetation or crop residues. Soil peds at the soil surface are broken apart by raindrops, depending on their water stability; the primary particles are dispersed. The splashing of rain drops and the presence of water causes the smallest particles to be "washed" into the spaces between the larger particles, forming a dense surface layer or "skin". The infiltration of rain water into the soil surface may be reduced a thousand fold or more. Thus, surface crust formation increases runoff and erosion on sloping land.

When the surface crust dries, the crust becomes very hard and may inhibit seedling emergence (*see* Fig. 2-17). Crust hardness increases with increased clay content

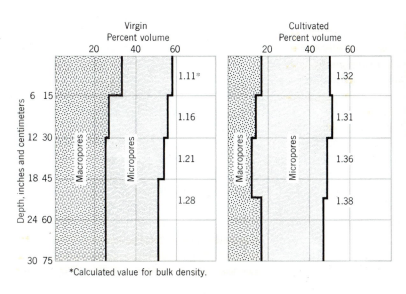

Figure 2-16
Effects of 90 years of cultivation on pore space and bulk density of Houston soils (Vertisols). Macropore and total pore space decreased and mircopore space increased. (Data from Laws and Evans, 1949.)

*Calculated value for bulk density.

Figure 2-17
Soil crusts consist of dense surface layer that may inhibit seedling emergence when dry, and greatly reduce water infiltration and increase runoff water on sloping land when wet. (Courtesy of Dr. L. S. Robertson.)

of the soil and soil drying. High organic matter content reduces crust formation and strength, if the organic matter increases the water stability of the soil peds.

Tilth and Tillage

Tilth is the physical condition of the soil as related to its ease of tillage, its fitness as a seedbed, and its impedence to seedling emergence and root penetration. Tilth is related to structural conditions, presence or absence of pressure pans, and also to soil moisture content and aeration. The effect of tillage on soil tilth is importantly related to soil moisture content at the time of tillage. This is especially true when tillage of wet clay soils creates large massive chunks that are difficult to break down into a good seedbed. On the other hand, the plowing of dry clay soils may also create large clods that are hard to break down. As shown earlier, in the long run,

tillage causes aggregate deterioration and reduced porosity. Farmers need to exercise considerable judgment about both the kind of tillage and the timing of tillage operations to minimize the detrimental effects, and to maximize the beneficial effects, of tillage on soil tilth.

Traffic and Soil Compaction

The changes in soil pore space resulting from compaction are the same whether compaction is the result of long-time cropping or of traffic involving tractor tires, animal hooves, or shoes. These changes are important enough to warrant their summarization as an aid to a further discussion of effects of tillage and traffic on soils.

Soil compaction results in:

1. Decrease in total pore space.
2. Decrease in macropore space.
3. Increase in micropore space.

The decrease in total pore space is associated with an increase in bulk density.

Compaction Layers or Pressure Pans

Compaction at the bottom of the furrow frequently occurs during plowing because of the running of tractor wheels in the furrows that were made by the previous pass over the field. Subsequent tillage is usually too shallow to break up the compacted soil; compaction increases with time. This type of compaction produces a layer with high bulk density and lower porosity at the bottom of the plow layer, and it is appropriately termed a plow sole or pressure pan.

Pressure pans are a problem on sandy soils that have a clay content insufficient to cause enough shrinking and swelling, via wetting and drying, to naturally break up the compacted layer. Bulk densities as high as 1.9 grams per centimeter have been found in pressure pans of sandy coastal plain soils of the southeastern United States. Cotton root extension was inhibited when the bulk density exceeded 1.6 grams per cubic centimeter.

Effect of Wheel Traffic on Soils and Crops

Up to 75 percent of the entire area of an alfalfa field may be run over by machinery wheels in a single harvest operation. Plant injury and soil-compaction potential is great, because 10 to 12 harvests per year are achieved where the crop is grown all year with irrigation. Wheel traffic damages plant crowns, and plants are weakened and more susceptible to disease infection. Root development is restricted. Alfalfa stands (plant density) and yields have been reduced by wheel traffic.

Potatoes are traditionally grown on sandy soils that permit easy development of well-shaped tubers. Mechanical impedance in compacted soils not only reduces tuber yield, but increases the amount of deformed tubers, which have reduced market value. In studies where bulk density was used as a measure of soil compaction, potato tuber yield was negatively correlated with bulk density, and the amount of deformed tubers was positively correlated with bulk density (*see* Fig. 2-18).

Most potatoes grown in the western states are irrigated. Soil compaction has also been observed in potato fields where irrigation water was allowed to reach excessive heights on the soil.

Effects of Recreational Traffic

Maintenance of plant cover in recreation areas is important to preserve the natural beauty and prevent the undesirable consequences of water runoff and erosion. Three campsites in the Montane zone of the Rocky Mountain National Park in Colorado were studied to determine effect of camping activities on soil properties. Soil in the areas where tents were pitched, and where the fireplaces and tables were located, had a bulk density of 1.60 grams per cubic centimeter compared to 1.03 grams per cubic centimeter in areas with little soil use at the Glacier Basin campground. This data suggests that camping

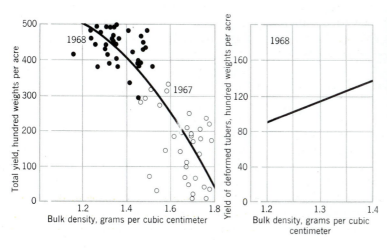

Figure 2-18
Relation of total potato tuber yield and yield of deformed tubers to soil bulk density (or soil compaction). (Adapted from Grimes, 1971. Courtesy of *The American Potato Journal* [1971], Vol. 48.)

activities can compact soil as much as tractors or other heavy machinery.

Snowmobile traffic on alfalfa fields significantly reduced the yield of forage at one out of four locations in a Wisconsin study. It appears that injury to alfalfa plants and reduced yields are importantly related to snow depth; there being less effect as the depth of snow increases. The large cross-sectional area of the snowmobile treads makes it likely that soil compaction will be minimal or nonexistent. More study is needed to determine the full effects of heavy traffic on public trails. It seems that moderate traffic on fields with a good snow cover has little effect on soil properties or dormant plants buried in the snow.

One of the most conspicuous, recent changes in many of the arid landscapes of the southwestern United States has been caused by the indiscriminate use of off-road vehicles. Studies done in the Mojave Desert have shown that one pass of a mo-

torcycle increases the bulk density of loamy sand soils from 1.52 to 1.60 grams per cubic centimeter, that 10 passes increases the bulk density to 1.68 grams per cubic centimeter. Parallel changes in porosity occurred with the greatest reduction in the largest pore spaces. Water infiltration was decreased and runoff and erosion were greatly accelerated. Recovery of natural vegetation and restoration of soils occur very slowly in deserts. It has been estimated that about 100 years will be required to restore bulk density, porosity, and infiltration capacity based on extrapolation of 51 years of data from an abandoned town in southern Nevada.

Heavy human traffic compacts soil on golf courses and lawns. Aeration and water infiltration can be increased by using coring aerators. Numerous small cores about 2 centimeters in diameter and 10 centimeters long are removed by a revolving drum and left lying on the ground (*see* Fig. 2-19).

Figure 2-19
Soil aerator used to increase aeration on a college campus in California. The insert shows detail of removal of small soil cores that are left lying on the surface of the lawn.

Effect of Logging Traffic on Soils and Tree Growth

There has been a rapid increase in the use of tractors to skid logs because of their maneuverability, speed, and economy. About 20 to 30 percent of the land area may be affected by logging. Tractor traffic can disturb or break shallow tree roots that are feeding just under the surface organic layers (*see* Fig. 1-11).

The permeability of soil to water was found to be 65 and 8 percent as great on cutover and logging road areas, respectively, in comparison with to undisturbed areas in southwestern Washington. The reductions in permeability increase water runoff and erosion and reduce the amount of water available for the growth of either transplanted seedlings or any remaining trees. Reduced growth and survival of chlorotic Douglas-fir seedlings on tractor roads in western Oregon was considered to be caused by poor soil aeration and low nitrogen supply.

Changes in bulk density and pore space of forest soils, because of logging, are similar to changes produced by long-time cultivation. After logging, and in the absence of further traffic, however, forest soils are gradually restored to their former condition. Extrapolation of 5 years' data suggests that restoration of logging trails takes 8 years; wheel-ruts require 12 years in northern Mississippi. In Oregon, however, soil compaction in tractor skid trails was still readily observable after 16 years.

Effects of Flooding and Puddling on Soil Physical Properties

Rice or paddy (wetland rice) is one of the world's three most important food crops; over 90 percent of it is grown in Asia. The paddy is grown mostly in small fields or paddies that have been leveled and bunded (enclosed with a ridge to retain water). The paddy fields are flooded and puddled before rice transplanting, and a unique physical soil environment is created that is discussed in this section.

Effect of Flooding

Flooding of dry soil causes water to enter aggregates and to compress the air in the pores, resulting in small explosions that break aggregates apart. The anaerobic conditions result in the reduction and dissolution of iron and manganese compounds, and the decomposition of organic binding materials. Aggregate stability is greatly reduced, and the aggregates that remain are easily crushed. The breakdown of soil aggregates and the clogging of pores with microbial wastes reduces soil permeability or hydraulic conductivity.

Effect of Puddling

Puddling is the tillage of water-saturated soil when water is standing on the field (*see* Fig. 2-20). Aggregates that are already weakened and broken down by flooding are worked into a uniform mud, which is essentially a two-phase system of solids and liquids. Human and animal foot traffic are effective in forming a pressure pan about 5 to 10 centimeters thick at the base of the puddled layer. The next result of puddling is increased bulk density, elimination of large pores, and increased capillary porosity. Soil stratification may occur in medium-textured soils, where sand settles first after puddling. This is followed by silt and clay to create a thin surface layer high in clay and low infiltration. These changes are desirable for

Figure 2-20
Puddling soil to prepare land for paddy (rice) transplanting.

paddy production, because soil permeability is reduced by a factor of 1,000; the amount of water needed to produce the crop is greatly reduced. Much of the rice is grown without a dependable source of irrigation water—during the monsoon season when rainfall can be erratic. Low soil permeability is, therefore, of great importance for maintaining standing water on the paddy during the growing season. Standing water promotes paddy growth (no water stress and an increased supply of nutrients) and the nearly 0 level of soil oxygen inhibits many of the weeds.

The pressure exerted by human feet during transplanting and weeding also contributes to pressure pan formation. Optimum pan formation occurs in fine loamy soils. Pressure pans do not form in sands; if formed in soils high in swelling clays, drying and cracking will break up the pan. Rice is the preferred dietary staple in much of Asia, where puddling and pressure pan formation make paddy production possible on a wide range of soils. In many landscapes, nearly all the land is paddy—even on steep slopes—if sufficient water is available. A pressure pan also makes many wetland soils accessible for

man, beasts, and machines due to improved trafficability. Where pressure pans have been destroyed by deep tillage, tractors have mired down and become inoperative.

In the United States, pressure pans are not created intentionally for rice production. Rice is grown on soils that naturally have low permeability. In Louisiana, most of the rice is grown on clay-pan soils. In Arkansas and Texas, soils with a high content of swelling clays are used.

Paddies are drained and allowed to dry before harvest. Where winters are dry, and no irrigation water is available, a dryland crop is frequently planted. Drying of puddled soil promotes aggregate formation; however, many times, large and hard clods are formed that are difficult to work into a good seedbed. The pressure pan creates a shallow root zone for the dryland crops that severely restricts rooting depth and reduces the supply of available water and nutrients. Pressure pans are retained from season to season and seriously inhibit the use of land for dryland crops.

Oxygen Relationships in Flooded Paddy Soils

The water standing on a paddy soil has an oxygen content that tends toward equilibrium with oxygen in the atmosphere. This oxygenated water layer supplies oxygen via diffusion to a thin, upper soil layer 1 millimeter to 1 centimeter thick (*see* Fig. 2-21). This thin upper layer retains the color of oxidized soil. The soil below the thin surface layer and above the pressure pan is oxygen deficient, and reducing conditions exist. Colors indicative of the presence of reduced iron and manganese occur; that is, grayish and black colors. The soil in the immediate vicinity of pad-

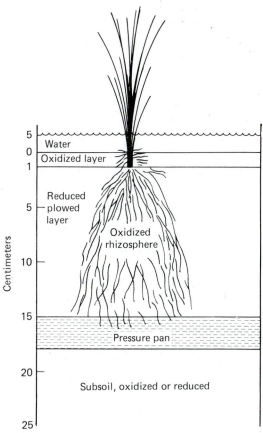

Figure 2-21
Oxidized and reduced zones in flooded paddy soil.

either oxidized or reduced. If the soil occurs in a depression and is naturally poorly drained, a gray subsoil that is indicative of reduced soil is likely. Many rice paddies are formed high on the landscape, provided there is a source of water, and they naturally have well-aerated subsoils remaining as such in paddies. Well-aerated subsoils are also favored by high sand content.

Soil Color

Color is the most obvious and easily determined soil property. The significance of soil color, however, is mainly its use as an indirect measure of other important soil characteristics that are difficult to accurately observe or measure, such as drainage. Thus, soil color, when used with other characteristics, is useful in making many important inferences regarding soil genesis and land use.

Factors Affecting Soil Color

Organic matter is a major coloring agent of soils that affects color depending on its nature, amount, and distribution in the soil profile. Raw peat is usually brown, whereas well-decomposed organic matter such as humus is black or nearly so. Organic matter is usually highest in the surface soil layer; in the temperate regions, darker surface soil color is commonly associated with a greater organic matter content. Colors tend to range from pale brown to black. However, black soils such as those of the Blacklands of Texas are not necessarily high in organic matter content. These soils also may be black or nearly black to a depth of 1 meter, even though there is a marked decrease in organic matter with increasing soil depth. Similarly, the black soils on the Deccan

dy roots is oxidized because of the excretion of oxygen by roots. This produces a thin layer of yellowish-red soil around the roots where ferrous iron has been oxidized to ferric iron and precipitated at the root surfaces. In some soils, where a high content of soluble iron exists, genuine pipes of iron oxide may form around the roots. As the soil dries and cracks form, iron migrates from the interior of the wet clods to the edge of the cracks, where it is oxidized to produce an iron mottling of the puddled layer.

The soil below the pressure pan may be

plateau in India may exist near reddish-brown soils that contain more organic matter. For a given amount of organic matter, the soils of cool regions have a darker color than soils of warm regions.

The red color in soils is generally produced by unhydrated and oxidized iron oxide (hematite). In the tropics, many soil horizons have a reddish-brown color due to the combined effects of iron oxide and organic matter. Red soils tend to occur on convex slopes, where underlying rocks are pervious. Long periods of weathering and good drainage are required to form the red soils that are common to the tropics. Some red soils, however, have inherited their color from their parent material.

Yellow color is produced by hydrated iron oxide (limonite); in subsoils, it is an indication of imperfect drainage, but not of water saturation. Iron is reduced in water-saturated soils low in oxygen, resulting in a gray color that may appear to be blue. Soil color tends to vary with topographic position, reflecting differences in water content and aeration that affect the organic matter accumulation, and the oxidation, reduction, and hydration of iron oxides.

Soils with gray-colored A horizons that overlie red- and yellow-colored B horizons are commonly found on the coastal plain of the southeastern United States, where red and yellow colors are related to soil drainage and topography. In general, red and yellow soil colors increase in abundance and intensity from cool regions to the equator.

Light gray or nearly white color is sometimes inherited from parent material such as marl or quartz sand, and where conditions are unfavorable for plant growth and the accumulation of organic matter in the soil. In arid and subhumid regions, surface soils may be white due to evaporation of water and soluble salt accumulation; white subsoil horizons may be due to limited leaching and calcium carbonate accumulation. Leaching of iron oxide from the E horizons of humid, region forest soils developed from quartz sand-parent material results in light gray horizons.

Determination of Soil Color

Soil colors are determined by comparing them to a color chart. The Munsell Color Chart consists of 175 color chips arranged systematically on seven charts according to hue, value, and chroma—the three variables that combine to give the colors. *Hue* refers to the dominant wave-length or color of the light. *Value,* sometimes called brilliance, refers to the total quantity of light. It increases from dark to light colors. *Chroma* is the relative purity of the dominant wave-length of light. It increases with decreasing proportions of white light.

The Munsell notation of color is a systematic numerical and letter designation of each of the three variable properties of color. The three properties are always given in this order: hue, value, and chroma. For example, in the Munsell notation 10YR 6/4, 10YR is the hue, 6 is the value, and 4 is the chroma. This color is a light yellowish-brown.

The Munsell notation for a given soil sample can be quickly determined by comparison of the sample with a standard set of color chips. The chips are mounted in a notebook with all the colors of a given hue on one page. The pages are arranged in the order of increasing or decreasing wave-length of the dominant color to facilitate the matching of the unknown soil color with the color of the standards.

Many soil horizons have a single dominant color. Horizons that are dry part of the year and wet part of the year tend to exhibit a mixture of two or more colors. These colors are intermediate between those of well-drained and poorly drained soils. When several colors are present in a spotted pattern, the word *mottled* is used to describe the condition. In these cases, several of the dominant colors may be recorded.

Significance of Soil Color

White soils usually have low native fertility, but they may respond to management. Broad generalizations between soil color and soil fertility are not valid, however, due to variations in clay mineralogy, texture, and organic matter content. Within a local region, there may be a color pattern related to organic matter content and soil drainage (*see* Fig. 2-22). Gray subsoil color may indicate a saturated soil during soil genesis, and that drainage is required for agriculture. Such soils are poor building sites, because basements tend to be wet and septic tank filter fields do not operate properly. Unmottled subsoils that are not gray are indicative of good aeration and drainage. Many landscape trees and plants have specific needs regarding soil moisture and aeration, and soil color can be a very useful guide for plant selection at a given site.

Soil Temperature

Below freezing, there is no biological activity, water does not move through the soil as a liquid, and, unless there is frost heaving, time stands still for the soil. Germination of seeds and root growth hardly occur in the range 0 to 5°C. A horizon as cold as 5°C acts as a *thermal pan* to most plant roots. Each plant species has its own temperature requirements. The chemical processes and activities of microorganisms, which convert plant nutrients into available forms, are also materially influenced by temperature. Freezing and thawing play a role in rock weathering, structure formation, and heaving of plant

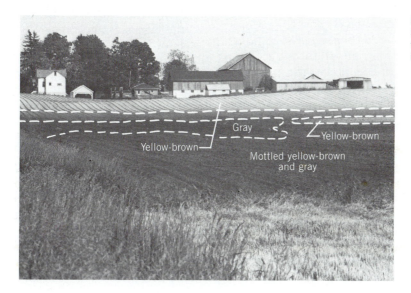

Figure 2-22
Subsoil colors and topographic relationships in a humid region.

Yellow-brown

Gray

Yellow-brown

Mottled yellow-brown and gray

roots. Temperature is thus seen to be an important soil property.

Heat Balance of a Soil

The heat balance of a soil consists of the gains and losses of heat energy. Solar radiation received at the soil surface is partly reflected back into the atmosphere and partly absorbed by the surface of the soil. A dark-colored soil and a light-colored quartz sand may absorb, respectively, about 80 and 30 percent of incoming solar radiation. The amount reflected is the *albedo;* it is less than 10 percent for water, about 20 percent for soils, and over 50 percent for snow. Of the total solar radiation available for the earth, about 34 percent is reflected back into space (albedo), 19 percent is absorbed by the atmosphere, and 47 percent is absorbed by the earth.

Absorbed heat is lost from the soil by: (1) evaporation of water, (2) reradiation back into the atmosphere as long-wave radiation, (3) heating of air above the soil, and (4) heating of the soil. In the long run, the gains and losses balance each other. For the short run, which is considered to be daytime or summer, heat gains exceed heat losses and soil temperatures increase. During the night and winter, the reverse is true.

Heat Capacity and Thermal Conductivity

The **specific heat** is the number of calories needed to increase the temperature of 1 gram of a substance 1°C. The specific heat of soil minerals and water are about 0.2 and 1.0, respectively. The *heat capacity* of a soil is the sum of the specific heat of each component multiplied by its mass. The heat capacity of the air component is

so small that it can be discounted. The heat capacity of a soil that contained 25 percent water (25 grams of water for every 100 grams of oven dry soil) would be approximately:

$$\frac{(100 \text{ grams})(0.2) + (25 \text{ grams})(1.0)}{100 \text{ grams} + 25 \text{ grams}}$$
$$= \frac{20 \text{ calories} + 25 \text{ calories}}{125 \text{ grams}}$$
$$= 0.36 \text{ calories/gram}$$

This calculation draws attention to the fact that water content plays a major role in determining the heat capacity of soil. In general, wet soils warm more slowly in the spring and cool more slowly in the fall than dry soils.

The water content of soils also affects soil temperature through its effect on thermal conductivity and rate of heat transfer. Heat transfer by conduction is more rapid through soil solids than soil air. As the water content of soil increases, its potential for heat transfer increases, because water occupies air space and the water conducts heat faster than air. Even though water increases heat transfer from the atmosphere or soil surface to the underlying soil, the high specific heat of water in the soils of temperate and cooler regions commonly reduces seed germination and early spring growth due to low soil temperature in the spring.

Location and Temperature

In the northern hemisphere, soils located on southern and southeastern slopes warm up more rapidly in the morning than those located on level or northern slopes. The reason is that they are more nearly perpendicular to the sun's rays; hence, a maximum amount of radiant energy strikes a given area. Soils with a southern or southeastern exposure often are selected for the growing of early vegetables and fruits. Landscape plants have a greatly different microenvironment in terms of light, temperature, and soil moisture supply when growing on the north side of a house, as compared to the south side in temperate regions.

Temperature decreases with an increase in elevation. Within the 40 miles from Fort Collins, Colorado to the top of the Rocky Mountains, the average July temperature changes as much as from Texas to North Dakota. The presence of bodies of water also modifies temperature; this is well illustrated by the location of the Hawaiian Islands in the Pacific Ocean. Temperature adjacent to the Great Lakes is modified so that the land warms slowly in the spring, delaying the blossoming of fruit trees. Delayed blossoming and an earlier date for the last killing frost in the spring make land near the lakes suitable for fruit production (*see* Fig. 2-23). A delay in the date of the first killing frost in the fall also lengthens the growing season.

Fluctuations in Soil Temperature

Soil temperatures vary in a characteristic manner on a daily and seasonal basis. Both fluctuations are greatest at the soil surface and decrease with increasing depth. A well-defined, seasonal time lag occurs, because the seasonal soil temperature is slow to change. Below about 3 meters, the temperature remains quite constant. Temperature fluctuations are greater at the soil-air interface than in the air above or soil below. Lethal temperatures for plant seedlings may occur at mid-day. Below 15 centimeters, there is little daily variation in soil temperature.

Snow acts as an insulator. A snow cover in the winter results in higher soil temperatures and a decrease in the depth of frost penetration. The insulation effect of both a snow cover and a layer of organic matter at the top of the soil in forests may result in little if any soil freezing. Melting of snow in the spring delays the warming up of the soil because of the high specific heat of the melt water. A layer of organic matter on top of the soil and a snow cover

Figure 2-23
Dates for last killing frost in spring and first killing frost in the fall at left and right, respectively, for Erie County, Pa. The longer growing season along Lake Erie favors use of land for fruits and vegetables. (Data from Taylor, 1960.)

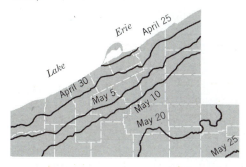
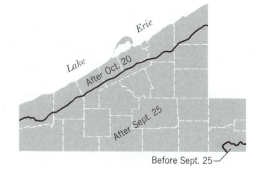

are thus seen to reduce fluctuations in soil temperature.

Control of Soil Temperature

As has been pointed out, removal of excess water from a soil will facilitate changes in soil temperature. By providing drainage, humans may exert some influence on the temperature relations of soils that are so situated they hold excessive quantities of water. By use of mulches and various shading devices, the amount of solar radiation absorbed by the soil, loss of heat energy from the soil by radiation, infiltration of water, and loss of water by evaporation can be altered.

Light-colored organic matter mulches: (1) reflect a large part of solar radiation, (2) retard heat loss by radiation, (3) increase infiltration of water, and (4) reduce evaporation of water from the surface of the soil. The net effect of a light-colored, organic matter mulch is to reduce soil temperature. In regions where summers are cool, the reduced soil temperature has been found to reduce crop yields. Dark-colored plastic mulches: (1) absorb most solar radiation, (2) reduce heat loss from the soil by radiation, and (3) reduce the evaporation of water from the surface of the soil. The net effect of black plastic mulches is to increase soil temperature in the soil under the mulch, when used. The higher soil temperature increases crop yields in regions with cool summers. Clear plastic mulches increase soil temperature because of the "greenhouse effect," and have proved more successful than opaque mulches when used with herbicides.

Cold air, being more dense than warm air, moves downslope and collects in low areas or depressions. Such areas have greater frost hazard. Land-forming and large electrically driven fans (wind machines) have been used to improve air drainage and to reduce frost hazard for fruit crops (*see* Fig. 2-24).

Permafrost

When the mean annual soil temperature is below 0°C, the depth of freezing in winter may exceed the depth of thawing in

Figure 2-24
Fruit land that has been reshaped to permit drainage of cold air onto the lake in the background. Spring frost hazard is reduced, and the smooth land results in more efficient operation of machinery.

summer. As a consequence, a layer of permanently frozen soil or grounds, called *permafrost*, may develop. Permafrost ranges from material that is essentially all ice to frozen soil, in which ice can not be seen; the soil appears ordinary, except that it is hard. When the only ice in permafrost is the bridges between contact points of soil grains, it is called *dry* permafrost.

The permafrost layer in a soil does not increase in thickness indefinitely, but it tends toward a maximum thickness balanced by the heat from the earth's interior and the depth of thawing in summer. The thickest permafrost reported is over 1,500 meters thick. In northern Alaska and Canada, permafrost is over 300 meters thick. The surface layer that freezes and thaws annually is the *active layer*, as shown in Fig. 2-25. The base of the active layer is the upper surface of the permafrost, or the permafrost table.

The active layer is used by plants. It is thickest where the mineral soil surface is bare and the maximum heating of the soil occurs in summer. The active layer is thinnest where a thick layer of organic matter, such as that produced by the growth of mosses and lichens, insulates the soil from the summer heat. Disturbances of the plant cover or insulation layer, such as road construction, increases permafrost melting and causes subsidence, or *thermokarst*. To prevent permafrost melting caused by road construction, a thick base layer, such as a 1-meter-thick layer of gravel, can be used to insulate the permafrost. Buildings are situated off the ground on pilings that are solidly set in permafrost to prevent melting of the permafrost by heat from the buildings.

Figure 2-25
Soil with permafrost in Northern Canada.

Active layer

Permafrost table

Permafrost

The soil is also shaded in summer and in winter; there is free movement of air under the buildings to permit maximum refreezing of the active layer.

The permafrost retards root and other biological activities brought on by low temperature. It also continually supplies water during the summer as the ice melts and inhibits the downward percolation of water. Therefore, landscapes made with permafrost are typically wet in summer; plant growth is meager. Organic matter content of soils, however, can be great because of the inhibitory effect of low temperature and water content on decom-

position processes. Frost action, or cryoturbation, commonly creates an irregular soil surface, as shown in Fig. 2-25.

Soil Temperature Regimes

Soil temperature, being an important soil property, is used to classify soils. Soil temperature classes, or regimes, are defined according to the mean annual soil temperature (MAST) in the root zone (arbitrarily set at 5 to 100 centimeters). Regime definitions and some characteristics and land uses associated with the regimes are given in Table 2-7. The use of soils for agricul-

Table 2-7

Definitions and Features of Soil Temperature Regimes

Temperature Regime	Mean Annual Temperature in Root Zone (5 to 100 cm)		Characteristics and Some Locations
	C	F	
Pergelic	<0	<32	Permafrost and ice wedges common. Tundra of northern Alaska and Canada and high elevations of the mid and northern Rocky Mountains.
Cryic	0–8	32–47	Cool to cold soils of the northern Great Plains of the United States and southern Canada where spring wheat is dominant crop. Forested regions of eastern Canada and New England.
Mesic	8–15	47–59	Midwestern and Great Plains regions where corn and winter wheat are common crops.
Thermic	15–22	59–72	Coastal plain of southeastern United States where temperatures are warm enough for cotton. Central Valley of California.
Hyperthermic	Over 22	Over 72	Citrus areas of Florida peninsula, Rio Grande Valley of Texas, and southern California, and at low elevations in Puerto Rico and Hawaii. Tropical climates and crops.

ture and forestry is related importantly to soil temperature because of the specific temperature requirements of plants.

Bibliography

Arkley, R. J., "Soil Survey of the Eastern Stanislaus Area. California," *U.S.D.A. and Cal. Agr. Exp. Sta.,* 1964.

Baver, L. D., W. H., Gardner, and W. R. Gardner, *Soil Physics,* 4th ed., Wiley, New York, 1972.

Bouyoucos, G. J., "Hydrometer Method Improved For Making Particle Size Analyses of Soils," *Agron. Jour., 54:*464–465, 1962.

Camp, C. R., and L. F. Lund, "Effect of Soil Compaction on Cotton Roots," *Crops and Soils,* November 1964:

Chesters, G., O. J. Attoe, and O. N. Allen, "Soil Aggregation in Relation to Various Soil Constituents," *Soil Sci. Soc. Am. Proc., 21:*272–277, 1957.

Childe, V. E., *Man Makes Himself,* Mentor, New York, 1951.

Cook, R. L., J. F. Davis, and M. G. Frakes, "An Analysis of Production Practices of Sugar Beet Farmers in Michigan—1958," *Ag. Exp. Sta. Quart. Bul., 42:*401–420, 1959.

Dickerson, B. P., "Soil Changes Resulting From Tree-Length Skidding," *Soil Sci. Soc. Am. Proc., 40:*965–966, 1976.

Dotzenko, A. D., N. T. Papamichos, and D. S. Romine, "Effect of Recreational Use on Soil and Moisture Conditions in Rocky Mountain National Park," *Jour. Soil Water Con., 22:*196–197, 1967.

Emerson, W. W., "The Structure of Soil Crumbs," *Jour. Soil Sci., 10:*235–244, 1959.

Erickson, A. E., and D. M. Van Doren, "The Relation of Plant Growth and Yield to Soil Oxygen Availability," *7th Int. Cong. Soil Sci., 3:*428–434, 1960.

Foster, R. C., "Polysaccharides in Soil Fabrics," *Science, 214:*665–667, Nov. 6, 1981.

Froehlich, H. A., "Soil Compaction from Logging Equipment: Effects on Growth of Young Ponderosa Pine," *Jour. Soil and Water Cons., 34:*276–278, 1979.

Grimes, Donald W., and James C. Bishop, "The Influence of Some Soil Physical Properties on Potato Yields and Grade Distribution," *Am. Pot. Jour., 48:*414–422, 1971.

Iverson, R. M., B. S. Hinckley, and R. M. Webb, "Physical Effects of Vehicular Disturbances on Arid Landscapes," *Science, 212:*915–916, May 22, 1981.

Khalifa, E. M., and S. W. Buol, "Studies of Clay Skins in a Cecil (Typic Hapludult) Soil: Effect on Plant Growth and Nutrient Uptake," *Soil Sci. Soc. Am. Proc., 33:*102–105, 1969.

Laws, W. Derby, and D. D. Evans, "The Effects of Long-Time Cultivation on Some Physical and Chemical Properties of Two Rendzina Soils," *Soil Sci. Soc. Am. Proc., 14:*15–19, 1949.

Moormann, F. R., and N. van Breemen, *Rice: Soil, Water, Land,* Int. Rice Res. Inst., Manila, 1978.

Page, J. B., and C. J. Willard, "Cropping Systems and Soil Properties," *Soil Sci. Soc. Am. Proc., 11:*81–88, 1946.

Retzer, J. L., "Soil Development in the Rocky Mountains," *Soil Sci. Soc. Am. Proc., 13:*446–448, 1948.

Russell, E. W., *Soil Conditions and Plant Growth,* 10th Ed., Longman, London, 1973.

Sanchez, P. A., *Properties and Management of Soils in the Tropics,* Wiley, New York, 1976.

Soil Survey Staff, *Soil Survey Manual,* USDA Handbook 18, Washington, D.C., 1951.

Soil Survey Staff, *Soil Taxonomy,* Agr. Handbook 436, USDA, Washington, D.C., 1975.

Steinbrenner, E. C., and S. P. Gessel, "The Effect of Tractor Logging on Physical Properties of Some Forest Soils in Southwestern Washington," *Soil Sci. Soc. Am. Proc., 19:*372–376, 1955.

Taylor, David C., "Soil Survey of Erie County, Pennsylvania," USDA and Pennsylvania State University, 1960.

Tedrow, J. C. F., *Soils of the Polar Landscapes,* Rutgers University Press, New Brunswick, N. J., 1977.

Ulrich, Rudolph, "Some Chemical Changes Accompanying Profile Formation of the Nearly Level Soils Developed from Peorian Loess in Southwestern Iowa," *Soil Sci. Soc. Am. Proc., 15:*324–329, 1950.

Van Doren, D. M., and G. B. Triplett, Jr., "Mulch and Tillage Relationships in Corn Culture," *Soil Sci. Soc. Am. Proc., 37:*766–769, 1973.

Walejko, R. N., et al., "Effect of Snowmobile Traffic on Alfalfa," *Jour. Soil Water Con., 28:*272–273, 1973.

Wascher, Herman L., et al., "Characteristics of Soils Associated with Glacial Tills in Northeastern Illinois," *Univ. Ill. Agr. Exp. Sta. Bul. 665,* 1960.

Wilde, S. A., "The Significance of Texture in Forestry and Its Determination by a Rapid Field Method," *Jour. For., 33:*503–508, 1935.

Wimer, D. C., and M. B. Harland, "The Cultivation of Corn," *Univ. Ill. Agr. Exp. Sta. Bul. 259,* 1925.

Youngberg, C. T., "The Influence of Soil Conditions, Following Tractor Logging on the Growth of Planted Douglas-Fir Seedlings," *Soil Sci. Soc. Am. Proc., 23:*76–78, 1959.

CHAPTER 3

SOIL WATER

Water is the most common substance on the earth; it is necessary for all life. The supply of fresh water on a long-time sustained basis is equal to the annual precipitation, which averages 66 centimeters for the world's land surface. The soil, located at the atmosphere-lithosphere interface, plays an important role in determining the amount of precipitation that runs off the land and the amount that enters the soil for storage and future use. Approximately 70 percent of the precipitation in the United States is evapo-transpired and returned to the atmosphere as vapor, with the soil playing a key role in water retention and storage. The remaining 30 percent of the precipitation represents the long-time annual supply of fresh water for use in homes, industry, and irrigated agriculture.

Only a tiny portion of the earth's water is active in the hydrologic cycle. Less than 3 percent of earth's water is fresh, and most of this exists in ice sheets and glaciers, as shown in Fig. 3-1. The water in the atmosphere, soils, lakes, and streams represents only 0.03 percent of the total water. The 0.005 percent in soils is of great importance, since it forms the water reservoir for land plants.

Fortunately, water is not easily destroyed. The earth has as much water now as it did thousands of years ago. However, the water is unevenly distributed by rainfall, changes form, moves from place to place, and can be polluted. For students

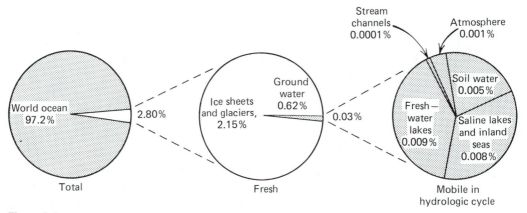

Figure 3-1
Distribution of the earth's water into its flow segments. (Adapted from *Elements of Physical Geography*, Strahler, A. N., and A. H. Strahler, Wiley, New York, 1976. Used by permission.)

interested in pollution of the environment, resource use, and agronomy, this chapter contains important concepts and principles that are essential for gaining an understanding of the soil's role in the hydrologic cycle and the intelligent management of water resources.

Energy Concept of Soil Water

As water cascades over a dam, the energy (ability to do work) of the water decreases. If water that has gone over a dam is returned to the reservoir, work will be required to lift the water back up into the reservoir and the energy level of the water will be restored. Water movement in soils and from soils into plant roots, like water cascading over a dam, is from regions of higher-energy water to regions of lower-energy water. Thus, "water runs downhill." For this reason it is necessary to consider the forces that determine the physical state or energy content of water to understand the behavior of water in soils and plants.

Forces That Control the Physical State of Soil Water

The change of water from a vapor to a liquid (condensation) is accompanied by a great reduction in the movement of the molecules and their energy. Energy is released as heat when water changes from vapor to liquid. The liberation of heat from the formation of raindrops is a major source of energy for storm systems. When raindrops fall on dry soil and are adsorbed on the surface of soil particles, a further reduction occurs in the motion and energy of the water molecules. The adsorbed water may still be in a liquid state, but the tendency of the water molecules to move has been further reduced. This change in energy can be explained by considering the forces operating between soil particles and water molecules.

Water molecules (H_2O) are electrically neutral; however, the electrical charge within the molecule is asymmetrically distributed. As a result, water molecules are strongly polar and attract each other by H bonding (*see* Fig. 3-2). Soil particles are

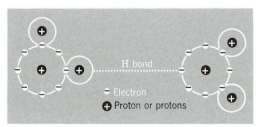

Figure 3-2
Schematic diagram of two water molecules. The sharing of electrons by oxygen and hydrogen produces water molecules that are negatively charged on the oxygen side and positively charged on the side where the proton of the hydrogen sticks out. Molecules attract each other by H bonding, the attraction of the proton of one water molecule for the negatively charged oxygen side of an adjacent water molecule.

also charged and have negative- and positive-charged sites. The strong attraction of soil for water molecules (adhesion) results in a spreading of water over the surface of the soil particles as a film when liquid water comes in contact with dry soil particles. The adsorption of water on the surface of soil particles produces (1) a reduction in the motion of the water molecules, (2) a reduction in the energy of the water, and (3) the release of heat associated with the transformation of water to a lower energy level. You can observe the release of heat, called *heat of wetting*, by adding water to oven dry clay soil and observe the increase in soil temperature.

When oven dry soil is exposed to a moist atmosphere, water molecules from the atmosphere will be strongly adsorbed to the soil particles because of these strong adhesive forces (*see* Fig. 3-3). This water is called *adhesion water* and will be several molecular layers thick. Adhesion water moves little, if at all, and some scientists believe that the innermost layers of water molecules exist in a crystalline state similar to the structure of ice. Adhesion water is not available to plants and is always present in the normal soil (even in the dust of the air), but the adhesion water can be removed by drying the soil in an oven.

You can demonstrate the existence of the forces important in holding water in soils by placing two clean microscope slides in water and bring them together so that their flat sides are flush against each other. Then, try to pull them apart. Your failure to pull the slides apart demonstrates the existence of attractive forces between glass and water molecules. You can also observe that the forces are effec-

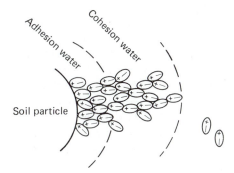

Figure 3-3
Schematic drawing of relationship between various forms of soil water. The gravitational water is beyond the sphere of cohesive forces, where gravity causes gravitational water to flow down and out of the soil (unless downward flow is inhibited).

tive only over a very short distance, since the glass slides must be very close together before a strong attraction develops between the slides.

Beyond the sphere of strong attraction of the soil particles, water molecules are held in the water film by cohesion (H bonding between water molecules). This outer film water is called *cohesion water* (*see* Fig. 3-3). Molecules of cohesion water, compared to adhesion water, are in greater motion, have greater energy, and move more readily. The water film (including adhesion and cohesion water) in soils may be as much as 15 to 20 molecular layers thick. Cohesion water also exists in soil micropores due to surface tension forces. This micropore water plus approximately the outer two thirds of film water can be considered available to plants; it constitutes the major source of water for plant growth (*see* Fig. 3-4).

Cohesion and adhesion water occupy the micropore space and are retained in soil by forces that exceed gravity. Macropore water, by contrast, is more weakly held and moves down and out of the soil unless an impermeable layer inhibits the water's downward movement. Macropore space water is called *gravitational* water because it tends to drain downward and out of the soil, because gravity attracts this water more strongly than the soil (*see* Fig. 3-3).

Energy and Pressure Relationships

We have just noted that water exists in the soil over a range of energy contents. This energy content of water can be expressed in terms of water pressure. Because it is much easier to determine the pressure of water, as contrasted to the energy content or level, we usually categorize water on the basis of pressure. For this reason we must understand the relationship between the energy content and the water pressure.

The hull of a submarine must be sturdily built to withstand the great pressure encountered in a dive far below the ocean's surface. The head of water above the submarine exerts pressure on the hull of the submarine. If the water pressure exerted against the submarine was directed against the blades of a turbine, the gravitational energy in the water could be used to generate electricity. The greater the water pressure, the greater the tendency of water to move and do work, and *the greater is the energy content of the water*. Thus, the existence of a relationship between the pressure of water and the energy of water is established. Our next consideration will be that of the water pressure relationships in saturated soils, which is analogous to the water pressure relationships in a beaker of water or any body of water.

Figure 3-4
Schematic drawing showing the relationship of adhesion and cohesion water with respect to soil particles and air-filled macropores.

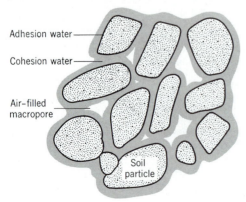

Adhesion water

Cohesion water

Air-filled macropore

Soil particle

Water Pressure in a Beaker or in Saturated Soil

The beaker in Fig. 3-5 has a bottom with an area of 100 square centimeters. The height of the water in the beaker is 20 centimeters. The water has a volume of 2,000 cubic centimeters, weighs 2,000 grams, and exerts a total force on the bottom of the beaker equal to 2,000 grams. The pressure of the water at the bottom of the beaker is equal to:

$$P = \frac{\text{force}}{\text{area}} = \frac{2,000 \text{ grams}}{100 \text{ square cm}}$$
$$= 20 \text{ grams per square cm}$$

The water pressure at the bottom of the beaker could also be expressed simply as equivalent to a column of water 20 centimeters high. This is analogous to saying that air pressure at sea level is equal to 1 atmosphere or a 1,033-centimeter water column, or 1,033 grams per square centimeter.

Figure 3-5
A beaker with a cross-sectional area of 100 square centimeters contains 2,000 grams of water when the water is 20 centimeters deep. At the water surface, water pressure is 0; the water pressure increases with depth to 20 grams per square centimeter at the bottom.

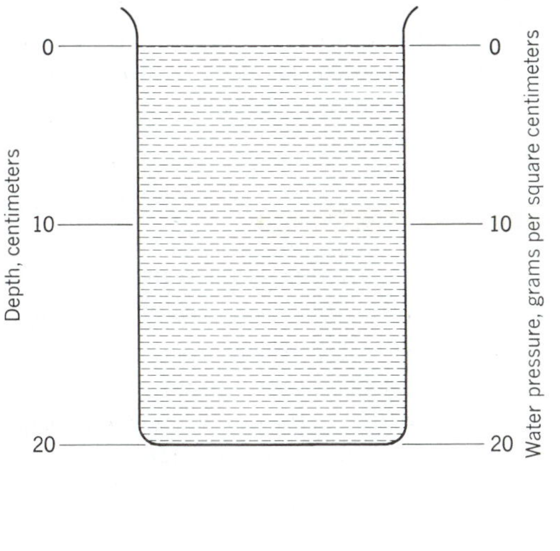

At the 10-centimeter depth, the water pressure is half of that at the 20-centimeter depth and is, therefore, 10 grams per square centimeter. The *water pressure* decreases with distance toward the surface and becomes zero at the free water surface (*see* Fig. 3-5).

Water Pressure and Energy Relationships in Unsaturated Soil

If the tip of a capillary tube is inserted in a beaker of water, the attraction between glass and water molecules (adhesion) causes water molecules to migrate up the interior wall of the capillary. The cohesive force between water molecules causes other water molecules to be drawn up the capillary. Now we need to ask, "What are the pressure relationships in the capillary tube?" and "Why is this knowledge important?"

We have already seen that the water pressure decreases from the bottom to the top of a beaker; at the top of the water surface, the water pressure is zero. Beginning at the water surface of the beaker and moving upward into the capillary tube, the water pressure continues to decrease. Thus, the water pressure in a capillary tube is less than zero or is *negative*. From Fig. 3-6, one can observe that the water pressure decreases in a capillary tube with height above the water in a beaker. At a height of 20 centimeters above the water surface in a beaker, the water pressure in a capillary tube is equal to −20 grams per square centimeter. Figure 3-6 also shows that at this same height above the water surface, the water pressure in *unsaturated* soil is the same, −20 grams per square centimeter at the 20-centimeter height. The two pressures are the same at any one height, and there is

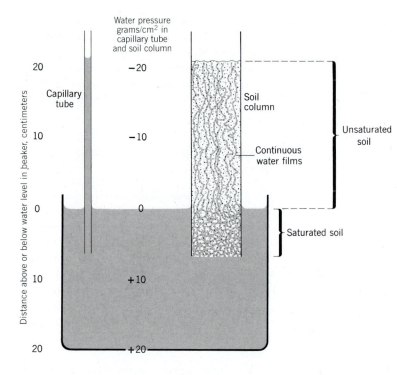

Figure 3-6
Water pressure in a capillary tube decreases with increasing distance above the surface of the water in the beaker. It is -20 grams per square centimeter at a height 20 centimeters above the water surface. Since the water column of the capillary tube is continuous through the beaker and up into the soil column, the water pressure in the soil 20 centimeters above the water level of the beaker is also -20 grams per square centimeter.

no net flow of water from the soil or capillary after an equilibrium condition is established.

Applying the considerations of water in a capillary tube to an unsaturated soil, we can make the following statements: (1) water in unsaturated soil has a *negative* pressure or is under *tension*, (2) the water pressure in unsaturated soil *decreases* with *increasing distance* above the surface of a water table, and (3) water in unsaturated soil, compared to saturated soil, has a *lower pressure* and *lower energy* level.

Summary Statement of Forms of Soil Water

Adhesion water in soils:

1. Is held by strong surface (electrical) forces existing between soil particles and water molecules.

2. Is mostly "crystalline," exhibits little or no movement, and possesses a low energy content.

3. Exists as a film on the surface of soil particles, several molecular layers thick.

4. Is unavailable to plants.

5. Exists on surfaces of dry soil particles that occur as dust in the air, but can be removed by drying the soil in an oven.

Cohesion water in soils:

1. Is held by the attraction of water molecules for each other through hydrogen bonding.

2. Exists in the liquid state in the water films around soil particles and in micropores, and approximates the soil solution.

3. Is the major source of water for plant growth.

4. Has greater energy than adhesion water.

Gravitational water in soils:

1. Exists in the macropores.

2. Has greater energy than cohesion water.

3. Moves freely through macropore spaces in response to gravity.

The Soil Water Potential

The energy status of soil water is expressed by the soil water potential, which is defined as *the amount of work that must be done per unit quantity of water in order to transport reversibly (without energy loss due to friction) and isothermally (without energy change due to temperature change) an infinitesmal quantity of water from a pool of pure water at specified elevation and at atmospheric pressure to the soil water at the point under consideration.*[1] The total water potential consists of several subpotentials. The total water potential is equal to:

$$\psi_T = \psi_M + \psi_g + \psi_p + \psi_\pi + \psi_\Omega$$

where ψ_M, ψ_g, ψ_p, ψ_π, and ψ_Ω are the matric, gravity, gas pressure, osmotic and overburden potentials, respectively.

The gravitational potential is due to the position of the water in a gravitational field. The gravitational potential is very important in water-saturated soils. It accounts primarily for the movement of water through saturated soils and the movement of water, in general, from high to low elevations. Usually, the soil water being considered is higher in elevation

[1]From *Glossary of Soil Science Terms*. Soil Science Society of American Madison, Wis., October, 1979.

than the reference pool of pure water. The gravitational potential represents the amount of work that must be done to move water from a reference pool at a low elevation to a point at a higher elevation in the soil. The gravitational potential has a positive sign.

We have noted that a dry soil adsorbs rain water with the release of heat. Thus, the movement of pure water from a pool onto soil particle surfaces, and into the small soil pores, occurs with a loss of energy in the water and the loss of that energy as heat. This results in a negative sign for the matric potential. When a moist soil is allowed to dry, the matric potential decreases as the soil dries (the matric potential becomes a larger negative number). The matric potential is the major factor affecting the movement of water from the soil into plant roots, seeds, and microorganisms.

The osmotic potential is created mainly by the adsorption of water molecules by ions from soluble salt. Normally, the salt content of soil is low, and the osmotic potential has little significance. In saline soils, however, the osmotic potential may control the movement of water from the soil into plant roots and microorganisms. Since the adsorption of water molecules by ions releases heat, the sign for osmotic potential is also negative.

Measurement and Expression of Soil Water Potential

The gravitational potential is due to the position of soil water relative to a reference pool of pure water at a lower elevation. Thus, if the soil water in question was 20 centimeters above the reference pool, the gravitational potential would be equal to a 20-centimeter water column or

20 grams/cm². Water potentials are normally expressed in bars, which are about equal to atmospheres (atmospheres × 1.013 = bars).[2] Expressed as bars, the gravitational potential is:

$$\text{bars} = \frac{20 \text{ cm}}{1{,}020 \text{ cm}} = 0.0196 \text{ bars}$$
$$= 0.02 \text{ bars}$$

The matric potential can be determined in several ways. As we noted in Fig. 3-6, water in soil 20 centimeters above the free water surface, where the matric potential is zero, has a pressure equal to a −20-centimeter water column. This translates into −0.02 bars matric potential.

Vacuum gauge potentiometers consist of a rigid plastic tube that has a fired clay-porous cup on one end and a vacuum gauge on the other. The potentiometer is filled with pure water and buried in the soil, so that the porous cup has good contact with the surrounding soil. Since the potential of pure water in the potentiometer is greater than water in dry soil, water will move from the potentiometer into the soil (assuming the soil is not saturated with water). At equilibrium, the vacuum gauge will record the matric potential (see Fig. 3-7). Vacuum gauge potentiometers work in the range 0 to −0.8 bars, which is a biologically important range for plant growth.

For matric potentials lower than −0.8 bars, a pressure chamber is used. Wet soil is placed on a porous ceramic plate with very fine pores that is confined in a pressure chamber. Air pressure is applied, and the air forces water through the fine pores of the ceramic plate. At equilibrium, when water is no longer leaving the soil, the air pressure applied is equated to the matric potential. This apparatus is very useful in developing soil-moisture characteristic curves that relate matric potential to the water content of soil. This technique can be used to determine the amount of water a soil can retain or hold for plant use, which is generally considered to be matric potential in the range of −0.3 to −15 bars (see Fig. 3-8).

---

[2]A bar is equal to 10^6 dynes; a dyne equals the force that imparts to a mass of 1 gram an acceleration of 1 cm/sec². A bar is also equal to a 1,020-centimeter water column or 1,020 grams per square centimeter.

Figure 3-7
Soil water potentiometers for measuring the matric potential. The one on the left has been removed from the soil to show the porous clay cup.

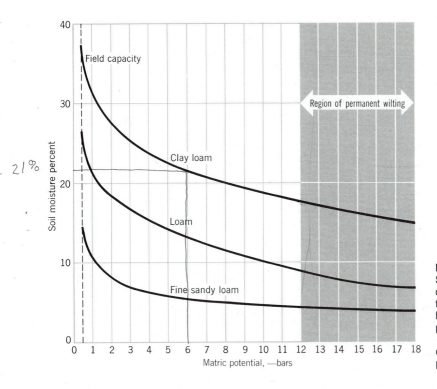

Figure 3-8
Soil moisture characteristic curves for three soils. The water between the matric potential of −0.3 and −15.0 bars is generally considered available for plant use.

Soil Water Movement

Soil water movement is important in the movement of water from the soil via evaporation and/or drainage (from wet to dry soil), and from the soil into plant roots (to the leaves and to the atmosphere). Water movement in soils occurs as a liquid flow in water-saturated soils, and as a liquid and vapor flow in unsaturated soils. In all cases, water movement is controlled by the rate of water flow known as the soil-hydraulic conductivity, and also by the driving force, which is the difference between the water potentials at two locations in the soil. Mathematically, we can summarize the description of soil water movement as follows:

where $V = kf$
V = volume of water flow
k = soil hydraulic conductivity
f = water potential difference or gradient

Saturated Water Flow

In saturated flow, all soil pore space is water filled; the water moves rapidly through the larger pores. The matric potential is nearly zero and the gravitational potential is the major force that is producing flow. Saturated water flow into and through a flooded field with water that is standing on the soil surface is analogous to water flow from a water tank on a roof top through a pipe to a faucet (tap) in a house. Water flow will be a function of the difference in elevation, between the water level in the tank and the faucet (gravitational water potential difference), and the size of the pipe (conductivity). Water flow in pipes and pores is related directly to the fourth power of the radius. Thus, water can flow in a large pore 10,000 times faster if the pore has a radius or diameter that is 10 times greater. Obviously, in saturated flow, the great bulk of the water is flowing through the largest pores. As a consequence, sands having a large number of macropores have a much greater saturated conductivity compared to clays. Since saturated flow is always in soils that are saturated, and all the pores are water filled, the hydraulic conductivity remains constant for a given soil unless the amount and size of the pores change.

Unsaturated Flow

If some of the water in a saturated soil is allowed to drain out, water quickly leaves the soil via the largest pores, and air is pulled into the soil. This rapidly moving water is moving mainly in response to gravitational potential differences. In the field, this rapidly moving water in unsaturated soil almost ceases within a day or so; then, the soil is at *field capacity*. Field capacity occurs when the soil retains the maximum amount of water with little or no further loss of water by drainage or loss of gravitational water.

Further water movement in unsaturated soil slows quickly as water movement becomes restricted to smaller pores and to soil particle surfaces where there is an air-water interface. The driving force is the matric potential gradient. As the matric potentials decrease, however, conductivity decreases very rapidly; it may change a million fold within the matric potential range, which is biologically important as shown in Fig. 3-9. When soils become slightly drier than their field capacity, the remaining water is quite immobile. It moves very slowly because of low conductivity.

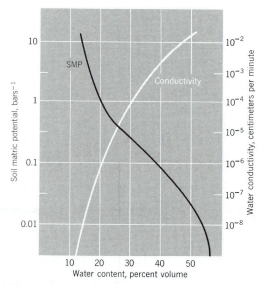

Figure 3-9
Soil matric potential and water conductivity as related to water content of silt loam soil. Soil matric potential and conductivity decrease very rapidly as the soil dries. (Data from Kunze, et al., 1968.)

In unsaturated soils with matric potentials near zero (moist soil), the conductivity of sands is much greater than clays. As the soil dries and the matric potential decreases several bars, water films between soil particles become discontinuous, which inhibits the movement of water. Water films become discontinuous in sands at higher matric potentials compared to clays, which causes clay soils to have a greater conductivity than sands when soils are quite dry. The differences in conductivity in moist soil versus dry soil of differing textures is shown in Fig. 3-10. When water is moving through unsaturated silt loam, and a dry sand layer is encountered, there is a sudden and drastic reduction in conductivity (dry sand has great film discontinuity). The water continues to move downward in the silt loam. Instead of continuing downward into the

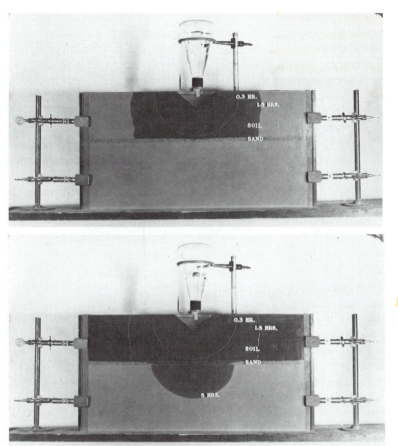

Figure 3-10
Photographs illustrating water movement in stratified soil where water is moving as unsaturated flow through a silt loam into a sand layer. In the upper photo, water movement into the sand is inhibited by the low conductivity of dry sand. The lower photo shows that after an elapsed time of 1.5 to 5.0 hours, the water potential just above the sand increased sufficiently to cause water movement into the sand. (Photos courtesy Dr. W. H. Gardner of Washington State University.)

sand, the water moves laterally into dry silt loam, which has greater conductivity than the dry sand. As water continues to move downward, the water potential just above the sand increases. Eventually, the driving force, matric potential difference, and the conductivity cause water to move into the sand (*see* Fig. 3-10). The soil above the sand layer is able to retain more water at field capacity than if the sand layer was absent. Many agricultural soils have an increased ability for retaining water for plants due to underlying horizons or layers with a coarser texture.

If a clay layer underlies a silt loam layer, when the water-moving front encounters the dry clay layer, the high conductivity of the dry clay causes water to continue to move, but at a greatly reduced rate because of the limited capacity of small pores for transmitting water. The water content of the soil just above the texture contact increases, because the water conductivity of moist silt loam is greater than that of moist clay.

Because soil conductivity decreases so rapidly with a decreasing matric potential (see Fig. 3-9), unsaturated flow is important in water that moves only a few centimeters to plant roots as they absorb water.

Basically, soil water is quite immobile or moves very slowly when matric potentials drop to less than −0.1 to −0.3 bars, therefore, water can not move from the lower part of the solum to the upper soil layer to supply plant needs. Instead, roots must ramify all the soil from which significant amounts of water are extracted. Just above the water tables (water-saturated soil), matric potentials are relatively high (near zero) and significant water can move upward via capillarity within the distance of about 1 meter.

Vapor Flow

When soils become so dry that unsaturated liquid ceases, soils are at or near air dry; further water movement is mainly *vapor flow*. The *driving force* is the difference in water potential expressed as vapor pressure. Vapor pressure, in essence, is caused by the rate of water evaporation in soil. It is affected mainly by temperature and matric potential. As the temperature increases, more water evaporates and the vapor pressure (pressure created by water vapor) increases. As the soil dries, however, water molecules are held more tightly to soil particles and soil drying reduces vapor pressure. In the summer, vapor flows in a direction from warm surface soil to cooler subsoil; vapor flow is reversed in winter. During the fall, with the transition to winter, vapor flow is frequently upward. It is encouraged by a low air temperature to create a slippery smear on the bare soil surface because the water vapor condenses on a cool soil surface. If one soil layer is moist and cold and another is dry and warm, the offsetting vapor movement tendencies may cancel each other resulting in no movement.

　　Soil conductivity in vapor flow is not affected by pore size, but increases with an increase in total porosity and pore space continuity. Water vapor movement is minor in most soils, but it becomes important when soils crack at the surface, thus allowing for deep soil drying.

Plant-Soil Water Relations

In the course of a summer day, it is not unusual for a plant to transpire an amount of water equal to many times its weight. Since the water that plants absorb from soils is not free flowing, but diffuses slowly into plant roots by osmosis, an enormous area of contact between roots and soil particles is required. The total length of the root system of a corn plant could easily equal the distance from New York to San Francisco and back. Our concern in this section centers around the ability of plants to satisfy their water requirements from soil.

Water Absorption and the Wilt Point

At field capacity, the matric potential is high, relatively speaking. Roots can easily absorb water. As roots absorb water by osmosis, water near the roots will move slowly in the direction of the root (*see* Fig. 3-11). We have already observed that as the soil becomes drier, the conductivity rapidly decreases and movement and uptake of water becomes slower. Eventually, if no additional water is added to the soil, the plant will absorb water slower than water is lost by transpiration. A water deficit is developed inside the plant and, eventually, wilting occurs (unless the plant has some special adaptation, such as found in many desert plants that can stop transpiration loss). To determine the wilt point, plants such as sunflowers or wheat are grown in soil until the plants wilt and

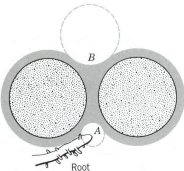

Figure 3-11
As the root absorbs moisture from the accumulation between two soil particles, the film curvature increases as is shown by the projected circles. Since the force that tends to move water into a given portion of the film varies inversely with the radius of the curvature $\left(p = \dfrac{2T}{\pi} \right)$, it follows that moisture will move to the feeding point of the root.

are unable to regain turgor when placed in a saturated atmosphere. A soil-water matric potential of about -15 bars has been found to correspond generally to the wilt point.

The water between field capacity and wilt point is considered to be the available water for plants. Assuming a field capacity of $-\frac{1}{3}$ bar matric potential for each of the soils (*see* Fig. 3-8), plants would have to exert the same amount of energy to remove water from each soil at field capacity (even though the water content of the clay loam is about three times greater than that of the fine sandy loam). This point brings out a most important fact: the ability of plants to remove water from soils is primarily related to soil water potential and not water content. It is the potential and not the water content that indicates when to irrigate. This shows the importance of potentiometers in irrigation agriculture. The wilt point, like the field capacity, is not a precise value. It var-

ies with soil, plant, and environmental conditions. High temperatures and strong winds could cause some plants to wilt with matric potential of -2 bars. The minimum moisture content in subsoils for wheat on the Great Plains has a potential that is less than -26 bars. For most soils and plants, the wilting range appears to be about -10 to -60 bars. By the time the water content of the soil has been reduced to the wilting range, the matric potential decreases rapidly, with little change in moisture content of soils. Thus, a plant that wilts on a *fine sandy loam* at -10 bars has extracted about as much water from the soil as a plant that wilts at -50 bars (*see* Fig. 3-8).

Other Soil Moisture Coefficients

The soil still contains water at the wilt point that is considered to be unavailable to plants. To remove the remaining water (excluding water of hydration), the soil is dried in an oven for 24 hours at 105 °C. The soil is then brought to the *oven dry* state. If oven dry soil is placed in a water-saturated atmosphere, water will be adsorbed by the soil. At equilibrium, the soil will contain an amount of water described as the *hygroscopic* coefficient. It has little relevancy for plant growth, but it is a qualitative measure of the surface area in the soil.

Water Content of Soils

The oven dry soil weight is easily and consistently determined, it is used to express the content of various soil components, such as the content of organic matter, clay, sulfur, and so on. The oven dry soil basis is also used to express the water content of soils. About 120 grams of moist loamy soil at field capacity would contain about 20 grams of water; consequently, it would lose 20 grams of water by oven drying. The percentage of water on an oven dry basis is:

$$\frac{20 \text{ grams of water}}{100 \text{ grams of oven dry soil}} \times 100 = 20\%$$

In considering the amount of rainfall, the water use by crops per day or week, or the application of irrigation water, it is useful to consider water content on a volume basis. When the oven dry density (bulk density) of the soil is the same as the density of water, the percentage of water on the oven dry basis is the same as on the volume basis. Most soils have a bulk density different than 1.0 gram per cubic centimeter. The percentage of water, on an oven dry basis, can be converted to the volume basis as follows:

% water on an oven dry basis

$\times \dfrac{\text{density of soil}}{\text{density of water}}$

= % water on a volume basis

A soil with 20 percent water on an oven

dry basis, and a bulk density of 1.25 grams per cubic centimeter, contains 25 percent water on a volume basis. This value is representative of many loamy soils; it means that at field capacity, about 25 percent of the soil volume is occupied by water (about 50 percent of the pore space) as shown in Fig. 3-12. A root zone that is 100 centimeters thick would contain water equal to a layer 25 centimeters thick.

About 60 percent of the water in loamy soils, at field capacity, is available to plants (−0.3 to −15.0 bars of water). Therefore, plants are able to extract about 15 centimeters of water from a 100-centimeter-thick root zone. Conversely, if the loamy soil is at the wilt point, the addition of 15 centimeters of rain or irrigation water will recharge the 100-centimeter-thick root zone to field capacity.

Effect of Texture on Available Water

The capacity of the soil to hold water is related to both surface area and pore space volume. Water-holding capacity is

Figure 3-12

Relationship of soil texture to available water-holding (between field capacity and wilt point) capacity of soils.

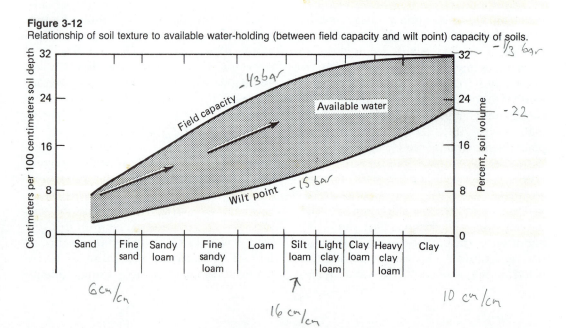

therefore related to structure as well as to texture. Fine-textured soils have the maximum *total* water-holding capacity, but maximum *available* water is held in medium-textured soils. Research has shown that available water in many soils is closely correlated with the content of silt and very fine sand (*see* Fig. 3-12). As the texture becomes finer, a smaller percentage of the water at field capacity is available, about 40 percent for clays.

It is generally known that sandy soils are more droughty than clayey soils. One reason is that the finer-textured soils are able to retain more available water. Another important difference between sands and clays is related to the differences in the slope of the soil-moisture characteristic curves shown in Fig. 3-8. The flatness of the curve for the fine sandy loam at water matric potential that is less than -4 bars means most of the available water in the sandy soil has relatively high potential, . . . it can be rapidly used by transpiring plants. Since more of the available water in the clay loam has a lower matric potential, plants use this water less rapidly, conserving the available water supply. A given amount of water was found to be more efficiently used or that it produced greater yields of wheat growing on a clay soil than on a sandy loam soil, in Saskatchewan (*see* Fig. 3-13). In areas of limited rainfall, sandy loam soils may out-produce clay soils due to the greater infiltration and lesser runoff on the sandy loam soils.

Many red tropical soils (Oxisols) with high iron oxide contents are rich in clay; however, the soils exhibit moisture characteristics of sands. That is, a modest amount of water is held at "high" potential. As discussed in Chapter 2, such soils may be composed of very stable sand-sized aggregates. The aggregates act as sand particles for water retention. Mois-

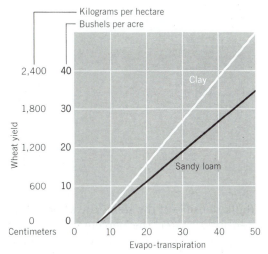

Figure 3-13
Effect of soil texture on wheat yields in Saskatchewan. A given amount of evapo-transpiration produced more wheat on the clay soil. (Data from Lehane and Staple, 1965. Reproduced courtesy of the *Can. Jour. of Soil Sci.,* June 1975, p. 213.)

ture within the aggregates is held in a matrix that is high in clay. It is mostly held at potentials less than the wilt point.

Water Potential and Plant Growth

We have already noted that at very high matric potential (near zero), the lack of air may limit plant growth. The rate of plant growth is at or near a maximum at field capacity, because there is adequate oxygen accompanied by rapid water absorption. As soil water is absorbed, the water films become thinner, the matric potential decreases, and the rate of water absorption decreases. Generally, the decreasing water potential between field capacity and wilt point is associated with a reduced rate of photosynthesis and growth. Corn yields in an Iowa experiment were reduced by 3 percent for each day of water stress in the last half of the growing season. Yields

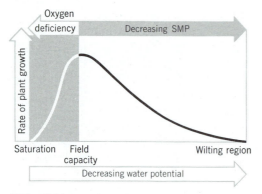

Figure 3-14
Generalized relationship between the soil-water
matric potential and plant growth.

were most reduced by stress during the
silking period. Thus, two important facets
of plant growth are associated with soil
water potential: lack of oxygen at high potential and slow rate of water uptake at
low potential (*see* Fig. 3-14). Forest tree
growth is commonly limited by low water
potential; in fact, it is very common for
plants to experience water stress due to
low soil water potential.

Pattern of Water Removal from Soils by Plants

When the water potential throughout the
root zone is high or near field capacity,
roots will absorb water most rapidly from
the upper part of the soil where oxygen is
the most abundant and near the base of
the plant. As the soil dries and the soil water potential in the surface soil layers decrease, water uptake will shift to deeper
soil layers, where the oxygen supply is less
but the soil is moist and the potential is
higher. In this way the root zone is progressively depleted of available soil moisture (in the absence of rain or irrigation
water) (*see* Fig. 3-15). When the upper soil
layers are rewetted by rain or irrigation
water, water absorption shifts back toward
the surface soil layers near the base of the
plant. This pattern of water use results in:
(1) more deeply penetrating roots in dry
years than in wet years, and (2) a greater
use of water from the upper soil layers
than from the lower soil layers. Crops that
have a long growing season and a deep

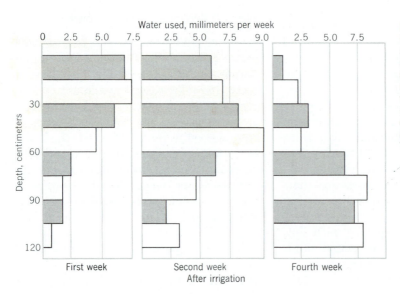

Figure 3-15
Pattern of water used from
soil by sugar beets during
a 4-week period following
irrigation. Water in the
upper soil layers was used
first. Then water was
removed from increasing
depths with time since
irrigation. (Data from
Taylor, 1957.)

root system, such as alfalfa, absorb a greater proportion of water below 30 centimeters (*see* Fig. 3-16).

Forests exert considerable influence on how water is used. Much of the precipitation is intercepted by the canopy. Pine and hardwood forests can intercept as much as $\frac{1}{2}$ centimeter of water, which is lost by evaporation. A similar amount may be retained in the unincorporated organic matter on the forest floor, so that rains must exceed about 1 centimeter before the upper mineral soil is wetted. Frequent rain in humid regions contributes to a large proportion of the soil water being absorbed by roots near the surface of the soil. This is related to the extensive root growth in forests in the upper soil horizons. Use of water by deep roots is minor but critical in enabling forest trees to survive drought.

Studies of beech trees in Ohio showed that considerable rainfall moved down along the branches and trunk. About five times more water was estimated to reach the soil near the trunk as compared to soil further away (450 versus 90 centimeters annually). The soil near the trunk was also more leached and acidic. The channeling of water from summer rains into the soil near the trunk results in a deeper penetration of water into the soil and a more efficient use of water by the trees.

Water Loss by Transpiration and Water Uptake

The amount of water transpired to produce 1 gram of dry matter was studied by Briggs and Shantz shortly after the turn of this century. They grew plants in large galvanized pots that had tight-fitting covers with openings only for the stems of the plants. They measured the amount of water used by the plants via weighing the pots. They also harvested the plants to de-

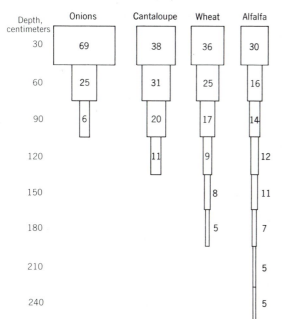

Figure 3-16
Percent of water used from each 30-centimeter layer of soil when produced with irrigation in Arizona. Onions are a shallow-rooted annual crop that were grown in the winter and used a total of 44 centimeters of water. Alfalfa, by contrast, is deep-rooted perennial crop that grew the entire year and used a total of 186 centimeters of water. (Data from *Arizona Agr. Exp. Sta. Tech. Bul.* 169, 1965.)

termine the quantity of dry matter produced. Some of their findings, presented in Table 3-1, show that plants commonly transpire 500 grams or more of water for each gram of dry matter produced. Furthermore, differences existed in the amount transpired between crops.

Because transpiration is simply the evaporation of moisture from plant surfaces, it is influenced by the same factors that affect the evaporation of water from any moist surface; exposure to direct sunlight, air temperature, humidity, wind movement, and atmospheric pressure are among the most important. Since these vary from year to year, so should the amount of water transpired (Table 3-1). Perhaps you have reduced the transpiration of plant cuttings by placing them in a plastic bag to prevent wilting and enabling the cuttings to get established.

The relationship between the atmosphere, the plant, and soil regarding water use by plants has resulted in the SPAC concept, which consists of the soil-plant-atmosphere-continuum. Loss of water by transpiration creates the major driving force for water uptake by roots. Low water potentials, created in leaves by loss of transpiration water, are transmitted to the xylem (water-conducting vessels) of the stem and eventually to the roots. Water potentials less than -4 or -5 bars are common in the xylem of trees, and they are sufficient enough to move water to the tops. Water potentials less than -80 bars have been found in desert plants.

Consumptive Water Use

Consumptive water use is that amount of water lost by evaporation from the soil and plants during the time a crop is grown. The quantity varies widely from less than 25 centimeters for a quickly maturing crop in a humid region to more than 175 centimeters for a long-season crop in an arid environment. Soybeans commonly require about 30 to 60 centimeters per season, which averages to about 0.4 to 0.5 centimeters per day.

Evaporation of water requires energy, making the climatic environment the major factor in determining the amount of water used. This is shown in Table 3-1, where the amount of water transpired to produce 1 gram of dry matter varied greatly between years. In fact, an actively

Table 3-1
Water Transpired by Plants

Grams of Water Transpired per Gram of Dry Plant Tissue Produced

Crop	1911	1912	1913	Greatest Variation	Average
Wheat	468	394	496	102	452.7
Oats	615	423	617	194	551.7
Corn	368	280	399	119	349.0
Sorghum	298	237	296	61	277.0
Alfalfa	1,068	657	834	411	853.0

From Briggs and Shantz, 1914.

growing crop that completely covers the ground may lose water more rapidly than a free water surface (open pan), if the soil water potential is high.

On any given day, a given amount of radiation is available to evaporate water. Therefore, many different kinds of crops require essentially the same amount of water when the cover is complete, they are green, and the soil water potential is the same. Great differences, however, exist in the total amount of water crops used, because they have different maturation periods or they grow at different seasons of the year (*see* Fig. 3-17).

Role of Water in Nutrient Absorption

Plant roots do not engulf and absorb the soil solution that contains nutrients the same as animals drink water containing soluble material. Instead, the water enters the roots as pure water without regard to the intake of any of the materials dis-

solved in it. The entrance of nutrients is entirely a separate process. Nutrients dissolved in the soil solution move with it, so that when water moves toward the roots to replace what has been taken up by plants, a supply of nutrients is moved near the roots. Although this action takes place over short distances only, the net result in the course of a growing season may add materially to the nutrient supply of plants.

Soil Moisture Regimes

The potential, available water supply for plants is governed mainly by the climate. The continental United States receives 75 centimeters of precipitation annually. About 20 percent runs off directly as overland flow to streams and 80 percent infiltrates the soil. An average of 7.5 centimeters, or 10 percent, percolates through the soil to the water table or to the underground water reservoir, the bulk of the water, or 70 percent, is retained by the soil and returned later to the atmosphere by transpiration and evaporation (evapo-transpiration).

The actual amount of water available to plants is affected by soil, as well as by climate. For example, wet soils exist in deserts where soils have impermeable layers and receive runon water from surrounding higher land or springs. Gravelly soils in humid regions may be droughty, because little water is retained. The soil property that expresses the order of soil moisture changes over time is the *soil moisture regime.* In addition to the supply of water for plants, the soil moisture regime expresses the availability of water for weathering and leaching, and whether the root zone lacks oxygen because of water saturation.

Figure 3-17
The seasonal distribution of water use by alfalfa and corn. The arrows indicate harvest dates. (From "Agricultural Water Use," D. E. Angus, *Adv. in Agronomy, II:* 20, 1959. Used by permission.)

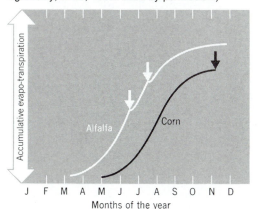

The Soil Moisture Control Section

Soil moisture regimes are based on moisture conditions in the *soil moisture control section*. The upper boundary of the moisture control section is the depth to which 2.5 centimeters of water will moisten dry soil (matric potential less than −15 bars, but not air dry) in 24 hours. The lower boundary is the depth of penetration of 7.5 centimeters of water in dry soil in 48 hours. These depths are exclusive of large cracks open at the soil surface. The moisture control section varies with texture and for many loamy soils, the moisture control section is located between depths of 20 to 60 centimeters (*see* Fig. 3-18).

Aquic Soil Moisture Regime

Soils with aquic (L. *aqua,* water) moisture regimes are wet; dissolved oxygen is virtually absent, because the soil is saturated. Very commonly, the level of ground water will fluctuate with the season. In some cases, as in tidal marshes, the water table is at or close to the soil surface all the time. Drainage is needed to grow plants that require an aerated root zone. Subsoil colors in mineral soils are frequently gray, indicating reducing conditions, or mottled, indicating alternating reducing and oxidizing conditions (*see* Fig. 3-19).

Aridic Soil Moisture Regime

The driest soils have aridic (L. *aridus,* dry) soil moisture regimes (also called *torric* meaning hot and dry). The moisture control section is dry in all parts more than half the growing season and is not moist (matric potential more than −15 bars) in some part for as long as 90 consecutive days during a growing season in most years. Most soils with aridic moisture regimes are in arid or desert regions with

Figure 3-19
Dark-colored A horizons, and gray and mottled subsoil horizons, are characteristic of soils with aquic-soil moisture regime.

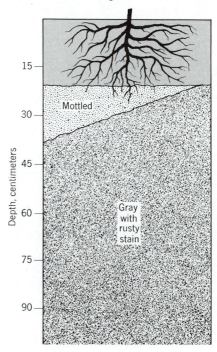

Figure 3-18
Approximate moisture control sections based on particle-size class. (Data from Ikawa, 1978.)

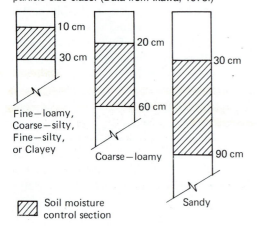

widely spaced shrubs and cacti. A crop can not be matured without irrigation.

Some climatic data and the soil water balance typical of an aridic soil moisture regime are given in Fig. 3-20. Rainfall is low most months and the potential evapotranspiration (PE) is very high in summer, relative to the precipitation. Therefore, little soil moisture recharge or water storage occurs, and the little water that is stored is quickly used in late winter or early spring. The result is a lack of water for plant growth most of the time and a large water deficit. Many of the native plants have an unusual capacity to endure a high degree of desiccation without serious injury. Grazing is the dominant land

use, and crop production requires irrigation.

Weathering occurs when soils are moist, but there is little or no leaching. Soluble salts commonly accumulate in a zone, marking the average depth of moisture penetration. Some soils with aridic moisture regimes exist in the semiarid regions, if they are shallow over rock or have very low infiltration rates.

Udic Soil Moisture Regime

Udic (L. *udus,* humid) means humid. Soils with udic moisture regimes have a moisture control section that is not dry in any part as much as 90 cumulative days in

Figure 3-20

Climatic data and water balance representing aridic, udic, ustic, and xeric soil moisture regimes. (Data from *Soil Taxonomy,* 1975.)

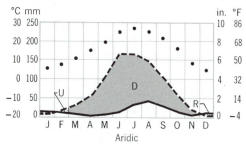

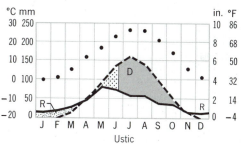

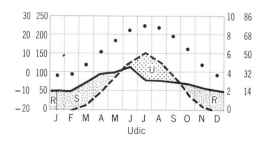

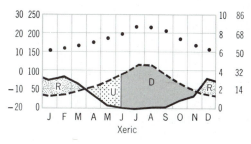

°C   Degrees centigrade
mm Millimeters
In.   Inches
°F   Fahrenheit

Precip. ——   U, Utilization
PE   ----   D, Deficit
Temp. • •   R, Recharge
                        S, Surplus

most years. Udic soil moisture regimes are common in soils of humid climates that have well-distributed rainfall. The amount of summer rainfall, plus stored soil water, is approximately equal to or exceeds the amount of evapo-transpiration (*see* Fig. 3-20). Forests and tall grass prairies are typical vegetation on udic soils. If precipitation exceeds the amount of evapo-transpiration each month, the moisture regime is called *perudic*.

In most years there is surplus water and leaching occurs. Data in Table 3-2 from a location in Ohio show that about one sixth of the precipitation percolated through the soil. Significant amounts of plant nutrients were removed from the soil and the soils developed acidity. Leaching losses of calcium and magnesium in Ohio each year were more than the amount required to produce an average crop of

wheat, but less than the amount contained in an average crop of alfalfa. Sufficient leaching to produce soil acidity and low natural fertility is typical of forest soils with udic moisture regimes.

Short droughts may occur occasionally. Irrigation is not widespread. It is used for specialty crops and for special benefits, as frost control. Udic soils are common from the east coast to the western boundaries of Minnesota, Iowa, Missouri, Arkansas, and Louisiana.

Ustic Soil Moisture Regime

The ustic (L. *ustis*, burnt, implying dryness) soil moisture regime is intermediate between aridic and udic regimes. The concept is one of limited available soil moisture; however, soil moisture is available for significant plant growth when

Table 3-2

Plant Nutrient Losses in Lysimeter Percolates on Kenne Silt Loam During a Period of High Precipitation (1950) and Low Precipitation (1953), as Compared with the 16-Year Average (1940–1955), by Practice

Practice and Period	Total Precipitation, cm	Percolation, cm	Nutrients Percolated Per Hectare kg[a]					
			Ca	Mg	K	N	Mn	S
Conservation								
1950 (high precipitation)	120	32	63	37	14	6.8	0.78	99
1953 (low precipitation)	72	9	7	1	3	0.8	0.16	10
16-year average (1940–1955)	94	16	36	21	12	4.2	0.36	38
Poor								
1950 (high precipitation)	120	34	37	23	25	3.5	1.07	62
1953 (low precipitation)	72	13	6	1	4	1.4	0.23	15
16-year average (1940–1955)	94	18	25	15	16	5.0	0.45	25

From Harrold and Dreibelbis, 1958.
[a]Multiply by 0.892 to convert to pounds per acre. Kilograms per hectare can be translated to pounds per acre with little or no significant change in the interpretation.

Figure 3-21
Grape production on soils with a xeric soil moisture regime in the Central Valley of California. Irrigation water in this case is being applied in March, because the root zone was not completely recharged by winter rains.

other conditions are favorable for growth. Compared to the aridic regime, significantly more water storage occurs between fall and spring, and summer rainfall is greater (*see* Fig. 3-20). The water deficit is much less than in the aridic regime and much greater than in the udic regime. Ustic soil moisture regimes are common on the Great Plains, where water is available for wheat production in winter, spring, and early summer in most years. Droughts are not uncommon. Sorghum is grown, because it can interrupt growth when water is lacking and grow again if more rainfall occurs. Corn requires irrigation. Wheat and sorghum are the major dry land crops and grazing is important. Surplus water is rare and soils are unleached. Native vegetation was mainly mid, short, and bunch grasses. Many areas in Asia, with a monsoon climate, have ustic soil moisture regimes.

Xeric Soil Moisture Regime

Soils in areas with Mediterranean climates typically have xeric (Gr. *xeros*, dry) soil moisture regimes. Winters are cool and moist and summers are hot and dry. The precipitation occurs in the cool months, when evapo-transpiration is low and very effective for weathering and leaching (*see* Fig. 3-20). Surplus water may occur. Pasture and crops are well supplied with moisture in the winter, and the landscape is green. A large water deficit occurs in summer, hilly grasslands are brown. Xeric soils of the Central Valley of California are used for a wide variety of crops including vines, fruits, nuts, vegetables, seeds, and agricultural crops (*see* Fig. 3-21). Xeric soils of the Palouse in Washington and Oregon are used for winter wheat.

Bibliography

Angus, D. E., "Agricultural Water Use," in *Advances in Agronomy*, vol. 11, Academic, New York, 1959, pp. 19–35.

Baver, L. D., W. H. Gardner, and W. R. Gardner, *Soil Physics*, 4th ed., Wiley, New York, 1972.

Blumrich, C., "The Browning of America," *Newsweek*, 26–30, Feb. 23, 1981.

Briggs, L. J., and H. L. Shantz, "Relative Water Requirements of Plants," *Agr. Res., 3*:1–65, 1914.

Erie, L. J., Orrin F. French, and Karl Harris,

"Consumptive Use of Water by Crops in Arizona," *Arizona Agr. Exp. Sta. Tech. Bull. 169,* 1965.

Gardner, W. H., "How Water Moves in the Soil," *Crops and Soils,* pp. 7–12, November 1968.

Gersper, P. L., and N. Halowaychuk, "Effects of Stemflow Water on a Miami Soil Under a Beech Tree: I. Morphological and Physical Properties," *Soil Sci. Soc. Am. Proc., 34:*779–786, 1970.

Harrold, L. L., and F. R. Dreibelbis, "Evaluation of Agricultural Hydrology of by Monolith Lysimeters 1944–1955," *USDA Tech. Bull.,* 1179, 1958.

Ikawa, H., "Occurrence and Significance of Climatic Parameters in the Soil Taxonomy," in *Soil Resource Data for Agricultural Development,* Hawaii Agr. Exp. Sta., 1978.

Jacobs, H. S., and L. V. Withee, "Soil Moisture Tension: Presentation for Beginning Soil Students," *Agron. Jour.,* 57:639–642, 1965.

Kunze, R. J., G. Uehara, and K. Graham, "Factors Important in the Calculation of Hydraulic Conductivity," *Soil Sci, Soc. Am. Proc., 32:*760–765, 1968.

Lehane, J. J., and W. J. Staple, "Influence of Soil Texture, Depth of Soil Moisture Storage, and Rainfall Distribution on Wheat Yields in Southwestern Saskatchewan," *Can. Jour. Soil Sci., 45:*207–219, 1965.

Musgrave, G. W., "How Much of the Rain Enters the Soil?," in *Water,* USDA Yearbook, Washington, D.C., 1955, pp. 151–159.

Scholander, P. F., H. T. Hammel, E. D. Bradstreet, and E. A. Hemmingsen, "Sap Pressure in Vascular Plants," *Science, 148:*339–346, 1965.

Sharma, M. L., and G. Uehara, "Influence of Soil Structure on Water Relations in Low Humic Latosols: I. Water Retention," *Soil Sci. Soc. Am. Proc., 32:*765–770, 1968.

Soil Survey Staff, *Soil Taxonomy,* Agr. Handbook 436, USDA, Washington, D.C., 1975.

Strahler, A. N., and A. H. Strahler, *Elements of Physical Geography,* Wiley, New York, 1976.

Taylor, S. A., "Use of Moisture by Plants," in *Soil,* USDA Yearbook, Washington, D.C., 1957, pp. 61–66.

Taylor, S. A., and G. L. Ashcroft, *Physical Edaphology,* Freeman, San Francisco, 1972.

United States Department of Agriculture, *Water,* USDA Yearbook, Washington, D.C., 1955.

Voight, G. K. "Distribution of Rainfall Under Forest Stands," *Forest Science, 6:*2–10, 1960.

CHAPTER 4

SOIL WATER MANAGEMENT

In a very real sense, the story of water is the story of mankind. Civilization and cities emerged along the rivers of the Near East. The world's oldest known dam in Egypt is over 5,000 years old. It was used to store water for drinking and irrigation, and perhaps to control flood waters. As the world's population and the need for food and fiber production increase, water management becomes more important. The rapidly increasing need of water in some urban areas is causing serious competition for available water between agriculture and industry.

Basically there are three approaches to water management for increasing food and forest production: (1) conservation of natural precipitation in subhumid and arid regions, (2) removal of excess water from wet lands, and (3) adding water to supplement the amount of natural precipitation.

Water Conservation

Water conservation is important where large water deficits occur in soils with aridic, ustic, and xeric soil moisture regimes. Most wheat is produced on these soils. Techniques for water conservation aim at increasing the amount of water that enters the soil and making better use of this water.

Effect of Surface Soil Conditions on Infiltration

Infiltration is the movement of water into the soil surface. The nature of the pores and water content are the most important factors determining the amount of precipitation that infiltrates and the amount that runs off. High infiltration rates, therefore, not only increase the amount of water stored in the soil for plant use, they also reduce flood threats and erosion resulting from runoff.

Raindrop impact on bare soil (*see* Fig. 4-1) breaks up soil aggregates and causes the average pore size in the surface soil to decrease; this decreases infiltration. Infiltration is also decreased by overgrazing, deforestation, and soil compaction resulting from traffic. The presence of a vegetative cover that absorbs raindrop impact is effective in maintaining a high infiltration rate in a given situation.

The presence of crop residues or other organic matter on the surface of the soil has the same effect as a living vegetative cover. At Hays, Kansas, the infiltration during a storm was 2.9 centimeters, where the stubble of a previous wheat crop had been left on the land compared to 1.8 centimeters where the stubble had been burned. Standing stubble is also effective for trapping snow. Efforts to trap snow are used extensively in the newly developed virgin lands of Kazakhstan in the Soviet Union, where natural precipitation is adequate for about one good crop of wheat every 4 years. Narrow strips of plants are grown perpendicular to the winds to trap snow.

Many farmers modify the soil surface with contour tillage and terraces that hold water on the land for a longer time, which produces a greater opportunity for infiltration (*see* Fig. 4-2).

Figure 4-1
Raindrops falling on bare soil. *(a)* A drop just before striking the soil. *(b)* Just after the drop has struck the soil. *(c)* Scattering of soil particles in all directions. The scattered particles form a compact layer that reduces infiltration. (Photo courtesy of USDA.)

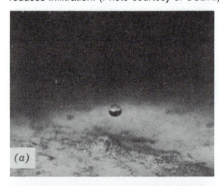

Effect of Internal Soil Properties on Infiltration

Typically, when water infiltrates in the field, the immediate surface of the soil is water saturated. After infiltrating, the wa-

Figure 4-2
Lister furrows on the countour hold water on the land to increase infiltration on this Oklahoma field. Runoff and erosion are reduced. (Photo courtesy of USDA.)

ter moves downward as unsaturated flow that is dependent on water potential gradients and soil conductivity. Soils with a high content of expanding clay, as in the Blacklands of Texas, develop large cracks in the dry season that permit water from intense storms to move quickly as saturated flow deep into the dry soil without runoff. When these soils become wet in the rainy season however, infiltration approaches zero and nearly all the rainfall runs off.

Many Great Plains soils have well-developed argillic horizons that limit the downward movement of water. It would seem that the use of deep tillage to disrupt these horizons would increase infiltration. Numerous studies were conducted from 1909 to 1916 at 12 different locations. Results at most locations showed that there was no increase in wheat yields as a result of deep tillage. Deep tillage may be effective in some cases, but this is uncommon. When one considers the extra cost, it is unlikely that any increases in yield would be able to pay the extra cost for deep tillage on the Great Plains.

Summer Fallowing for Water Conservation

Many soils have ustic soil moisture regimes in the wheat-growing areas of the western United States and Canada. They receive too little water to produce a profitable crop every year. Summer fallowing is used to increase soil water storage so that a profitable crop can be produced every other year or so. After wheat harvesting, the land is left fallow (no crop is grown) to accumulate soil moisture. Weeds are controlled by cultivation to prevent water loss by transpiration. The land in these areas has a characteristic pattern of alternating strips of bare fallow land and land used for wheat production (see Fig. 4-3). An explanation of how water storage occurs in fallow land in dry regions depends on a consideration of both evaporation of water from the soil surface and water movement within the soil.

Any rain on dry soil immediately saturates and infiltrates the soil surface, and the excess water moves downward as unsaturated flow. The depth of water penetration depends on the texture and the amount of water that infiltrates. Each cen-

Figure 4-3
Characteristic pattern of alternate fallow and wheat strips where land is summer fallowed for wheat production.

timeter of water that infiltrates will be able to moisten about 6 to 8 centimeters of loamy soil near the wilt point (*see* Fig. 4-4). Water will evaporate from the soil surface after the rainfall. Some water in moist soil below will migrate upward as an unsaturated flow of liquid water to keep the soil surface moist. This phase of soil drying is characterized by a rapid water loss and is represented by horizontal part of the curve shown in Fig. 4-5. As water below the surface migrates upward and is lost by evaporation, hydraulic conductivity decreases rapidly. After about 1 day of dry, sunny weather the conductivity of the upper surface soil becomes so small, and water moves upward so slowly, that the soil surface dries. A sharp boundary is created between the dry surface soil layer and the moist soil below. Drying of the soil surface causes a sharp decline in water loss from the soil (*see* Fig. 4-5).

Once the surface soil dries, water can move from the underlying moist soil to

— Moist soil with about
− 0.3 bar matric potential

__ Boundary between moist
and dry soil

— Dry soil

Figure 4-4
The moist upper soil layer has been recharged by a recent rain. Note the sharp boundary between the moist upper and dry lower layers. If another rain occurs, the soil will be recharged to a greater depth.

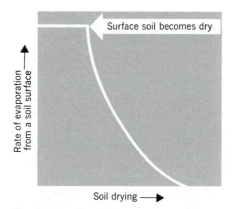

Figure 4-5
Evaporation of water from a bare soil surface decreases sharply when the surface soil becomes dry, thus conserving the moisture remaining in the soil (Adapted from Evans and Lemon, 1957.)

and through the dry thin surface layer only as vapor flow. This vapor movement is so slow as to be largely discounted (unless cracks exist). This produces a "capping" feature, which has a great significance for water conservation. Once the surface of the soil has become dry after a rain (with or without cultivation), the wa-

ter in the underlying moist layers is largely protected from loss by evaporation, when the time considered is a few weeks or months. If another rain occurs, the soil may be moistened to a greater depth but, when the surface of the soil again becomes dry, the water is again trapped in the soil. Repetition of this sequence of events during the fallow period progressively increases the amount of moisture stored in the soil. A sequence of events during a fallow period is illustrated in Fig. 4-6. Many studies have shown a close correlation between water stored at planting time and grain yield.

The fallowing system is not 100 percent efficient, because the self-mulching effect does not work perfectly; some runoff and evaporation occur. A good estimate is that about 25 percent of the rainfall during the fallow period will become stored in the soil for use in crop production. This extra quantity of water, however, has a great effect on yields (Table 4-1). Yields over 1,345 or 2,690 kilograms per hectare were significantly increased by summer fallowing. Fallowing was also more effec-

Figure 4-6
Changes in soil water during a summer fallow period for wheat production.

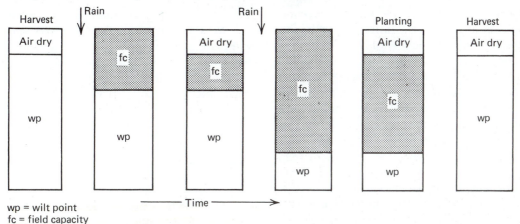

wp = wilt point
fc = field capacity

Table 4-1

Percentage Distribution of Wheat Yield Categories at Two Locations in the Great Plains and One Location in the Columbia River Basin

Yield Category	Wheat After Wheat, %			Wheat on Fallowed Land, %		
	Akron, Colo.	Hays, Kans.	Pendleton, Ore.	Akron, Colo.	Hays, Kan.	Pendleton, Ore.
Under 336 kg/ha (5 bu/A)[a]	44	33	0	15	17	0
Under 1,345 kg/ha (20 bu/A)	88	62	100	61	36	0
Over 1,345 kg/ha (20 bu/A)	12	38	0	39	64	100
Over 2,690 kg/ha (40 bu/A)	0	5	0	5	12	83

From Mathews, O. R., [a] A = acre. 1951

tive at Pendleton, Oregon, where maximum rainfall occurs in winter (xeric soil moisture regime), as compared to Akron and Hays, where maximum rainfall occurs in the summer (ustic soil moisture regime). Stored soil moisture has been shown to be as effective as precipitation during the growing season for wheat production (*see* Fig. 4-7).

Saline Seep Due to Fallowing

Most of the world's spring wheat is grown in areas where winters are severe, such as the plains areas of the U.S.S.R., Canada and United States. Low temperature results in low evapo-transpiration, which increases the efficiency of water storage during the fallow period. In fact, if a fallow period is followed by a wet year, surplus water may occur and percolate downward and out of the root zone. Where surplus water encounters an impermeable layer and saturates the soil, water moves laterally; it may appear as a spring or seep at a lower elevation. These seep areas remain wet and, as water evaporates, soluble salt accumulates at the soil surface to create *saline seep* areas. The osmotic effect of soluble salts reduces soil water potential

and water uptake; wheat yields are reduced. In some cases, the soils are too saline for plant growth.

Factors that encourage saline seep development include frequent fallowing during periods of above normal rainfall on sandy soils with a low water retention capacity. Any practice that increases water infiltration increases saline seep development, including the use of snow fences to trap snow or construction of ponds to collect runoff water. Control of saline seep depends on reducing the likelihood that surplus water will occur. Management practices to control saline seep include: (1) reduced fallowing frequency, (2) use of fertilizers to increase plant growth and water use, (3) and growth of deep-rooted perennial crops, such as alfalfa, to remove soil water. When alfalfa is grown and has dried out the soil, it grows very slowly, which indicates it is time to begin wheat production.

Effect of Fertilizers on Water Use Efficiency

Plants growing in a medium that contains relatively small quantities of nutrients appear to grow slower and to transpire more

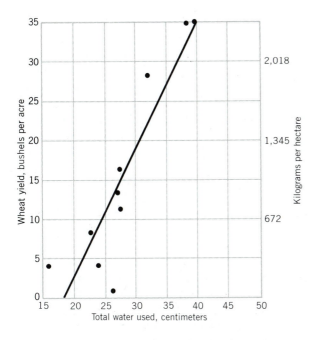

Figure 4-7
Wheat yield and moisture use (stored soil water plus precipitation) at Edgely, N.D. At least 17 centimeters of water was needed for some grain production; each additional 2.5 centimeters of water used increased grain yield by 270 kilograms per hectare (4 bushels per acre). (Data from Mathews, 1940.)

water per gram of plant tissue produced than those growing in a medium that contains an abundance of plant nutrients. Since it has been shown that water loss is mainly dependent on the environment, any management practice that increases the rate of plant growth will tend to result in more dry matter produced per unit of water used. The data in Table 4-2 show that the yield of oats was increased from 86 to 145 kilograms per hectare for each 2.5 centimeters of water used by the addition of fertilizer. In humid regions, it has been commonly observed that fertilized crops are more drought resistant. This may be explained on the basis that increased top growth results in increased root growth and root penetration so that the total water consumed is greater (Fig. 4-8).

Table 4-2
Oats Production per 2.5 Centimeters of Water Used

Year	Low Nitrogen		High Nitrogen	
	Kilograms per Hectare	Bushels per Acre	Kilograms per Hectare	Bushels per Acre
1949	75	2.1	158	4.4
1950	97	2.7	133	3.7
Average	86	2.4	145	4.0

From Hanks and Tanner, 1952.

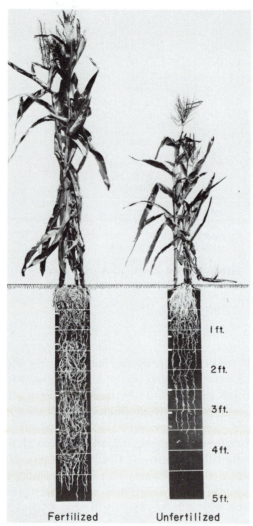

Figure 4-8
Fertilizer increased the corn yield nearly 4-fold on the Cisne soil of southern Illinois. Fertilizer use resulted in a deeper rooting depth and a greater use of soil water (and nutrients). (Photo courtesy of Dr. J. Fehrenbacher, University of Illinois.)

Fertilizer used in a subhumid region may occasionally decrease yields. If fertilizer causes a crop to grow faster early in the season, the greater leaf area at an earlier date results in greater loss of water by transpiration. If no rains occur, or no irrigation water is applied, plants could run out of water near harvest and a serious reduction in grain yield could occur. Work in Kansas has shown this to be the case for grain sorghum. Forage, when fertilized, does not depend on a supply of water to the end of the growing period; it can be produced the same as in the humid region, when fertilized, with less water per unit of forage produced. The forage will stop growing when the water is exhausted, after having used the water efficiently as long as it lasted.

This concept may have far-reaching consequences in the future for crop production. Allocation of water to those situations resulting in the most efficient use of the water, in terms of crop yield per unit of water consumed, may be required if supplies become sufficiently limited.

Soil Drainage

About 1200 A.D., farmers in the Netherlands became engaged in an interesting water control problem. Small patches of fertile soil affected by tide and flood waters, were enclosed by dikes and drained. Windmills provided the energy to lift gravitational water from drainage ditches into higher canals for return to the sea. Now, 800 years later, about one third of the Netherlands is protected by dikes and kept dry by pumps and canals. The use of drainage systems to remove water from soil is important throughout the world, where the water table is close enough to the surface so that saturated soil occurs within the root zone of plants, foundations of engineering structures, or septic tank drain fields. We will discuss the principles underlying these problems in this section.

Properties and Distribution of Aquic Soils

Aquic soils have aquic soil moisture regimes and develop under the influence of water saturation. As a result, aquic soils have dark-colored surface horizons high in organic matter content, are commonly on level land with minimal erosion hazard, and are frequently fine textured with high native fertility. Aquic soils comprise some of the best agricultural soils after drainage (*see* Fig. 4-9).

Small areas of wet or aquic soils are widely distributed locally. Extensive areas occur on the coastal plain of the southeastern United States because of high precipitation and elevation near sea level. Other major areas include the Mississippi River Valley, the lake plains in the Midwest, the recently glaciated, nearly level plains of northern Iowa and southern Minnesota, and the Red River Valley between Minnesota and North Dakota.

Effect of the Water Table on Air and Water Content of Soil

If a well is dug and the lower part fills with water, the top of the water in the well is the top of the water table. Water does not rise in a well above the water table. In the adjacent soil, however, capillarity causes water to move upward in soil pores, and a water-saturated soil zone is created above the water table unless the soil is so gravelly and stony that capillarity does not occur. The water-saturated zone above the water table is the *capillary fringe* (*see* Fig. 4-10).

Soils typically have a range in pore sizes. Some pores will be so large that they are air filled above the capillary fringe. Pores will be water filled, depending on pore sizes. This creates a zone above the capillary fringe that has a decreasing wa-

Figure 4-9
Large, lateral subsurface drain emptying into a surface drainage ditch on a lake plain in the Corn Belt of the United States.

ter content and an increasing air content, with distance upward from the capillary fringe. This capillary affected zone, plus capillary fringe, generally range in thickness from several centimeters to several meters. For all practical purposes, soil more than about 2 meters above the water table is unaffected significantly by the water table. This means that most plants, even in humid regions, do not use water from the water table. In most cases, the soil above the capillary fringe will have a water content governed by the balance between infiltration and evapo-transpiration and percolation. Soils with aridic moisture regimes will have a permanently dry zone between the root zone and the water table. These relationships are shown in Fig. 4-10.

Benefits of Soil Drainage

Root tips are regions of rapid cell division and elongation; they have a high oxygen requirement. Typically, roots of most crop plants do not penetrate water-saturated soil because of oxygen deficiency. The major purpose of drainage in agri-

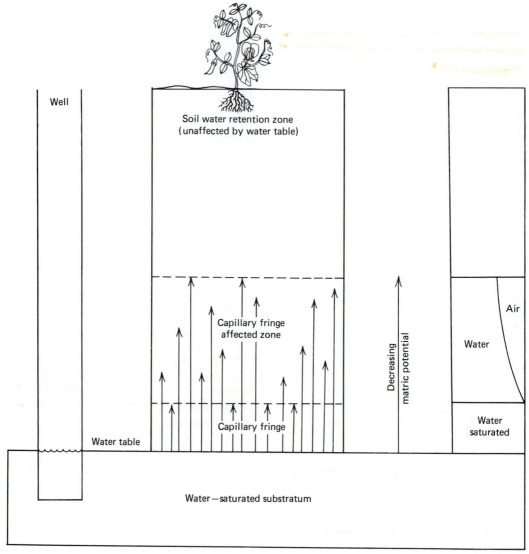

Figure 4-10
Generalized relationship of the water table to the soil moisture zones, and the air and water content of soil above the water table. The capillary fringe is water saturated.

culture and forestry is lowering the water table to increase the depth of rooting.

Some of the most phenomenal increases in growth, resulting from drainage of aquic soils, occur in forests. Drainage on the Atlantic Coastal Plain increased the growth of pine from 80 percent to 1,300 percent (Table 4-3). There are few, if any, management practices as effective as drainage for increasing tree growth. Ad-

Table 4-3

Effect of Drainage on Mean Annual Growth of Pine on Aquic Soils of the Atlantic Coastal Plain

| Species | Ages | Mean Annual Growth[a] | | Percent Increase Over Undrained |
		Drained	Undrained	
Planted loblolly	0–17	17.9	1.28	1,298
Natural pond	0–22	0.74	0.41	80
Natural slash	19–22	9.0	4.9	84
Planted slash	0–5	14.6	5.7	156
Planted slash	0–5	13.3	5.7	133
Planted loblolly	0–5	7.6	1.3	585
Planted loblolly	0–5	4.9	1.3	277
Planted loblolly	0–13	4.3	0.54	696

Adapted from Terry and Hughes, 1975.

[a]Cubic meters per hectare.

ditional benefits of drainage in forests are easier logging, less soil disturbance during logging, and easier site preparation for the next crop.

Drainage of wet soils also increases the length of the growing season in regions where low soil temperature and frost hazard restrict plant growth. Drainage lowers the water content in the spring, causing soils to warm more rapidly. The higher, early soil temperature is associated with drier soil that can be tilled earlier so crops can be established sooner. Seed germination is more rapid and roots grow faster. The net result is a greater potential for plant growth.

Surface Drainage

Surface drainage is the collection and removal of water from the surface of the soil. Two conditions favoring the use of surface drainage are (1) low areas that receive water from surrounding higher land and (2) impermeable soils that have insufficient capacity to dispose of the excess water by movement downward through the soil profile. Surface drainage is a good choice on the Sharkey clay soils along the lower Mississippi River. The soil is very slowly permeable and receives water from surrounding higher land. Sugar cane is widely grown and sensitive to poor aeration. The ridges on which the sugar cane is planted also provide improved soil aeration for roots.

It is difficult to irrigate without applying excess water in many instances. Surface drainage ditches are used to dispose of excess water to prevent saturation of soil on the low end of irrigated fields. Surface drainage is also widely used along highways and in urban areas for water control. Rapid removal of surface water from low areas of golf courses is essential to permit golfing soon after a rain.

Subsurface Drainage

Ditches can be quickly and inexpensively made to remove gravitational water. Drainage ditches, however, require periodic cleaning and are inconvenient for the use of machinery. There are also many situations where ditches are not satisfactory, as in the case of water removal around the walls of the basement of a building (*see* Fig. 4-11). The drainage tiles shown in Fig. 4-11 are made of fired clay and are laid side by side, with a small crack between adjacent tiles. When the soil surrounding the tile is saturated with water, water seeps into the tile laid on a grade. The water eventually reaches an outlet where it is disposed.

Drain tile are installed in fields with trenching machines. Perforated plastic tubing for small-diameter laterals is less expensive than ceramic tile and is gaining in popularity. Some drainage machines

Drainage tile

Figure 4-11
Clay tile lain at the base of the foundation of the walls of a basement. Cracks between adjacent tile are covered with a durable black material to keep soil from entering the tile.

carry rolls of plastic drainage tubing that is automatically laid at the bottom of the trenches as the machine moves across the field. Unless some unusual condition prevails, the tile or plastic tubing should be laid at least 1 meter deep to have a sufficiently low water table between the drains to permit crops to develop an adequate root system (*see* Fig. 4-12).

Drainage in the Soil of Container Grown Plants

Most of us grow plants in containers that we use to decorate rooms. One of the most common problems of growing indoor plants is poor soil aeration caused by overwatering. Perhaps you wonder how this could be true, since most flower pots have a large drainage hole in the bottom. We can use our knowledge of changes in air and water content of soil above the water table to help us understand this problem.

Consider a flower pot filled with soil, with a range in sizes of pore spaces, that is water saturated. Then, water is allowed to drain out of the large hole in the bottom of the pot. Gravitational water will exit rapidly from the largest soil pores at the top of the pot and air will be pulled into these pores. Over time, the soil will become progressively unsaturated from the top downward. In a few minutes, the

loss of gravitational water may essentially stop. At this time, some of the soil at the bottom may be water saturated, if all the pores in the soil are sufficiently small to attract water more strongly than gravity. Then, the air and water content of the soil, above the saturated soil at the bottom, is analogous to soil above the saturated capillary fringe (compare Fig. 4-13 with Fig. 4-10).

If all soil pores are very small, all the soil in the flower pot would likely remain water saturated after watering. A good soil for container grown plants has many large pores that can not retain water against gravity. After wetting and drainage, these pores will be air filled, even at the bottom of the flower pot, in the same way the space in a well that is above the water table is totally filled with air. Roots in tall containers will have a better-aerated environment than roots in shallow containers, if the water and air content vary as shown in Fig. 4-13.

Special attention must be given to soil used for container grown plants. Ordinary loamy garden soils are unsatisfactory. A mix of one-half fine sand and one-half sphagnum peat moss, by volume, is recommended by the University of California to produce a mix with excellent physical properties. If you want to prepare a mix using a garden or lawn soil, add only a small amount (perhaps 10 per-

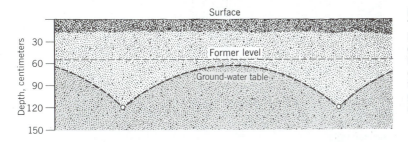

Figure 4-12
The effect of tile lines in lowering the water table or ground-water level. The benefit of drainage is first evident immediately over the tile lines, and it gradually spreads to the soil area between them.

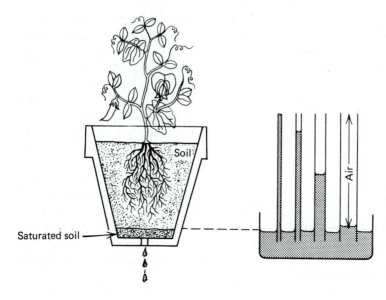

Figure 4-13
Illustration of conditions in a flower pot after a thorough watering and drainage. At the bottom of the pot, the soil is saturated. Soil-air content increases and water content decreases with distance upward above the water-saturated layer.

cent soil by volume) to some medium or fine sand. When equal amounts of soil and sand are mixed together, the small particles fill the spaces between the sand particles, producing a low-porosity mix with few large aeration pores (*see* Fig. 4-14).

Stratified soils interfere with water movement. Coarse material like gravel, with very small water conductivity when moist or dry, should not be placed in the bottom of holes dug for transplanting trees and shrubs, because a water-saturated zone will be formed above the

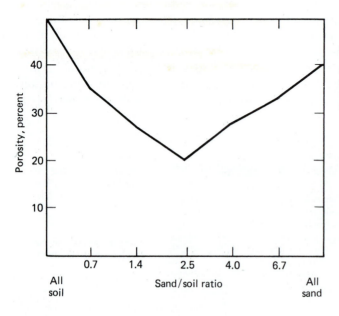

Figure 4-14
Effect of different mixtures of soil and sand on porosity. Lowest porosity occured with a sand/soil volume ratio of 2:5. (Based on data from Spomer, 1975.)

gravel layer in wet seasons. Blending of soil in the container with the surrounding soil will encourage better water movement and root extension from the container grown soil into the surrounding soil. This effect of a coarser layer underneath a finer layer was illustrated in Fig. 3-10.

Drainage of Septic Tank Effluent Through Soils

People who reside beyond the limits of municipal sewer lines have used septic-tank sewage disposal systems for many years. The recent rapid expansion of rural residential areas and the development of summer homes near lakes and rivers has greatly increased the role of soil in waste disposal. The sewage enters a septic tank where solid material is digested, and the liquid effluent flows out of the top of the septic tank and into the tile lines of a filter field (*see* Fig. 4-15). The seepage lines are laid in a bed of gravel from which the effluent seeps into and drains through the soil.

A major factor influencing the suitability of soil for filter-field use is permeability or *saturated* hydraulic conductivity. Soils with low hydraulic conductivity are unsuited for septic drain fields. Sands may have a too high hydraulic conductivity, because the sewage effluent may move so rapidly that disease organisms are not destroyed before shallow water supplies become contaminated.

Septic tank filter fields should never be installed in poorly drained soils, because the soil will become saturated with water; there is a danger that disease organisms in the effluent will contaminate work or play areas. The basements of homes constructed on poorly drained soils are likely to become flooded.

Irrigation

Irrigation is an ancient agricultural practice that was used 7,000 years ago in Mesopotamia. Other ancient, notable irrigation systems were located in Egypt, China, Mexico, and Peru. Today, about 11 percent of the world's cropland is irrigated. Some of the world's densest populations

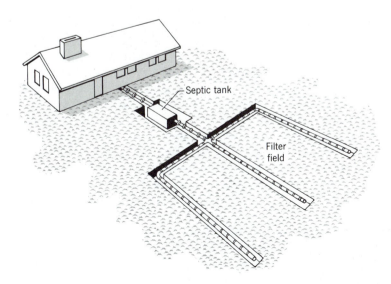

Figure 4-15
The layout for a septic tank and the filter field.

are supported by producing crops on irrigated land, as in the United Arab Republic (Egypt), where 100 percent of the cropland is irrigated. Other countries that have large percentages of irrigated cropland include: (1) Peru, 75 percent, (2) Japan, 60 percent, (3) Iraq, 45 percent, and (4) Mexico, 41 percent. Over 12 percent of the cropland, or about 24,000,000 hectares (61,000,000 acres), are irrigated in United States.

Nearly two thirds of the world's population lives in diet deficient countries having less than half of the world's arable land but with three quarters of the irrigated land.[1]

Thus, we can see the great importance of irrigation in the world today and for the future.

Water Sources

Most irrigation water is surface water resulting from rain and melting snow. Many rivers in the world have their headwaters in mountains; they flow through arid or semiarid regions. Examples include the Indus River, which starts in the Himalaya Mountains, and the numerous rivers on the western slope of the Andes Mountains, which flow through the desert of Peru to the Pacific Ocean. Much of the water in these rivers comes from the melting snow in the high mountains. In fact, the extent of the snow pack is measured to obtain information on streamflow and the amount of water that will be available the next season for irrigation (*see* Fig. 4-16).

Many farmers have their own source of irrigation water, because they use well wa-

Figure 4-16
Melting snow from the Rocky Mountains supplies much of the water for this irrigation reservoir in the United States.

ter. About 20 percent of the water currently used in the United States comes from irrigation wells.

Selecting Land for Irrigation

In choosing land for irrigation, a careful examination should be made of the soil to determine: (1) texture of soil to a depth of 1 to 2 meters, (2) presence of an impermeable stratum or of gravel within a depth of 2 meters, (3) accumulation of soluble salts in injurious quantities, (4) slope and evenness of the soil surface, and (5) behavior of the soil under irrigation. A desirable soil is readily permeable to water and, yet, is moisture retentive. Infiltration rates should be in the range of 0.5 to 8 centimeters per hour. It is good if the soil will absorb sufficient moisture in 24 hours to wet it to a depth of 50 or 80 centimeters. Some soils are so slowly permeable that they become wet to a depth of only 30 centimeters or less in 24 hours. Some soils are so coarse textured that little available moisture is retained. On the other hand, where fine- or medium-textured soil materials are underlain by coarser sands or gravels, greater retention of water in the upper layer occurs than if the

[1]The White House, *The World Food Problem*, A Report of the President's Science Advisory Committee, May 1967, p. 442.

soil was medium or fine textured throughout. The importance of soil properties in placing irrigated soils into land capability classes is illustrated in Table 4-4.

The land surface should be comparatively smooth, because the cost of leveling land is high (*see* Fig. 4-17). A uniform slope of 2 to 4 meters to the kilometer is desirable, although much steeper slopes are in use. Land cut up by ravines, gullies, or buffalo wallows or covered with sand dunes and hummocks should be avoided if possible.

Methods of Applying Irrigation Water

Choice of the various methods of applying irrigation water is influenced by a consideration of: (1) seasonal rainfall, (2) slope and general nature of the soil surface, (3) supply of water and how it is delivered, (4) crop rotation, and (5) infiltration rate. The methods of distributing water can be classified as surface, subsurface, sprinkler, and drip or trickle. A brief discussion of each of these methods is given.

Figure 4-17
Planned land with proper grade to give uniform distribution of water down the rows. Gated pipe is being used to distribute the water.

Table 4-4
Guide For Placing Irrigated Soils into Land Capability Classes (The Guide Shows Soil Properties of Importance in Irrigation)

| Land Capability Class | Surface Texture | Available Water-Holding Capacity, cm | | | Soil Permeability | Effective Depth, cm | Salinity or Sodium Hazard |
		Coarse Fragments in Surface, %	Surface 30 cm	Soil Profile to 150-cm depth			
I	Sandy loam to clay loam or silty clay loam	<15	>4	>20	Slow to medium	>100	None
II	Loamy sand to silty clay loam or clay loam	15–35	2.5–4	12–20	Slow to rapid	75–100	Slight
III	Sand to clay	>35	2.0–2.5	10–12	Slow to rapid	50–75	Moderate
IV	Sand to clay	No limits	<2	<10	Slow to rapid	<50	Severe
V, VI, VII, VIII	Sand to clay	No limits	No limits	No limits	Slow to rapid	<50	No limits

Adapted from *Soil Conservation Service*, USDA, Phoenix, Arizona State Guide, by Donald Post.

Surface Irrigation. Surface irrigation distributes water down rows or into basins and similar areas that are surrounded by ridges or bunds. Flooding of basins and similar areas is used for pastures, orchards, cereal crops such as wheat and rice (paddy), and the like. Crops commonly irrigated by furrow irrigation include row crops, such as potatoes, sugar beets, corn, grain sorghum, sugar cane, cotton, vegetables, and fruit trees. Furrows are made across the field, leading down the slope. Water is let into the upper end of the furrow from a "head ditch" or pipeline running across the end of the field. Siphon tubes are commonly used to transfer the water from the head ditch into the furrows (*see* Fig. 4-18). Gated pipe is also used extensively (*see* Fig. 4-17).

Subirrigation. Subirrigation is irrigation by water movement upward from a free water surface that is some distance below the soil surface. In arid regions, where almost all of the water used to grow crops is from irrigation, subirrigation would cause serious salt accumulation problems in the upper part of the soil. Subirrigation works best where natural rainfall removes any salts that may accumulate. In many poorly drained areas, as in the Sand Hills of Nebraska, there are areas where subirrigation is a natural occurrence. Artificial subirrigation is practiced in the Netherlands, where tile drainage systems in polders are used in wet seasons for drainage and in periods of drought for subirrigation. Subirrigation also works well on the nearly level Florida coastal plain. The sandy soils have high

Figure 4-18
Use of siphon tubes to transfer water from the head ditch into furrows. (Photo courtesy of USDA.)

saturated conductivity, and the natural water table is 1 meter or less within the soil surface much of the time. Subirrigation, as compared to other systems, is inefficient in use of water and is adapted only for special situations.

Sprinkler Irrigation. Everyone is familiar with the sprinklers used to water or irrigate lawns. Sprinkler systems are versatile and have special advantages where there are very high infiltration rates or the topography prevents proper leveling of the land for surface distribution of water. The rate of water application can also be carefully controlled. The portable nature of many sprinkler systems makes them ideally suited for use where irrigation water is used to supplement the nat-

ural rainfall. When connected to a soil-moisture measuring apparatus, sprinkler systems can be made automatic.

Large self-propelled sprinkler systems, with a center pivot, have been developed in recent years. Such systems can irrigate most of the land in a quarter-section (65 hectares or 160 acres) as shown in Fig. 4-19. If you have flown over the western part of the United States, you have probably seen large green circular irrigated areas produced by pivot sprinklers.

Sprinkler irrigation modifies the plant environment by completely wetting the soil and leaves. Reductions in temperature reduce water stress in plants. A very small amount of water applied with sprinklers has been observed to reduce midday surface soil temperature as much as 12°C

Figure 4-19
Circular patterns made by self-propelled pivot sprinklers in Holt County, Nebraska. A unit is operating in the front-left quarter section. (Photo courtesy of USDA-SCS.)

(22°F) (*see* Fig. 4-20). The high, specific heat of water makes sprinkling an effective means of reducing frost hazard. Sprinkler irrigation has been used for frost protection in such diverse situations as strawberries in Michigan and grapes in California.

Drip Irrigation. ==Drip irrigation is the frequent or almost continuous application of "small" amounts of water to localized areas of soil in a field.== Plastic hoses about 1 or 2 centimeters in diameter, with emitters, are placed down the rows or around trees (*see* Fig. 4-21). Only a small amount of the root zone is wetted, but roots in the localized, moistened areas can rapidly ab-

Figure 4-21
A section of plastic tubing and emitter. Rate of water application can be adjusted from less than 1 gallon to about 4 gallons per hour per emitter.

sorb the water, which has high matric potential. A major advantage of drip irrigation is the large reduction in the total amount of water used. Other advantages include a more uniform soil-water potential during the growing season and adaptability of the system to very steep land where other methods are unsuited.

Rate and Timing of Irrigation

An ideal application of water would be a sufficient quantity to bring the soil (to the depth of the root zone of the crop) up to its field capacity. More water may result in the waterlogging of a portion of the subsoil or in the loss of water by drainage. If much water percolates below the root zone, it may accumulate under low areas, thus raising the water table unless suitable drainage facilities are provided. On the other hand, unless enough water is applied to result in appreciable drainage, it is difficult to remove excess salts from soils in which there is a tendency for salt to accumulate.

An accepted generalization is that it is

Figure 4-20
Temperature profile above and within irrigated and nonirrigated muck soil at 2:00 P.M., June 9, 1964 at East Lansing, Michigan. (Data from Brink and Carolus, 1965.)

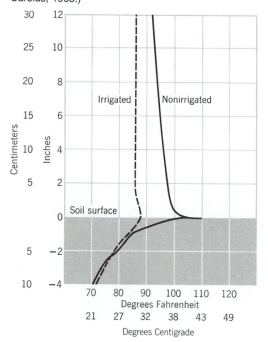

time to irrigate when 50 to 60 percent of the available water in the root zone has been used. Potentiometers are used to measure the matric potential in various parts of the root zone; they are used regularly to time the application of water. Computer programs are available for irrigation scheduling, based on type of soil and crop and weather information. Farmers commonly irrigate according to the appearance of the crop. Crops such as beans, cotton, and peanuts develop a dark-green color under moisture stress. Wilting symptoms are also used; however, in most cases, crops should be irrigated before marked wilting occurs. When irrigation water is in short supply, it may be better to curtail water use early in the season (delay irrigation), so that sufficient water is available during the reproductive stages of plant growth (when water stress is most injurious to crop yields).

Water Quality

Water in the form of rain and snow is quite pure. By the time the water has reached farm fields, however, the water has picked up soluble materials from the soils and rocks over and through which the water moved. The four characteristics that appear to be the most important in determining water quality are: (1) total salt concentration, (2) relative proportion of sodium relative to other cations, (3) concentration of boron and other elements that might be toxic to plants, and (4) the bicarbonate concentration.

Total Salt Concentration.
The total salt concentration or salinity can be expressed in terms of the electrical conductivity of the water; it is easily and precisely determined. Conductivity is expressed as mi-

cromhos per centimeter[2]. One micromho per centimeter is equal approximately to 750 parts per million, 10 milliequivalents per liter, and 0.01 normality. Four water quality classes have been established, ranging from 100 to 250 micromhos per centimeter for Class One, with low salinity hazard, to Class Four with salinity greater than 2,250 micromhos per centimeter and a very high salinity hazard. The conductivity of the four classes are given along the bottom scale of the diagram in Fig. 4-22.

The four classes of water have the following general characteristics. Low-salinity water (C1) can be used for irrigation with most crops, with little likelihood that soil salinity will develop. Some leaching is required, but this occurs under normal irrigation practices, except in soils with extremely low permeability.

Medium-salinity water (C2) can be used if a moderate amount of leaching occurs. Plants with moderate salt tolerance can be grown in most cases without special practices for salinity control.

High-salinity water (C3) can not be used on soils with restricted drainage. Even with adequate drainage, special management for salinity control may be required. Plants with good salt tolerance should be selected.

Very high-salinity water (C4) is not suitable for irrigation under ordinary conditions, but may be used occasionally under very special circumstances. The soils must be permeable, drainage must be adequate,

[2]The mho is a unit of conductivity that is the reciprocal of the ohm, which is a measure of resistance. A millimho is 10^{-3} mho and a micromho is 10^{-6} mho. The conversion of millimhos per centimeter to siemens per meter in the International System of Units (SI) is:

millimhos per cm $\times$ 0.1 = siemens per meter.

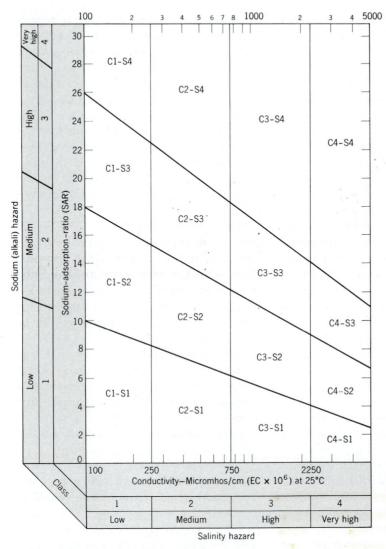

Figure 4-22
Diagram for the classification of irrigation waters. (Data from *USDA Agriculture Handbook,* 60, 1954.)

irrigation water must be applied in excess to provide considerable leaching, and very salt-tolerant crops should be selected.

Sodium Adsorption Ratio. The relative proportion of sodium, relative to other cations, is expressed by the sodium adsorption ratio (SAR). The SAR is calculated as follows:

$$SAR = \frac{Na^+}{\sqrt{\dfrac{(Ca^{2+} + Mg^{2+})}{2}}}$$

where the concentrations of the cations are given in milliequivalents per liter. The classification of irrigation waters with respect to SAR is based primarily on the effect of exchangeable sodium on the physical condition of the soil. Sodium-sensitive plants may, however, suffer injury as a result of sodium accumulation in plant tissues, when exchangeable sodium values are lower than those effective in causing deterioration of the physical condition of the soil. The SAR values of the four sodium hazard classes (S1, S2, S3, and S4) are given along the vertical scale of the diagram in Fig. 4-22.

Low-sodium water (S1) can be used for irrigation, on almost all soils, with little danger of the development of harmful levels of exchangeable sodium. However, sodium-sensitive crops such as stone-fruit trees and avocados may accumulate injurious concentrations of sodium.

Medium-sodium water (S2) will present an appreciable sodium hazard in fine-textured soils having high cation exchange capacity (especially under low-leaching conditions), unless gypsum is present in the soil. This water may be used on coarse-textured or organic soils with good permeability.

High-sodium water (S3) may produce harmful levels of exchangeable sodium in most soils and will require special soil management, good drainage, high leaching, and organic matter additions. Gypsiferous (high-calcium sulfate) soils may not develop harmful levels of exchangeable sodium from such waters. Chemical amendments may be required for replacement of exchangeable sodium, except that amendments may not be feasible with waters of very high salinity.

Very high-sodium water (S4) is generally unsatisfactory for irrigation purposes except at low and perhaps medium salinity, where the solution of calcium from the soil or the use of gypsum or other amendments may make the use of these waters feasible.

Sometimes, the irrigation water may dissolve sufficient calcium from calcareous soils to decrease the sodium hazard appreciably. This should be taken into account in the use of C1–S3 and C1–S4 waters. For calcareous soils with high pH values, or for noncalcareous soils, the sodium status of waters in classes C1–S3, C1–S4, and C2–S4 may be improved by the addition of gypsum to the water. Similarly, it may be beneficial to add gypsum to the soil periodically when C2–S3 and C3–S2 waters are used.

Boron Concentration. Boron is essential to normal plant growth, but the amount needed is very small. Boron is toxic to certain plants, and the concentration that will injure these sensitive plants is often approximately the amount required for the normal growth of very tolerant plants. The occurrence of boron in toxic concentrations in certain irrigation waters makes it necessary to consider this element in assessing the water quality. Permissible limit of boron in several classes of irrigation water is given in Table 4-5 and the relative tolerance of some plants is given in Table 4-6.

Bicarbonate Concentration. In waters containing high concentrations of bicarbonate ion (HCO_3^-), there is a tendency for calcium and magnesium to precipitate as carbonates, as the soil solution becomes more concentrated. This reaction does not go to completion under ordinary circumstances, but insofar as it does proceed, the concentrations of calcium and magnesium are reduced and the relative proportion of sodium is increased.

In making an estimate of the quality of water, the effect of salts on both soil and plant must be considered. Various factors such as drainage, soil texture, and the kind of clay minerals present influence

Table 4-5

Permissible Limits of Boron for Several Classes
of Irrigation Waters in Parts per Million

Boron Class	Sensitive Crops	Semitolerant Crops	Tolerant Crops
1	<0.33	<0.67	<1.00
2	0.33–0.67	0.67–1.33	1.00–2.00
3	0.67–1.00	1.33–2.00	2.00–3.00
4	1.00–1.25	2.00–2.50	3.00–3.75
5	>1.25	>2.50	>3.75

From *Agriculture Handbook* 60, USDA, 1954.

the effects on soil. The ultimate effect on a plant is the result of these two and other factors that operate simultaneously. Consequently, there can be no method of interpretation that is absolutely accurate under all conditions. The schemes for interpretation are accordingly based chiefly on experience.

Salt Accumulation and Plant Response

Soil salinity from the application of irrigation water is not a problem in regions where rainfall, at some time during the year, leaches soils. Most irrigation, however, is in low rainfall regions where leaching is limited and salt tends to accumulate in irrigated soils. The demise of some early civilizations, such as Mesopotamia, coincided with salt accumulation in irrigated lands. Crop production today is seriously limited by soil salinity. More than 50 percent of the world's irrigated land has developed salinity and/or drainage problems (*see* Fig. 4-23).

Saline soils contain sufficient salt to impair plant growth. The salts are mainly chlorides, carbonates and sulfates of sodium, potassium, calcium, and magnesium. The best method for assessing soil salinity is measurement of the electrical conductivity of the saturated soil extract (EC_e). The procedure involves preparing a water-saturated soil paste by stirring (during the addition of distilled water), until a characteristic endpoint is reached. A suction filter is used to obtain a sufficient amount of the extract for making the conductivity measurement.

The saturated extract method has two advantages over other methods. First, the conductivity measurement of the saturated extract is directly related to the field moisture range that is important to plants. In the field, the moisture content of the soil fluctuates between a lower limit, rep-

Figure 4-23

Extensive irrigated areas of saline soil occur on the Gangetic Plain in India where the soil moisture regime is ustic.

Table 4-6
Relative Tolerance of Plants to Boron (In each Group, the Plants First
Named are Considered as Being More Tolerant, and the Last Named
More Sensitive)

Tolerant	Semitolerant	Sensitive
Athel *(Tamarix aphylla)*	Sunflower (native)	Pecan
Asparagus	Potato	Black walnut
Palm *(Phoenix canariensis)*	Acala cotton	Persian (English) walnut
Date palm *(P. dactylifera)*	Pima cotton	Jerusalem artichoke
Sugar beet	Tomato	Navy bean
Mangel	Sweetpea	American elm
Garden beet	Radish	Plum
Alfalfa	Field pea	Pear
Gladiolus	Ragged Robin rose	Apple
Broadbean	Olive	Grape (Sultanina and
Onion	Barley	Malaga)
Turnip	Wheat	Kadota fig
Cabbage	Corn	Persimmon
Lettuce	Milo	Cherry
Carrot	Oat	Peach
	Zinnia	Apricot
	Pumpkin	Thornless blackberry
	Bell pepper	Orange
	Sweet potato	Avocado
	Lima bean	Grapefruit
		Lemon

From *Agriculture Handbook 60,* USDA, 1954.

resented by the permanent-wilting percentage, and the upper, wet end of the available range, which is approximately two times the wilting percentage. Measurements on soils indicate that over a considerable texture range, the saturation percentage *(SP)* is approximately equal to four times the -15 bar-percentage *(FBP),* which (in turn) closely approximates the wilting percentage. The soluble-salt concentration in the saturation extract, therefore, tends to be about 50 percent of the concentration of the soil solution at the upper end of the field-moisture range and about 25 percent the concentration that the soil solution would

have at the lower, dry end of the field-moisture range. The salt-dilution effect that occurs in fine-textured soils, because of their higher moisture retention, is thus automatically taken into account. For this reason, the conductivity of the saturation extract *(EC$_e$)* can be used directly in appraising the effect of soil salinity on plant growth.

The second advantage of the saturated extract method is that the differences in the effectiveness of different ions in different salts is accounted for. If the salt content was expressed as tons of salt per hectare or acre, there would be no account of the fact that different ions have different atomic (or molecular) weights and different solubilities or activities.

Salinity effects on plants, as measured by the conductivity of the saturated extract as millimhos per centimeter (EC_e) at 25°C, are as follows:

1. 0–1: Salinity effects mostly negligible.
2. 2–4: Yields of very sensitive crops may be restricted.
3. 4–8: Yields of many crops restricted.
4. 8–16: Only tolerant crops yield satisfactorily.
5. Over 16: Only a few very tolerant crops yield satisfactorily.

The salt tolerance of crops in Table 4-7 are arranged according to major crop divisions. Within each group, the crops are listed in the order of decreasing salt tolerance; however, a difference of two or three places in a column may not be significant. The EC_e values given represent the salinity level at which a 10-, 25-, and 50-percent decrease in yield may be expected, as compared to yields on nonsaline soil under comparable growing conditions.

Table 4-7

Salt Tolerance of Crops Expressed as the EC_e at 25°C for Yield Reductions of 10, 25, and 50 Percent, as Compared to Growth on Normal Soils

Crop	10%	25%	50%
Field crops			
Barley	12	16	18
Sugar beet	10	13	16
Cotton	10	12	16
Safflower	8	11	14
Wheat	7	10	14
Sorghum	6	9	12

Table 4-7 (continued)

Salt Tolerance of Crops Expressed as the EC_e at 25°C for Yield Reductions of 10, 25, and 50 Percent, as Compared to Growth on Normal Soils

Crop	10%	25%	50%
Soybean	5.5	7	9
Sesbania	4	5.5	9
Rice (paddy)	5	6	8
Corn	5	6	7
Broadbean	3.5	4.5	6.5
Flax	3	4.5	6.5
Beans	1.5	2	3.5
Vegetable crops			
Beet	8	10	12
Spinach	5.5	7	8
Tomato	4	6.5	8
Broccoli	4	6	8
Cabbage	2.5	4	7
Potato	2.5	4	6
Corn	2.5	4	6
Sweet potato	2.5	3.5	6
Lettuce	2	3	5
Bell pepper	2	3	5
Onion	2	3.5	4
Carrot	1.5	2.5	4
Beans	1.5	2	3.5
Forage Crops			
Bermuda grass	13	16	18
Tall wheat grass	11	15	18
Crested wheat grass	6	11	18
	7	10.5	14.5
Tall fescue	8	11	13.5
Barley hay	8	10	13
Perennial rye	8	10	13
Harding grass	6	8	10
Birdsfoot trefoil	4	7	11
Beardless wild rye	3	5	8
Alfalfa	2.5	4.5	8
Orchard grass	2	3.5	6.5
Meadow foxtail	2	2.5	4
Clovers, alsike and red			
Fruit crops			
Date palm	8		16

Table 4-7 (continued)
Salt Tolerance of Crops Expressed as the EC_e at 25°C for Yield Reductions of 10, 25, and 50 Percent, as Compared to Growth on Normal Soils

Crop	10%	25%	50%
Pomegranate			
Fig	4.6		9
Olive			
Grape (Thompson)	4		8
Muskmelon	3.5		No data
Orange, grapefruit, lemon	2.5		5
Apple, pear	2.5		5
Plum, prune, peach, apricot, almond	2.5		5
Boysenberry, blackberry, raspberry	1.5–2.5		4
Avocado	2		4
Strawberry	1.5		3

Adapted from *Agr. Information Bull. Nos. 283 and 292* and *Western Fertilizer Handbook*, 1975.

Many studies of the effects of salt on plants included a number of varieties. Significant differences in variety have been found for cotton, barley, and smooth brome, while the differences in variety were of no consequence for green beans, lettuce, onions, and carrots.

The effect of the salts on plants is mainly indirect; that is, the effect of the salt on the osmotic water potential, and the resultant reduced uptake of water by germinating seeds and roots. Thus, an increase in salinity produces the same effect on water uptake as that produced by decreased matric potential. In fact, the effects appear to be additive. For example, if a plant wilts when the matric potential is −20 bars, and the osmotic potential of the soil solution is −1 bar, the plant will wilt if the matric potential is −1 and the osmotic potential is −20 bars. The additive effects of the matric and osmotic potentials on plant growth are shown in Fig. 4-24.

Salinity Control and Leaching Requirement

For permanent agriculture, there must be a favorable salt balance. The salt added to soils in irrigated water must be balanced by the removal of salt by leaching. The fraction of the irrigation water that must be leached through the root zone to control soil salinity at some specific level is the leaching requirement (LR). Assuming that the goal is to irrigate a productive soil with no change in salinity, and where soil salinity will not be affected by leaching

Figure 4-24
Growth of bean plants as influenced by total stress. Data show the additive effects of matric and osmotic stress on plant growth. (Data from *USDA Agriculture Handbook*, 60, 1954.)

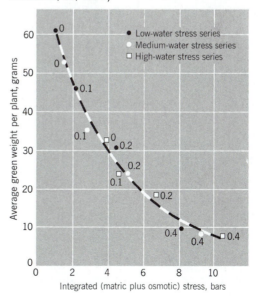

from rainfall or other factors, the leaching requirement is the ratio of the depth of drainage water (D_{dw}) to the depth of irrigation water (D_{iw}):

$$LR = \frac{D_{dw}}{D_{iw}}$$

LR, under the conditions specified (no change in soil salinity over time), is also equal to:

$$LR = \frac{EC_{iw}}{EC_{dw}}$$

If the electrical conductivity of the drainage water that can be tolerated is 8 millimhos per centimeter:

$$LR = \frac{EC_{iw}}{8}$$

For irrigation water with an electrical conductivity (EC) of 1, 2, and 3 millimhos per centimeter, LR = 0.13, 0.25, and 0.38, respectively. Of the water applied for irrigation to moisten the soil, 13, 25, and 38 percent represents the amount of water that must be leached through the soil. Under these conditions, the salt in the irrigation water is equal to the salt in the drainage water, and the salt content of soil due to irrigation is unchanged. When the irrigation need is 10 centimeters, the total water that needs to be applied is calculated as follows (when the LR is 25 percent):

$$LR = 10 \text{ cm} + 0.25 \ (10) \text{ cm} = 12.5 \text{ cm}$$

From the illustration, the drainage water would contain five times greater concentration of salt than original irrigation water. The need for leaching soils to control salt highlights a major disadvantage of drip irrigation in that the method is not suited for application of large amounts of water.

Leaching of salts on upland soils, as on the high plains of Texas, and other upland irrigated areas, results in salt being deposited below the root zone. Most of the irrigated land in arid regions is in basins or alluvial valleys, where long-continued leaching of salts results in a buildup of water tables. When the water table rises to 1 to 2 meters of the surface, water moves upward by capillary at a rate sufficient to deposit salt on top of the soil from the evaporation of water (*see* Fig. 4-25). It is essential in these cases that subsurface drainage systems be installed to remove drainage water. Thus, we see that salinity control and maintenance of permanently irrigated agriculture in arid regions is dependent on drainage for salt removal. Archeological studies in Iraq showed that the Sumerians grew about half wheat and half barley in 3500 B.C. One thousand years later, only one sixth of the grain was the less salt-tolerant wheat; by 1700 B.C., the production of wheat was abandoned. This decline in wheat production and increase in the more salt-tolerant barley coincided with soil salinization. Records show widespread land abandonment, and that salt accumulation in soils was a fac-

Figure 4-25
Salt accumulation in an irrigated field from the upward movement of water from a water table near the soil surface.

tor in the demise of the Sumerian civilization.

Even though good irrigation practices are followed, there can be important differences in the salt content of soil that is immediately under the irrigation furrow and beds. Downward movement of water below the furrow leaches the soil and produces low salinity. At the same time, water moves to the top of the beds carrying salt and deposits salt as water evaporates. Important differences in salt concentrations are produced (*see* Fig. 4-26). Seeds are commonly planted on the shoulder of the beds, between the furrow and high point of the beds, to insure greater germination and growth. Research has shown that water uptake is greatest in the soil with the lowest salt content.

Effect of Irrigation on River Water Quality

About 60 percent of water diverted for irrigation is evaporated or consumed. The remainder appears as irrigation return flow(includes drainage water) and is returned to the rivers. As a consequence, the salt concentration of river waters is in-

creased. A 2- to 7-fold increase in salt concentration is common for many rivers. Along the Rio Grande River in New Mexico, water diverted at Percha Dam for Rincon Valley irrigation had an average 8 milliequivalents of salt per liter from 1954 to 1963. Salt content increased to 9 at Leasburg, 13 at American Diversion Dam, and 30 at the lower end of El Paso Valley (*see* Fig. 4-27). The Rio Grande River serves as a sink for the deposition of salt. The Salton Sea serves as a salt sink for the Imperial Valley in southern California.

Salts also appear in rivers from salting of roads, industries, and the like. In the rivers of the western United States, however, irrigation is responsible for large increases in salt. The Colorado River, one of the largest, experiences a 21-fold salt increase between Grand Lake in northwestern Colorado and the Imperial Dam. Large increases in salt concentrations have become harmful to freshwater fish. Salts decrease water quality for downstream users, and there is no inexpensive method to remove the salt at the present time.

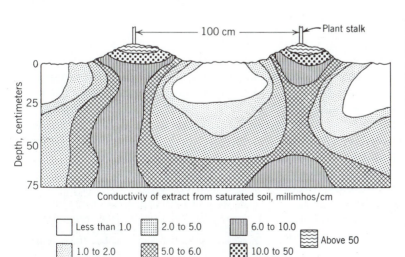

Figure 4-26
Salt distribution under furrow-irrigated cotton for soil initially salinized to 0.2 percent salt and irrigated with water of medium salinity. (From *USDA Agriculture Handbook*, 60, 1954.)

Conductivity of extract from saturated soil, millimhos/cm

Less than 1.0   2.0 to 5.0   6.0 to 10.0   Above 50
1.0 to 2.0   5.0 to 6.0   10.0 to 50

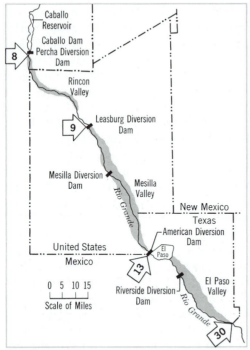

Figure 4-27
Total salt content of Rio Grande River water in milliequivalents per liter increased from 8 at Percha Dam to 30 at lower end of El Paso Valley. (Adapted from C. A. Bower, "Salinity of Drainage Waters," Agronomy Monograph #17, p. 481 (1974), by permission of the American Society of Agronomy.)

Nature and Management of Saline and Sodic Soils

The development of saline soils is a natural process in arid regions. Large areas of land have become saline via fallowing (saline seep) and irrigation. When sodium is an important component of salt, and a significant amount of exchangeable sodium (sodium absorbed on the negatively charged clay and humus particles) occurs, colloids disperse and soil develops peculiar physical properties. These soils are *sodic*. The quantity, proportion, and nature of salts that are present may vary in saline and sodic soils. This gives rise to

three kinds of soils: (1) saline, (2) saline-sodic, and (3) sodic soils. A consideration of the development, properties, and management of these soils follows.

Saline Soils. Saline soils contain sufficient soluble salt to impair plant growth and the conductivity of saturated soil extract is more than 4 millimhos per centimeter (0.4 siemens per meter). The salts are made up largely of chlorides, sulfates, and sometimes nitrates. Small quantities of bicarbonates may occur, but soluble carbonates are usually absent. Frequently, relative, insoluble salts such as calcium sulfate and calcium and magnesium carbonates are also present. The chief cations present are calcium, magnesium, and sodium; however, sodium seldom makes up more than 50 percent of the soluble cations, and it is not adsorbed to an appreciable extent on the colloidal fraction of soil.

The pH value of these soils is 8.5 or less, and the exchangeable sodium percentage is less than 15.[3] White crusts frequently accumulate on the soil surface, and streaks of salt are sometimes found within the soil. Saline soils have a favorable structure because the colloids are highly flocculated.

Good drainage is necessary for the reclamation process to remove the excess salts from the root zone. This can only be done by the application of sufficient water to wash them into the lower soil depths. Unless there is ample drainage, the addition of so much water will raise the water

---

[3]The exchangeable sodium percentage is the percentage of the negative charge on clay and humus that is neutralized by sodium (*see* Chap. 8). The SAR of a saturated soil paste is about equal to the exchangeable sodium percentage (ESP), and 15 percent exchangeable sodium is about equal to a SAR of 13.

table and, hence, lead to increased accumulation of salt in the surface soil instead of to a correction of the saline condition. Sufficient drainage should be provided to reduce the ground-water level well below the zone of root penetration. Preferably, the ground water should never be less than 2 to 3 meters below the soil surface.

With ample drainage provided, one may proceed to the leaching out of the salts. In fine-textured soils, the reclamation process will be slow; doubly so if the soil is underlain by dense clay subsoil. In fact, the presence of a dense clay layer makes it difficult to remove salts from even medium- or coarse-textured soils. From an economic standpoint, it is questionable if reclamation of soils with very deep clay subsoils is feasible.

Experiments have shown that leaching is all that is needed to reclaim saline soils that have adequate internal drainage. The addition of chemicals, plowing under of manure, or green-manuring crops are unnecessary. No specific directions can be given regarding the frequency of irrigation or the quantity of water to apply at each irrigation. The main points to observe are: (1) that the soil be kept moist so that soil solution will not become sufficiently concentrated to damage the growing crop, (2) that sufficient water be applied at each irrigation to result in some leaching of salts into the drainage water, and (3) that the soil of each irrigation check be carefully leveled so that water will enter the soil uniformly. Considerable research effort is being directed toward the development of crop varieties with a greater salt tolerance.

Sodic (Alkali) Soils. Sodic soils are nonsaline and have an exchangeable sodium percentage (ESP) of 15 or more, or the sodium adsorption ratio (SAR) of the saturated extract is 13 or more. The pH is usually in the range of 8.5 to 10.0. Sodium dissociates from colloids and small amounts of sodium carbonate form. Deflocculation of colloids occurs and, hence, a breakdown of soil structure. This creates a massive or puddled soil with very low infiltration of water into the soil and very low water conductivity within the soil. The soil is difficult to till and soil crusting may inhibit seedling emergence (*see* Fig. 4-28). Organic matter in the soil is highly dispersed and is distributed over the surface of the particles, giving a dark color, hence the term *black alkali,* which was formerly used to designate such soils. Sodic soils frequently occur in small irregular areas in regions of low rainfall and are referred to as "slick spots."

Sodic soils may develop as a result of irrigation. Because of the dispersed state of colloids, soils are difficult to till and are slowly permeable to water. After a long period of time the dispersed clay may migrate downward, forming a very dense layer with a prismatic or columnar structure. When this phenomenon occurs, several centimeters or inches of relatively

Figure 4-28
Dispersed sodic soil that forms large crusty clods, when dried, that inhibit seedling emergence. When wet, the soil has a very low infiltration and percolation rate.

coarse-textured soil may be left on the surface.

The soil solution of sodic soils contains only small amounts of calcium and magnesium, but larger quantities of sodium. The anions include sulfate, chloride, bicarbonate, and usually small quantities of normal carbonate. In some areas an appreciable amount of potassium salts is also present.

All that has been said concerning the need for drainage and the application of sufficient irrigation water to cause leaching is of as much, if not more, importance in the reclamation of sodic soils as in the treatment of saline soil. Although it has been shown that the application of ample irrigation water, coupled with good farming practices, will ultimately result in the removal of exchangeable sodium as well as soluble salt, the reclamation process may be materially hastened through the application of various chemicals. The basis of the treatments is the replacement of exchangeable sodium in the colloidal fraction by calcium and the conversion of the replaced sodium and any sodium occurring as the carbonate into neutral sodium sulfate (*see* Fig. 4-29). The desired changes may be brought about by application of considerable quantities of finely ground calcium sulfate (gypsum). The amount of gypsum to apply to replace the exchangeable sodium is the gypsum requirement. The method for its calculation is presented in Chapter 8. Ground sulfur, however, will accomplish the same results, but somewhat more slowly. The sulfur must first be oxidized in the soil. Then, it combines with water to make sulfuric acid. Other soluble sulfates, such as iron or aluminum, have also proved effective.

Figure 4-29

Illustration of the removal of exchangeable sodium by the addition of calcium sulfate. When irrigation water and calcium sulfate are applied to a sodic soil having dispersed particles and small pores, the calcium ions from the calcium sulfate replace the exchangeable sodium ions, which form sodium sulfate and leach out of the soil. Soil pH is lowered, colloids flocculate, larger pores develop, and soil permeability is increased. (Adapted from "Chemical Amendments for Improving Sodium Soils," *Agr. Inf. Bull.,* 195, USDA, 1959.)

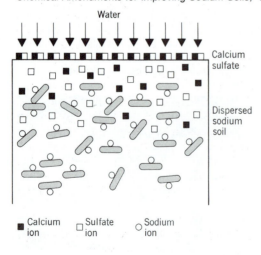

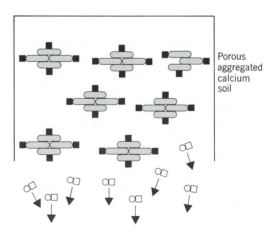

A supply of soluble calcium is needed to complete the reactions. Acid resulting from sulfur additions dissolves calcium carbonate ($CaCO_3$) that might be present in the soil to supply soluble calcium. Harmful exchangeable sodium can then be replaced by calcium, with a resulting improvement in the physical condition of the soil. The sodium is removed from the soil as sodium sulfate in the water that leaches through the soil for salinity control.

Saline-Sodic Soils. These soils are characterized by salinity and 15 or more percent exchangeable sodium. As long as a large quantity of soluble salts remains in the soil, the exchangeable sodium may not cause trouble; the soil pH seldom exceeds 8.5. If, however, the soluble salts are temporarily leached downward: (1) the pH goes above 8.5, (2) the sodium causes the colloids to disperse, and (3) a structure unfavorable for tillage, entry of water, and root development develops. The movement of the soluble salts upward into the surface soil may lower the pH and restore the colloids to a flocculated condition. The management of this group of soils is a problem until the excess soluble salts and the exchangeable sodium are removed from the zone of root growth. Unless calcium sulfate or another source of soluble calcium is present, the drainage and leaching of these soils convert them into sodic soils.

Bibliography

Allison, L. R., "Salinity in Relation to Irrigation," *Advances of Agronomy, 16:*139–180, 1964.

Baker, K. F., Ed., "The U.C. System for Producing Healthy Container-Grown Plants," Manual 23, Cal. Agr. Exp. Station, Berkeley, 1957.

Bernstein, L., "Salt Tolerance of Plants," *Agr. Information Bull.,* No. 283, USDA, Washington, D.C., 1964.

Bernstein, L., "Salt Tolerance of Fruit Crops," *Agr. Information Bull.,* No. 292, USDA, Washington, D.C., 1965.

Bower, C. A., "Salinity of Drainage Waters," in *Drainage for Agriculture,* Jan Van Schilfgaarde, Ed., Am. Soc. Agronomy, Madison, Wis., 1974, pp. 471–487.

Brink, C. Van Den, and R. L. Carolus, "Removal of Atmospheric Stresses from Plants by Overhead Sprinkler Irrigation," *Quart. Bull.,* Mich. Agr. Exp. Sta., *47:*358–363, 1965.

Cole, J. S., and O. R. Mathews, "Use of Water by Spring Wheat on the Great Plains," *USDA Bul.* 1004, 1923.

Criddle, W. D., and H. R. Haise, "Irrigation in Arid Regions," in *Soil,* USDA Yearbook, Washington, D.C., 1975, pp. 359–367, 1957.

Edminister, T. W., and R. C. Reeve, "Drainage Problems and Methods," in *Soil,* USDA Yearbook, Washington, D.C., pp. 379–385, 1957.

Epstein, E., J. D. Norlyn, D. W. Rush, R. W. Kingsbury, D. B. Kelly, G. A. Cunningham, and A. F. Wrona. "Saline Culture of Crops: A Genetic Approach," *Science, 210:*399–404, 1980.

Erickson, A. E., C. M. Hansen, and A. J. M. Smucker, "The Influence of Subsurface Asphalt Barriers on the Water Properties and the Productivity of Sand Soils," *9th Int. Cong. Soil Sci. Trans., 1:*331–337, 1968.

Evans, C. E., and E. R. Lemon, "Conserving Soil Moisture," in *Soil,* USDA Yearbook, Washington D.C., 1957, pp. 340–359.

Ferguson, H., and T. Bateridge, "Salt Status of Glacial Till Soils of North-Central Montana as Affected by the Crop-Fallow System of Dry Farming," *Soil Sci. Soc. Am. Proc., 46:*807–810, 1982.

Hanks, R. J., and C. B. Tanner, "Water Consumption by Plants as Influenced by Soil Fertility," *Agron. Jour., 44:*99, 1952.

Jacobsen, T., and R. M. Adams, "Salt and Silt

in Ancient Mesopotamian Agriculture," *Science, 128:*1251–1258, 1958.

Law, J. P., and J. L. Witherow, "Irrigation Residues," *Jour. Soil and Water Con., 26:*54–56, 1971.

Massee, T. W., and J. W. Cary, "Potential for Reducing Evaporation During Summer Fallow," *Jour. Soil Water Cons., 33:*126–129, 1978.

Mathews, O. R., "Place of Summer Fallow in the Agriculture of the Western States," USDA *Cir. 886,* 1951.

Richards, L. A., and S. J. Richards, "Soil Moisture," in *Soil,* USDA Yearbook, Washington, D.C. 1957, pp. 49–60.

Soil Improvement Comm. Cal. Fert. Assoc., *Western Fertilizer Handbook,* 5th Ed., Interstate, Dansville, Cal., 1975.

Spomer, L. A., "The Inside Story on Your Soil Mix," *Amer. Veg. Grower,* pp. 50–51, June 1975.

Staple, W. J., "Dryland Agriculture and Water Conservation," in H. L. Hamilton, Ed., *Research on Water,* ASA Spec. Pub. No. 4, Soil Sci. Soc. Am., 1964, pp. 15–30.

Taylor, S. A., "Use of Moisture by Plants," in *Soil,* USDA Yearbook, Washington, D.C., 1957, pp. 61–66.

Terry, T. A., and J. H. Hughes, "The Effects of Intensive Management on Planted Lobolly Pine (*Pinus taeda* L.) Growth on Poorly Drained Soils of the Atlantic Coastal Plain," Proc. Fourth North Am. Forest Soils Conf., University of Laval, Quebec, Canada, 1975, pp. 351–377.

The President's Science Advisory Committee Panel on World Food Supply, *The World Food Problem,* U.S. Govt. Printing Office, Washington, D.C., 1967.

Thorne, W., and H. B. Peterson, "Salinity in United States Waters," in *Agriculture and the Quality of Our Environment,* Am. Assoc. Adv. Sci., Washington, D.C., 1967.

Thorne, D. W., and M. D. Thorne, *Soil, Water and Crop Production,* AVI Publishing Company, Westport, Conn., 1979.

United States Salinity Laboratory Staff, "Diagnosis and Improvement of Saline and Alkali Soils," *USDA Agr. Handbook 60,* Washington, D.C., 1969.

Williamson, R. E., et al., "Effect of Water Table Depth and Flooding on Yield of Millet," *Agron. Jour., 61:*312, 1969.

CHAPTER 5

SOIL ECOLOGY

The soil is the home of innumerable forms of plant, animal, and microbial life. Some of the fascination and mystery of this underworld has been described by Peter Farb:

> We live on the rooftops of a hidden world. Beneath the soil surface lies a land of fascination, and also of mysteries, for much of man's wonder about life itself has been connected with the soil. It is populated by strange creatures who have found ways to survive in a world without sunlight, an empire whose boundaries are fixed by earthen walls.[1]

Life in the soil is amazingly diverse, ranging from microscopic single-celled organisms to large burrowing animals. As is the case with organisms above the ground, there are well-defined food chains and competition for survival. The study of the relationships of these organisms in the soil environment is *soil ecology.* We will discuss diverse activities, such as decomposition of organic matter, pesticide degradation by microorganisms, and predation and earth-moving activities of animals. A major theme of this chapter is the soil as a source of nutrients for living organisms and the role of soil organisms in nutrient cycling.

---

[1]*Living Earth,* Peter Farb, Harper and Brothers Publishers, 1959.

The Ecosystem

The sum total of life on earth, together with the global environments, constitutes the *escosphere*. The ecosphere, in turn, is composed of numerous self-sustaining communities of organisms and their inorganic environments and resources, called *ecosystems*. Each ecosystem has its own unique combination of living organisms and abiotic resources that function to maintain a continuous flow of energy and nutrients. All ecosystems have two types of organisms based on carbon source. *Autotrophs* use inorganic carbon, principally from CO_2, and are the *producers*. *Heterotrophs* use organic carbon and are the *consumers* and *decomposers*.

Autotrophs and heterotrophs are divided into groups based on energy source. Phototypes obtain energy from the sun, and chemotypes obtain energy from the oxidation of inorganic elements and compounds. Three groups that are most important in soils are photoautotrophs, chemoautotrophs, and chemoheterotrophs. Higher plants and many algae forms are photoautotrophs. Chemoautotrophs include nitrifying and sulfur-oxidizing bacteria. Animals, protozoa, fungi, and most bacteria are chemoheterotrophs.

Primary Producers

The major primary producers are vascular plants that use solar energy to fix carbon from carbon dioxide in photosynthesis. The tops of plants provide food for consumers and decomposers above the soil-atmosphere interface. Roots, tubers, and other underground organs provide food for consumers and decomposers within soil. A very small amount of photosynthesis occurs at or near the surface of some soils by algae, which are the dom-inant plants in aquatic ecosystems. A small amount of inorganic carbon is fixed by chemotrophic bacteria by using the energy of chemical bonds. Thus, the productivity of a terrestrial ecosystem is basically a measure of the net photosynthesis (photosynthesis less respiration) of vascular plants.

Tropical rain forests have high productivity and deserts low. Agricultural ecosystems are unique in that people determine the primary producers. In the case of monoculture, there is usually one major plant, as in the case of sugar cane, corn, or pineapples. Productivity of agricultural ecosystems is highly variable depending on soil, climate, and the inputs of labor, capital, and management. Biomass produced by the producers is food for consumers and decomposers, including humans.

Consumers and Decomposers

It is characteristic that ecosystems have many species, but that only a few species are common. In a meadow, for example, several grasses may account for over 90 percent of primary production. In turn, most primary production is consumed by a few animals, such as mice and rabbits. The fox and a few birds may represent the bulk of secondary consumers feeding on mice and rabbits (*see* Fig. 5-1).

About 1 gram of animal biomass is produced for every 10 grams of plant material consumed. In the transformation of plant material into animal biomass, considerable carbon is returned to the atmosphere as carbon dioxide from respiration; some energy is dissipated as heat. Most of the original carbon and most of the nutrients, however, appear in the dung or feces. The result is that fecal material is a good source of nutrients and en-

Figure 5-1
Examples of primary producers, consumers, and decomposers. Examples below the soil-atmosphere interface include (from left to right) centipede eating a springtail, bacteria decomposing soil organic matter, springtail feeding on soil organic matter, earthworm feeding on leaves, and nematode consuming plant roots.

ergy. It is not uncommon for primary consumers to recycle part of their feces. Rabbits eat feces when other food is in short supply. Some animals appear to recycle part of their feces instinctively as a survival mechanism or as a means to improve their nutrition. Recycling manure through animals is being studied as a means of economically producing animal products.

Primary consumers become food for secondary consumers, and there may be additional levels of consumers. Eventually all consumers die and are added to soil along with fecal material and unused primary production. These materials serve as food for soil-dwelling consumers and decomposers (*see* Fig. 5-1). Ultimately, all the carbon fixed in photosynthesis is returned to the atmosphere as CO_2, and the energy is lost as heat. The nutrients originally absorbed by producers are released for use in another cycle of growth and decay. The result is that the major function of soil organisms in the ecosystem centers around the flow or cycling of energy and nutrients.

We have just seen that the soil is the dumping ground for all unused primary production and wastes associated with plant and animal life. The great bulk of soil-organic matter is dead, as shown in Table 5-1. Soil is the "stomach" of the earth and fulfills the dictum: "Ashes to ashes and dust to dust." Without consumers and decomposers to release fixed carbon, the atmosphere would be depleted of carbon dioxide, life would cease, and the cycle would stop.

Microorganisms as Decomposers

Life within the soil is analogous to life above the soil. Roots, tubers, and other underground organs are parts of primary producers. There are consumers and decomposers that are interrelated by food chains. Perhaps the major difference between the ecology above and below the soil-atmosphere interface is that above the

Table 5-1

Estimates of Amount of Organic Matter and Proportions, Dry Weight, and Number of Living Organisms in a Hectare of Soil to a Depth of 15 Centimeters in a Humid Temperate Region

Item	Dry Weight		Estimated Number of Individuals
	%	kg/ha	
Organic matter, live and dead	6	120,000	—
Dead organic matter	5.28	105,400	—
Roots of higher plants	0.5	10,000	—
Microorganisms (protists)			
Bacteria	0.10	2,600	2×10^{18}
Fungi	0.10	2,000	8×10^{16}
Actinomycetes	0.01	220	6×10^{17}
Algae	0.0005	10	3×10^{14}
Protozoa	0.005	100	7×10^{16}
Nonarthropod animals			
Nematodes	0.001	20	2.5×10^{9}
Earthworms (and potworms)	0.005	100	7×10^{3}
Arthropod animals			
Springtails (Collembola)	0.0001	2	4×10^{5}
Mites (Acarine)	0.0001	2	4×10^{5}
Millipedes and centipedes (Myriapoda)	0.001	20	1×10^{3}
Harvestman (Opiliones)	0.00005	1	2.5×10^{4}
Ants (Hymenoptera)	0.0002	5	5×10^{6}
Diplopoda, Chilopods, Symphyla	0.0011	25	3.8×10^{7}
Diptera, Coleoptera, Lepidoptera	0.0015	35	5×10^{7}
Crustacea (Isopods, crayfish)	0.0005	10	4×10^{17}
Vertebrate animals			
Mice, voles, moles	0.0005	10	4×10^{5}
Rabbits, squirrels, gophers	0.0006	12	10
Foxes, badgers, bear, deer	0.0005	10	<1
Birds	0.0005	10	100

Adapted and reprinted by permission from *Soil Genesis and Classification* by S. W. Buol, F. D. Hole, and R. J. McCracken, copyright© 1972 by Iowa State University Press, Ames, Iowa, 50010.

interface, animals play the dominant role as *consumers,* and below the interface, microorganisms play the dominant role as *decomposers.* These decomposers are mainly single celled and microscopic. The dominant decomposer role of microorganisms in the soil is complemented by the activity of many, small animal consumers.

Some Characteristics of Microorganisms

Earliest life on earth consisted of microorganisms—both autotrophic and heterotrophic. The cycling of energy and nutrients was established before higher or vascular plants and animals evolved. No wonder that microorganisms play the major role as the ultimate decomposers.

They secrete enzymes that *digest organic matter outside the cell* and absorb the soluble end-products of digestion. These enzymes and processes are no different than those that occur in the digestive system of animals.

At the root-hair level the higher plants (vascular) and microorganisms have much in common. Both absorb soluble nutrients from the same soil solution and use energy to accumulate nutrients against a concentration gradient. Both are affected by the water potential gradients and water conductivities, are inhibited by the same salts, and compete for soil oxygen. Vascular plants and microorganisms are competitors for soil-growth factors, but they also depend on each other in the continuous cycling of energy and nutrients.

The major distinguishing feature of microorganisms is their relatively simple biological organization. Many are unicellular, and even the multicellular organisms lack differentiation into cell types and tissues characteristic of plants and animals. They are members of the protist kingdom. The words protists and microorganisms are used interchangeably. Protists are divided in the lower and higher protists based on degree of complexity.

Lower Protists

Lower protists include blue-green algae and bacteria. Bacteria are single-celled, the smallest living organisms, and exceed all other soil organisms in numbers and kinds. A gram of fertile soil may contain over 1,000,000,000 bacteria. The most common soil bacteria are rod-shaped, a micron (1/25,000 of an inch) or less in diameter, and up to a few microns long (*see* Fig. 5-2). Researchers have estimated that the live weight of bacteria per acre may

Figure 5-2
(*Left*) Colony of bacteria on sand grain magnified about 6,000 times (*Right*) Rod-shaped bacteria magnified 20,000 times (*Left* from "Stereoscan Electron Microscopy of Soil Microorganisms," T. R. G. Gray, *Science,* 155:1668, Fig. 1, March 31, 1967. Copyright © 1967 by the American Association for the Advancement of Science. *Right* courtesy of Dr. S. Flegler of Michigan State University.)

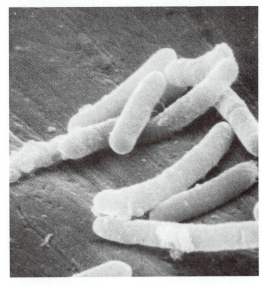

exceed 2000 kilograms per hectare (pounds per acre) (Table 5-1).

Most soil bacteria are chemoheterotrophs that are both dependent on preformed carbon and are nonphotosynthetic. These organisms play a major role in the cycling of energy and nutrients. Some bacteria, far fewer in number although very important for the growth of higher plants, are chemoautotrophs. Their carbon is derived from carbon dioxide and their energy from the oxidation of elements and compounds. Examples are bacteria that oxidize reduced nitrogen compounds to nitrate and oxidize sulfur to sulfate.

Most soil bacteria require oxygen from the soil air and are classified as *aerobes*. Some aerobic bacteria can adapt to living in the presence or absence of oxygen; they are *facultative aerobes*. Other bacteria can not live in the presence of oxygen and are *anaerobes*. Soil bacteria also differ considerably in their nutrition and in their response to environmental conditions. Consequently, the kinds and abundance of bacteria depend both on the available nutrients present and on the soil environmental conditions.

Bacteria normally reproduce by binary fission. Some divide as often as every 20 minutes and may multiply very rapidly under favorable conditions. It has been calculated that if a single bacterium divided every hour and every subsequent bacterium did the same, 17,000,000 cells would be produced in 1 day. A mass the size of the earth would be produced in 6 days. Such rapid growth rates cannot be maintained for long because nutrients and other growth factors become exhausted and waste products accumulate.

Blue-green algae are mainly filamentous, but they have a simplistic cell structure like bacteria. Actually, they have been classified as bacteria by some microbiologists.[2] Blue-green algae are aquatic photoautotrophs that thrive where light and moisture are favorable. They play an important role in flooded rice fields, where they fix atmospheric nitrogen and release O_2 from photosynthesis. The nitrogen eventually becomes available for the rice, and the excreted oxygen is used by rice roots. Some blue-green algae are among the first colonizers of rocks and fresh parent materials as a component of lichens.

Fungi—The Effective Lignin Decomposers

Fungi are heterotrophs that vary greatly in size and structure, from single-celled yeasts to molds and mushrooms. Fungi typically grow from spores by a threadlike structure that may or may not have crosswalls. Individual threads are *hypha*, and a mass of extensive threads is the *mycelium*. The mycelium is the working structure that absorbs nutrients, continues to grow, and eventually produces special hyphae that produce reproductive spores. The average diameter of hyphae is about 5 microns or about 5 to 10 times the diameter of a typical bacterium. Fungi have an advantage over bacteria in that fungi can invade and penetrate organic materials (*see* Fig. 5-3).

It is difficult to determine accurately the number of fungi per gram of soil, since mycelium are easily fragmented. It has been observed that a gram of soil commonly contains 10 to 100 meters of hyphae per gram. On the basis of the amount of filament, researchers have concluded that live weight of fungal tissue ex-

[2]There is some overlapping in the classification of protists. Blue-green algae have the simpler structure of lower protists similar to bacteria. Most algae are slightly more complex and are higher protists.

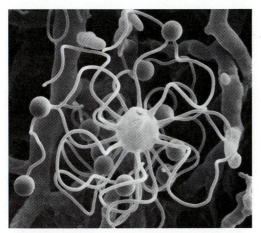

Figure 5-3
A soil fungi showing mycelium and reproductive structure with spores. (Photo courtesy of Michigan State University Pesticide Research Electron Microscope Laboratory.)

ceeds or equals that of the bacterial tissue in most soils.

All of us have seen mold mycelia growing on bread, clothing, or leather goods. Some mold colonies growing on plant leaves produce a white-cottonish appearance called *downy mildew disease.* Many fungi have morphological features that resemble higher plants. *Rhizopus,* a common mold in soil and on bread, has root-like absorptive structures called *rhizoids,* which penetrate the substrate on which the mold is growing. Hyphae elongating over the substrate are called *stolons.* Stalk- or stem-like hyphae originating from the stolons bear spore cases. Unlike higher or vascular plants, however, fungi have no specialized xylem or phloem-conducting tissue.

Fungi are important in all soils. Their tolerance of acidity makes them particularly important in acid forest soils. The woody residues of the forest floor provide an abundance of food for certain fungi

that are effective decomposers of lignin (*see* Fig. 5-4). In the Sierra Nevada Mountains, there is a large mushroom fungus called the "train wrecker" fungus because it grows abundantly on railroad ties unless they are creosoted.

Yeasts are fungi, are found in soils only to a limited extent, and are believed to be of no great importance in soil development or the growth of higher plants.

Actinomycetes—the "Fungi-like" Bacteria

Actinomycetes occupy a position between bacteria and fungi from the morphological viewpoint. They are frequently spoken of as *ray fungi* or *thread bacteria.* The actinomyces resemble bacteria in that they have the same cell structure and are about the same size in cross-section. They resemble filamentous fungi in that they produce a branched filamentous network. Many of these organisms reproduce by means of spores, and these spores appear to be very much like bacterial cells.

These organisms are present in great abundance in soil. They make up as much as 50 percent of the colonies that develop

Figure 5-4
Mushroom fungi (*Agaricales*) growing on soil, wood, and bark. The mushroom-fruiting bodies form after the hyphae have accumulated sufficient food and the moisture and temperature are favorable.

on plates containing artificial media inoculated with a soil extract. The numbers of actinomycetes may vary between 1,000,000 and 36,000,000 per gram of soil. In actual weight of live substance per acre, they may exceed bacteria, but as a rule, they will not equal fungus tissue.

Algae—Chlorophyllous Protists

Algae exhibit great diversity in form and size, ranging from single-celled organisms, with a diameter about 5 to 10 times greater than that of a bacterium to kelps of the ocean that are over 30 meters in length. Although algae are the most important plants living in water, algae are of only minor importance in most soils. The most common soil algae are single celled or small filaments. Algae are universally distributed in the surface layer of soils wherever moisture and light are favorable. A few algae are found below the soil surface, in the absence of light, and appear to function heterotrophically.

Protozoa—Aquatic Protists

Protozoa are single-celled protists that exhibit great diversity. Soil protozoa live in the films of water-surrounding soil particles and, in a sense, are aquatic organisms. When soil dries out, food supplies become short, or conditions are harmful, protozoa encyst; they become active again when conditions become favorable. Soil protozoa are largely predators, feeding on soil bacteria, although some protozoa also feed on fungi, algae, or dead organic matter. Although protozoa are very numerous in soils, they appear to have only a minor affect on organic matter decomposition and bacterial activity.

Vertical Distribution of Microbiota in the Soil

The surface of the soil is the interface between the lithosphere and the atmosphere. At or near this interface, the quantity of living matter is greater than at any region above or below. As a consequence, the A horizon contains more organic debris, which serves as food for microorganisms, than the B or C horizons. Although other factors influence the activity and numbers of microorganisms besides nutrient and energy supplies, the greatest number of microbes, as a rule, occurs in the A horizon or surface layers (*see* Fig. 5-5).

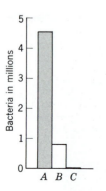

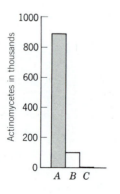

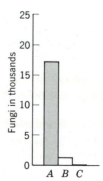

Figure 5-5
Distribution of microorganisms in the A, B, and C horizons of a cultivated Prairie (Mollisol) soil. All values refer to the number of organisms per gram of air dry soil. (Data from *Iowa Res. Bul.* 132.)

Soil Animals as Consumers and Decomposers

Perhaps higher plants could grow and provide us with food, and microorganisms could recycle all the nutrients without the aid of animals. Soil animals, however, are numerous (*see* Table 5-1) and play an important role in organic matter decomposition. Nematodes and white grubs are primary consumers that feed largely on the roots of primary producers (vascular plants). Secondary and tertiary consumers include predatory animals like mites, centipedes, spiders, ants, and moles. A food chain in the soil could consist of plant roots, nematodes, mites, and centipedes. Most of the soil animals are consumers that feed on dead and decaying refuse; they include springtails, sowbugs, millipedes, mites, slugs, earthworms, and various soil insects. Soil animals can be considered both consumers and decomposers, because they eat or ingest organic matter and some decomposition of organic matter occurs in digestion.

Nematodes—Parasitic Consumers

Nematodes are worms that are mainly microscopic in size and are the most abundant animals in soils. On the basis of their food requirements, three groups are distinguished: (1) those that feed on decaying organic matter, (2) those that feed on earthworms, other nematodes, plant parasites, bacteria, protozoa, and the like, and (3) those that infest the roots of higher plants, passing a part of their life cycle embedded therein. Nematodes are sometimes called eel worms; they are round or spindle shaped and usually have a pointed posterior.

Under a 10-power hand lens they appear as transparent thread-like worms (*see* Fig. 5-6). Nematodes lose water readily through their cuticles; they live mainly in the water films that surround soil particles or in plant roots. When soils dry, or other conditions become unfavorable, they encyst. They become active again when conditions are favorable.

Parasitic nematodes are considered the most important in agriculture. Many plants are attacked, including tomatoes, peas, carrots, alfalfa, turfgrass, ornamentals, and fruit trees. Parasitic nematodes have a needle-like anterior end (stylet) used to pierce plant cells and suck out the contents. Host plants respond in numerous ways; for example, they develop galls or knots or deformed roots.

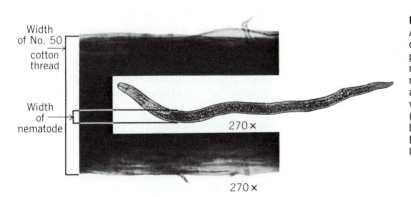

Width of No. 50 cotton thread

Width of nematode

270×

270×

Figure 5-6
A nematode and a piece of ordinary cotton thread photographed at the same magnification (270). The nematode is 1/15 as thick as the thread and is not visible to the naked eye. (Photo courtesy of H. H. Lyon, Plant Pathology Department, Cornell University.)

Investigations indicate that nematode damage is much more extensive than originally thought. Nematodes are a very serious problem for pineapple production in Hawaii, where the soil for new plantings is routinely fumigated to control nematodes. Nematodes may also become serious pests in greenhouse soils, unless special care is taken to avoid infestation. Not only do nematodes injure the plant roots themselves, but by puncturing the plant, they also prepare an entrance for other parasites. The main role of nematodes in soils appears to be economic, as consumer parasites.

Earthworms—Consumers and Soil Mixers

Perhaps the best-known group of larger animals inhabiting the soil is the common earthworm, of which there are several species. These organisms prefer a moist environment, with an abundance of organic matter and a plentiful supply of available calcium. Consequently, earthworms are found most abundantly, as a rule, in fine-textured soils that are both high in organic matter and not strongly acid. They occur only sparingly in acidic sandy soils that are low in organic matter.

Obviously, the number and activity of the earthworms vary greatly from one location to another; as with other soil organisms, figures indicating numbers are merely suggestive. The number of earthworms in the plowed layer of an acre may range from a few hundred or even less to more than 1,000,000. It has been estimated that the weight of earthworms commonly present is 200 to 1,000 kilograms per hectare (200 to 1,000 pounds per acre).

The common earthworm, *Lumbricus terrestris*, was imported into the United States from Europe. *Lumbricus terrestris*

makes a shallow burrow and forages on plant material at night. Some of the plant material is dragged into the burrow. Other kinds of earthworms exist by ingesting organic matter that exists in the soil. Excrement or castings are deposited both on and in the soil. The intimate mixing of soil materials, the creation of channels, and the production of castings leaves the soil more open and porous. Channels left open at the soil surface increase water infiltration.

Earthworms normally avoid water-saturated soil. If they emerge during the day, when it is raining, they are killed by ultraviolet radiation unless they quickly find protection. Casts deposited on the surface of lawns are objectionable. On the other hand, earthworms feed on thatch and help to prevent thatch buildup by turfgrass. Arable lands in the temperate climates are commonly low in earthworms. Their direct effect on plant growth is minimal, and attempts to increase plant growth by increasing earthworm activities in soils have been disappointing. Earthworms do not convert eroded soils low in organic matter into productive soils.

Arthropod Consumers and Decomposers

A high proportion of soil animals are arthropods. They have an exoskeleton and jointed legs. Most have a kind of heart and blood system and, usually, a well-organized nervous system. The most abundant are springtails and mites. Other arthropods include spiders, insects (including larvae), centipedes, millipedes, wool lice, snails, and slugs. Several of the most important will be discussed.

Springtails—Consumers with a Spring in Their Tail. Springtails are primitive insects less than 1 millimeter ($\frac{1}{25}$ inch) long.

They have a spring-like appendage under their posterior end that permits them to "flit to and fro" (*see* Fig. 5-7). They are very abundant (*see* Table 5-1) and are distributed worldwide. They live in the macropores of the litter layers and feed largely on dead plant and animal tissue, feces or dung, humus, and fungal mycelia. Water is lost rapidly through the cuticle, so they are restricted to moist soil layers. Those in the lower litter layers are the most primitive and are without eyes and pigment. Springtails are in the order *Collembola* and are commonly called Collembola. Their enemies includes mites, small beetles, centipedes, and small spiders.

Mites—Vegetarian and Carnivorous Consumers (Arachnids).

My experience with mites has been largely with red spider mites that bother plants, and with ear mites that bother dogs. Ticks are blood sucking mites. Mites are the most abundant air-breathing soil animals. They commonly have a sac-like body with protruding appendages, and are related to spiders (*see* Fig. 5-7).

Most mites feed on dead organic debris of all kinds, such as fungal hyphae and spores. Some mites are predaceous consumers and feed on nematodes, insect eggs, and other small animals, such as springtails. Activities of mites include breakup and decomposition of organic materials, movement of organic material to deeper soil layers, and maintenance of pore spaces (runways).

Vegetarian Millipedes and Carnivorous Centipedes (Myriapods).

Millipedes and centipedes are elongate, fairly large soil animals with many pairs of legs. They are common in forests, and overturning almost any stone or log will send them running for cover. Millipedes have many pairs of legs and are vegetarians or decomposers; they feed mainly on dead organic matter (saproghagous). Some browse on fungal mycelia. Centipedes typically have fewer pairs of legs compared to millipedes and are mainly carnivorous consumers. Centipedes will attack and consume almost any animal of any size they can master. Because they are fairly

Figure 5-7
Two of the most common soil animals. *(Left)* Springtail (Collembola) on top of pin head and magnified 50 to 100 times. *(Right)* Similar photo of a mite. (Photo courtesy of Michigan State University Pesticide Research Electron Microscope Laboratory.)

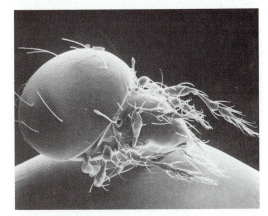

large compared to mites and springtails, their numbers in soils are small (*see* Fig. 5-8).

White Grubs—Consumers of Roots.

In some cases, the insect larvae play a more important role in soils than adults, as in the case of the white grub. White grubs are larvae of the familiar, brown May beetle or June bug. The grubs are round, white, about 2 to 3 centimeters long, and curl into a C shape when disturbed. The head is black, with three pairs of legs just behind the head. They feed mainly on grass roots, causing dead spots in lawns. A wide variety of other plants are also attacked, making white grubs an important agricultural pest. Moles feed on insect larvae and earthworms, resulting in greater likelihood of mole damage in lawns when white grubs are present.

Ants and termites are important soil insects that will be considered later in relation to earth-moving activities.

Interdependence of Microorganisms and Animals in Organic Matter Decomposition

Microorganisms and fauna work together as a team. When a leaf falls on the forest floor, both microorganisms and animals attack the leaf. Holes made in the leaf by springtails and mites facilitate the entrance of microorganisms inside the leaf. Soil animals ingest bacteria when they feed, and the bacteria continue to function in the digestive tract of small animals. The excrement of animals is attacked by both the microbes and the fauna. The entire decomposing mass, along with mineral soil particles, may be ingested by earthworms, thereby producing an intimate mixing of organic and mineral matter. The net result is the humification of organic matter, with both microorganisms and fauna playing important roles. In the process, the major role of animals is fragmentation and mixing, which greatly increase the surface area and prepare the organic matter for microorganisms. Major credit for the mineralization and recycling of mineral elements goes to microorganisms.

Nutrient Cycling

Nutrient cycling is the exchange of nutrient elements between the living and nonliving parts of the ecosystem. Two broad processes are involved. *Immobilization* is the uptake of inorganic nutrient ions by organisms. *Mineralization* is the conversion of nutrients in organic matter into inorganic ions, principally by microbial decomposers. Nutrient cycling conserves the nutrient supply and results in repeated use of nutrients. The net effect is greater ecosystem productivity than if nutrient cycling did not occur.

Nutrient Cycling Processes

Fig. 5-9 shows that the organic matter which is added to the soil consists of a variety of compounds. These include fats, carbohydrates, proteins, and lignins. Incorporation of these organic compounds into the soil stimulates, to the greatest ex-

Figure 5-8
A carnivorous centipede.

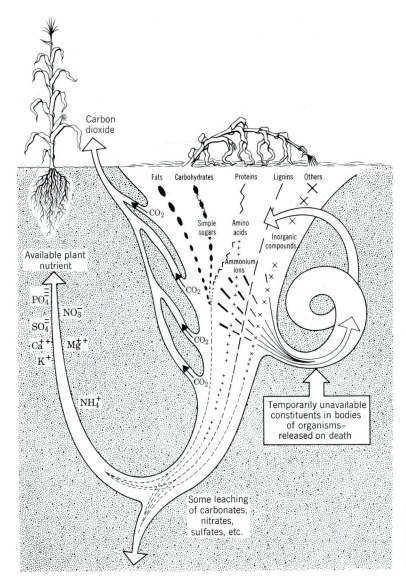

Carbon dioxide

Fats   Carbohydrates   Proteins   Lignins   Others

CO_2

Simple sugars

Amino acids

Inorganic compounds

Ammonium ions

Available plant nutrient

$\overline{PO_4}$

NO_3^-

$\overline{SO_4}$

Ca^{++} Mg^{++}

K^+

CO_2

CO_2

CO_2

NH_4^+

Temporarily unavailable constituents in bodies of organisms— released on death

Some leaching of carbonates, nitrates, sulfates, etc.

Figure 5-9
Schematic diagram of organic matter decomposition and nutrient cycling. (Courtesy of Dr. B. Sabey.)

tent, the organisms that are benefited the most. As decomposition proceeds, the most easily digested materials disappear first. All groups can effectively break down and use carbohydrates and proteins, but fungi are the most effective in decomposing lignin.

While digesting the plant residues, microbes use some of the carbon, energy,

and other nutrients for their own growth. In time, the synthesized tissue dies and becomes the substrate for further decomposition. Fig. 5-9 indicates this by the subcycle, where constituents in the living organisms are temporarily unavailable or *immobilized*. The immobilization of nutrients refers to the use and incorporation of nutrients into living matter by both microbes and higher plants. Immobilized nutrients are again mineralized when the organisms die. In time, even the most resistant materials succumb to the enzymatic attack of microbes. The net effect is the release of energy as heat, the formation of carbon dioxide and water, and the appearance of nitrogen as ammonium (NH_4^+), sulfur as sulfate ($SO_4^=$), phosphorus as phosphate (PO_4^-3), and many other nutrients as simple metallic ions (Ca^{++}, Mg^{++}, K^+). Most of these forms are available to living organisms for another cycle of growth.

As soon as some of the ions or elements released in organic matter decomposition appear, other specialized organisms oxidize some of them. These transformations are beneficial to the extent that the oxidized forms are more readily used by higher plants. This is the case for the oxidation of sulfur to sulfate. The oxidation of iron and manganese makes them less soluble and, therefore, less available to higher plants.

When soils become anaerobic, still other microbes become active. The oxidized forms of plant nutrients are reduced. Sulfate is reduced to hydrogen sulfide and insoluble mineral sulfides. Perhaps, one of the most important reactions in anaerobic soil is the reduction of nitrate, which results in the formation of nitrogen gas that escapes from the soil (denitrification). Loss of nitrogen by denitrification normally involves about 10 percent of the nitrogen turnover per year in well-aerated soils and more than this in poorly aerated soils.

A Case Study of Nutrient Cycling

One approach to the study of nutrient cycling is to select a small watershed and measure changes in nutrients over time.

Our example involves a 15 hectare (38 acre) watershed in the White Mountain National Forest in New Hampshire. For a mature forest, the amount of calcium (and other nutrients) taken up from the soil approximates the amount of nutrients returned to the soil in leaves, dead wood, and the like. For calcium (Ca) in New Hampshire, this was 49 kilograms per hectare or 44 pounds per acre taken up and returned to soil annually (*see* Fig. 5-10). About 9 kilograms of available calcium were added to the system by mineral weathering and 3 kilograms were added in the precipitation per hectare, annually. These additions were balanced by losses due to leaching and runoff. This means that 80 percent of the calcium in the cycle was recycled or reused by the forest each year. Plants were able to take up 49 kilograms of calcium each year, while only 12 kilograms of new calcium was being made available. The researchers concluded that northern hardwood forests have a remarkable ability to hold and circulate nutrients.

The data in Table 5-2 show that 70 to 86 percent of four major nutrients absorbed from the soil were recycled in the vegetation each year. Without such extensive recycling, the productivity of the forests would be lower. Nutrient cycling accounts for the existence of enormous forests on some very infertile soils. In this situation, most of the nutrients are in soil organic matter.

Effect of Deforestation on Nutrient Cycling

We have established that the litter layers or surface soil horizons enriched with organic matter are an important nutrient reservoir. Furthermore, vegetation immobilizes the mineralized nutrients and keeps them in the cycle. The question

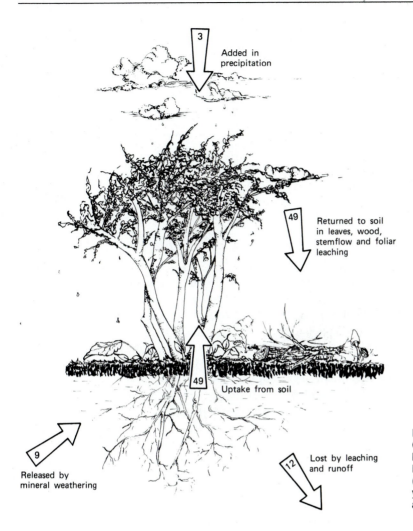

3
Added in
precipitation

49
Returned to soil
in leaves, wood,
stemflow and foliar
leaching

49
Uptake from soil

9
Released by
mineral weathering

12
Lost by leaching
and runoff

Figure 5-10
Calcium cycling in
hardwood forest in New
Hampshire in kilograms
(kilos) per hectare per
year. (Data from Bormann
and Likens, 1970.)

now is "What is the effect of deforestation on nutrient cycling?"

Researchers in the White Mountain National Forest cut all the vegetation on a watershed and measured the changes in nutrient cycling. The soil was not disturbed to minimize nutrient loss by erosion. Herbicides were applied to prevent nutrient immobilization by plants. Deforestation occurred in early 1966, and an increased loss of nutrients in the water leaving the watershed was detected in May. The losses of nitrogen from the watershed became greater than the amount of nitrogen normally immobilized by the vegetation. The concentration of nitrate in the *stream water* leaving the watershed was increased manyfold. Reestablishment of the vegetation resulted in an immobilization of nutrients and a reduced loss of nutrients from the watershed by mid-1968. In time, the nitrogen concentration

Table 5-2

Annual Uptake, Retention, and Return of Nutrients to the Soil in a Beech Forest

	Nutrients							
	Kilograms per Hectare				Pounds per Acre			
	N	P	K	Ca	N	P	K	Ca
Uptake from soil	50	12	14	96	45	11	13	86
Stored in wood or lost from soil	10	2	4	13	9	2	4	12
Returned to soil in litter	40	10	10	83	36	9	9	74
Percent recycled	80	82	70	86	80	82	70	86

Stoeckler, J. H., and H. F. Arneman, "Fertilizers in Forestry," *Adv. in Agron. 12:*127–195, 1960. By permission of author and Academic Press.

of the stream waters should equal the concentration that existed before deforestation.

Clear-cutting of forests not only removes the trees that immobilize nutrients, but it may also expose the soil to increased erosion. Particulate losses of nutrients are greatly increased as the surface soil is eroded away. Streams are enriched with nutrients and soil particles. As in the case of the deforestation experiment, the reestablishment of vegetation quickly reduces both nutrient leaching and erosion losses and causes a return to normalcy within several years.

Effect of Crop Harvesting on Nutrient Cycling

The grain and stalks or stover of a corn (maize) crop may contain over 200 kilograms of nitrogen per hectare (or 200 pounds per acre). It is not uncommon for farmers in the Corn Belt of the United States to add this amount of nitrogen to the soil annually to maintain high yields. According to the data in Table 5-2, by contrast, only 10 kilograms of nitrogen

per hectare (9 pounds per acre) are needed to maintain a beech forest. Naturally occurring nitrogen fixation processes and precipitation can readily supply this amount of nitrogen.

Basically, food production represents nutrient harvesting. The natural nutrient-cycling process can not provide enough nutrients to grow crops with yields high enough to feed the world's population. Yields of grain in many of the world's agricultural systems stabilize at about 500 to 600 kilograms per hectare (8 to 10 bushels per acre) without the addition of nutrients in manure or fertilizer.

Role of Manure Use in the Maintenance of Soil Nutrients

The early settlers in America found it difficult to establish a prosperous agriculture in the sandy, infertile soils (low in organic matter) along the eastern seaboard. You may recall that the Indian Squanto showed the Pilgrims how to fertilize corn with fish to increase the nutrient supply. Production of plantation crops and their shipment to Europe put a heavy nutrient drain on soils. Land abandonment and exploitation of new land further west was widespread as early as the 1700s. Subsequently, a permanent and prosperous agriculture was established in the east, and in states further west, based in part on the judicious use of animal manure.

Nutrients in feed, which are not used to build body tissue or milk, are excreted in the manure or feces. On the average, farm animals return 75 to 80 percent of the nitrogen, 80 percent of the phosphorus, 85 to 90 percent of the potassium, and about 50 percent of the organic matter in manure from consumed feed (*see* Fig. 5-11). Crops can be harvested with only a modest drain on the nutrients in the cycle, if the nutrients in the manure are returned to the land. Careful collection and management of the manure and the use of legumes to fix nitrogen, along with lime to reduce soil acidity, were the most important practices that resulted in the properous agriculture of dairy states such as New York and Wisconsin in the nineteenth century.

Nutrient Transference Between Ecosystems

The world does not gain or lose nutrients, practically speaking. Nutrients exist in many forms, are unevenly distributed, and are moved from place to place. The nutrients that leave one ecosystem are added to some other ecosystem. Some nutrients are moved about as dust in the air to create a general movement of nutrients from deserts to more humid regions. Certainly, some nutrients will find their way to the mouth of some river. Here, the nutrients will support aquatic plants (primary producers) and serve as food for fish (consumers). Many ocean surface waters, however, are so low in nutrients as to be called a "biological desert." Good fisheries tend to occur at the mouths of rivers that bring in nutrients. Construction of dams on the Nile River has reduced fishing at the river's mouth and in the adjacent Mediterranean Sea because of reduction in transference of nutrients from the headwaters. Flooding and deposition of silt accounts for the generally high fertility of alluvial soils in the Nile Valley and of other alluvial soils found in the world. Nutrients are transferred from headwaters to floodplains more rapidly than they are removed by leaching, and other methods.

The transference of nutrients from terrestrial ecosystems to aquatic ecosystems is a natural phenomenon. In the larger geologic cycle, calcium in soil minerals weathers into an available form that is subject to plant use and removal from soils by leaching. In humid regions, calcium is gradually moved to the oceans. After calcium

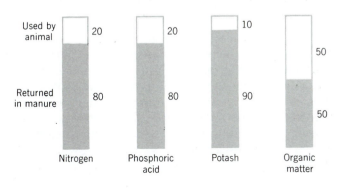

Figure 5-11
Average proportion of plant nutrients and organic matter in feed consumed by animals and excreted by manure.

reaches the ocean, it may be cycled many times among the primary producers, consumers, and decomposers before calcium is precipitated in the form of limestone. Uplift of limestone and initiation of a new cycle of soil formation frees calcium to make another trip through the larger geologic cycle.

The creation of cities resulted in movement of nutrients (in food) from farms to cities. In the Orient, there has been a considerable transfer of nutrients in wastes back to the land. The high population density of China, in the early 1900s and today, has been made possible by efficient transference of nutrients from the land to cities, or villages, and back to the land. In the United States, nutrients are moved to the cities; however, the nutrients are mainly disposed of in some water course and are not returned to the land. Fertilizers are used to make up the difference. As nutrients in fertilizers become more expensive, there is more interest in returning the nutrients back to the land for reuse.

A key point is that food production depends on nutrients in both terrestrial and aquatic ecosystems. In nature, there is considerable conservation and reuse of nutrients. The large shipments of grain from the United States and Canada to other countries represent a massive movement of nutrients to foreign countries.

Relation of Soil Organisms to Higher Plants

The role of soil organisms in nutrient cycling has been stressed; we have seen that some organisms are parasites feeding on plants. In this section, we will consider the specialized roles of microorganisms living near, or adjacent to, plant roots.

The Rhizosphere

Plant roots leak or exude a large number of organic substances. Sloughing of root caps also provides much new organic matter. These substances are food for organisms, and they cause a zone of intense biological activity near the roots in the area called the *rhizosphere*. Many kinds of organisms inhabit the rhizosphere; the bacteria are benefited most. Bacterial colonies may form a continuous film around the root. Roots supply microorganisms in the rhizosphere with food.

Fungus Roots or Mycorrhizae

Fungi infect the roots of most plants. Fortunately, most of the fungi form a symbiotic relationship—a relationship that is mutually beneficial. After mycorrhizal spores germinate, hyphae invade rootlets, growing both inside and outside the rootlet (long or major roots are seldom invaded). Fungal hyphae on the exterior of roots serve as an extension of roots for water and nutrient absorption. These fungus roots are called *mycorrhizae*.

There are two types of hyphae: ectotrophic and endotrophic. Ectotrophic hyphae exist between the epidermal root cells, using pectin and other carbohydrates for food. Continued growth of hyphae outside the root may result in the formation of a sheath, or mantle, that completely surrounds the root (*see* Fig. 5-12).

There are literally thousands of species that belong in the group of fungi that form ectomycorrhizae and produce mushroom or puffball-fruiting bodies (Basidiomycetes). The fungi shown in Fig. 5-4 form mycorrhizae. They are associated typically with trees and, thus, benefit plants, as illustrated in Fig. 5-13.

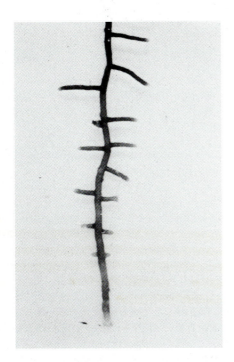

Anatomy of an Ectotrophic Mycorrhiza

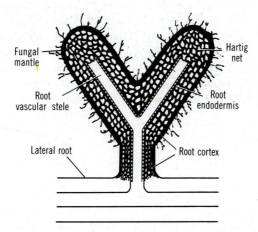

Figure 5-12
Ectomycorrhiza. *(Top)* Uninfected pine root. *(Bottom)* Infected root with fungal mantle. The Hartig net is the hyphae between the root cells. (Photo Courtesy of Dr. D. H. Marx, University of Georgia.)

Figure 5-13
Effect of mycorrhizal fungi on the growth of 6-month-old Monterey pine seedlings: *(a)* fertile Prairie (grassland) soil, *(b)* fertile Prairie soil plus 0.2 percent by weight of Plainfield sand from a forest, and *(c)* Plainfield sand alone. (Photo courtesy of the late Dr. S. A. Wilde, University Wisconsin.)

Endotrophic mycorrhizae are more abundant than ectomycorrhizae; they benefit most field and vegetable crop plants. Hyphae invade roots and ramify within and between cells, usually avoiding invasion of the very center of the root. Coils of hyphae or branched hyphal structures, formed within root cells, are called *arbuscles*. Swellings that form on the hyphae and contain oil are *vesicles*. These structures form the basis for referring to endomycorrhizae as *vesicular-arbuscular* (VA) mycorrhizae.

The host plant provides the fungus with food. Benefits to the host include:

1. Increasing effective root surface with increased effectiveness in absorption of nutrients (particularly phosphorus) and water.
2. Rootlets functioning longer.
3. Increasing heat and drought tolerance.
4. Rendering soil nutrients more available.

5. Deterring infection by disease organisms. This last benefit only occurs in the case of ectomycorrhizae.

Nitrogen Fixation

We live in a sea of nitrogen, because the atmosphere is 79 percent nitrogen. In spite of this, nitrogen is perhaps the most limiting nutrient for worldwide plant growth. Nitrogen in the air exists as an inert gas (N_2) and, as such, is generally unavailable to biological organisms. There are some species of bacteria, algae, and actinomycetes that can absorb nitrogen gas from the air and convert it into ammonia, which plants can use. This process is *nitrogen fixation*.

Some of the (nitrogen-fixing) bacteria are free living or nonsymbiotic. Many of the important nitrogen fixers are symbiotic and live in association with a host plant. The symbiotic, nitrogen-fixing bacteria, called *Rhizobia*, infect a root; the plant responds by the formation of a nodule (*see* Fig. 5-14). The host plant supplies bacteria with food and, in return, the host plant is benefited by the fixed nitrogen. Many plants, both legumes and nonlegumes, are involved in symbiotic nitrogen fixation. Symbiotic nitrogen fixation is the major means by which inert nitrogen in the atmosphere is transferred to soils for use by plant roots and microorganisms.

Blue-green algae (Cyanobacteria) are nonsymbiotic nitrogen fixers which live, generally, in aquatic ecosystems; they are close relatives of bacteria. Blue-green algae are unique in that they are autotrophs and can fix nitrogen nonsymbiotically. The limited growth of algae in soils means that they play a very minor role in adding nitrogen to most terrestrial ecosystems. Algae fix important amounts of ni-

 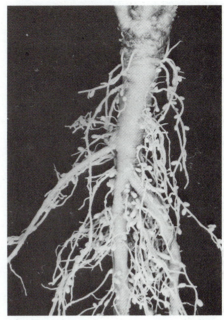

Figure 5-14
Symbiotic nitrogen fixing nodules on the roots of soybeans on left and sweet clover on the right. (Photo courtesy of Nitragin Co.)

trogen in the water of rice paddies and contribute importantly to the yields of rice. Whereas bacteria play the dominant role in transferring nitrogen from the atmosphere to the soil in terrestrial ecosystems, algae play the dominant role in transferring nitrogen from the air into the waters of aquatic ecosystems.

Production of Disease

Soil frequently contains a rather large number of organisms that cause diseases either in plants or animals. Some of these organisms live in soil only temporarily, and others use it as a permanent habitat. Soil may harbor organisms that cause bacterial diseases, such as wilt of tomatoes and potatoes, soft rots of a number of

vegetables, leaf spots, and galls. Some of the most destructive parasites are the disease-causing fungi, such as those that cause damping-off of seedlings, cabbage yellows, mildews, blights, certain rusts, wilt diseases, scab, dry rot of potatoes, and many others. Certain species of *Actinomycetes* may cause diseases, such as scab in potatoes and sugar beets and pox in sweet potatoes. The catastrophic potato famine in Ireland from 1845 to 1846 was caused by a fungus that produced potato blight.

Soil used for bedding and greenhouse plants is routinely treated with heat or chemicals to kill plant pathogens. Weed seeds are also killed. Fumigating field soils is a common treatment for controlling nematodes. Heat treatment does not sterilize soil, since many bacteria and fungi are not killed.

Competition for Nutrients

Microorganisms or protists have some nutrient needs in common with higher plants, absorb nutrients by similar processes, and compete for the same nutrient supply. We have noticed, however, that microorganisms may contribute to a greater available nutrient supply. The major cause of competition appears to be for nitrogen. When large quantities of decomposable organic matter are added to soils that are low in available nitrogen, microorganism growth may be stimulated to such an extent that serious competition for nitrogen may occur. This problem will be discussed in more detail in the next chapter.

Soil Organisms and Environmental Quality

During the billions of years of organic evolution, organisms evolved that could decompose all compounds formed directly or indirectly from photosynthesis. Today, human beings have become important contributors of compounds to the environment. The problems of pesticide degradation, disposal of sewage effluent, and contamination of soils with oil spills will be considered as they relate to environmental quality.

Pesticide Degradation in Soils

Pesticides include those substances used to control or to eradicate insects, disease, organisms, and weeds. One of the first and most successful pesticides was DDT, used to kill mosquitoes for malaria control. Today, about 40 years later, there is evidence to indicate that some DDT exists in the cells of all living animals. This has dramatized the resistance of DDT to biodegradation. Evidence supports the view that the structure of DDT is different from most naturally occurring compounds and, as a consequence, few soil organisms have developed enzyme systems that degrade DDT. In general, it is believed that pesticides with structures similar to those found in naturally occurring compounds are degradable; pesticides with new structures that are not naturally found are persistent. This is illustrated by comparing 2,4-D(2,4-dichlorophenoxyacetic acid) with 2,4,5-T(2,4,5-trichlorophenoxyacetic acid). The two compounds have very similar structures (*see* Fig. 5-15), except that 2,4,5-T has an extra chlorine at the meta position of the ring. Although 2,4-D is readily decomposed (in fact, some organisms can use 2,4-D as their only source of carbon), 2,4,5-T is very resistant. A chlorine on the meta position is metabolized with difficulty, if at all; consequently, the 2,4,5-T is a very persistent (or is a "hard") pesticide.

We have become very dependent on pesticides, and the likelihood of eliminating their use is remote. The challenge for

Figure 5-15
Structure representing 2,4-D on the left and 2,4,5-T on the right. The structures are very similar except for the additional Cl at the meta position of 2,4,5-T, which is metabolized with great difficulty, if at all, by soil microorganisms.

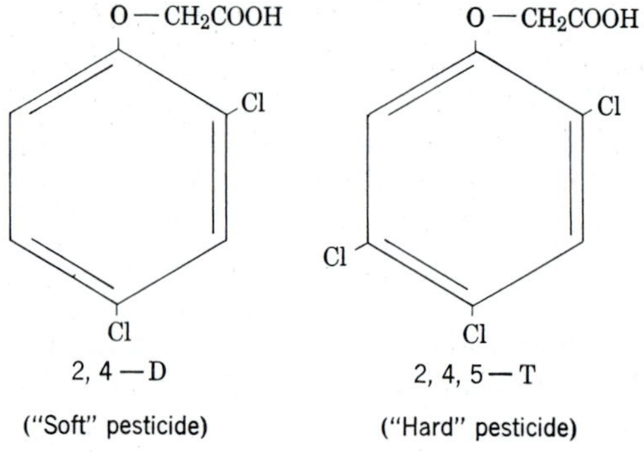

2, 4 —D

("Soft" pesticide)

2, 4, 5 — T

("Hard" pesticide)

scientists is the development of pesticides that can perform their useful function and disappear from the ecosystem without any undesirable side effects.

Soil as a Living Filter for Sewage Effluent Disposal

A relatively small community of 10,000 people may produce about 1,000,000 gallons or 4,000,000 liters of waste water or effluent per day from its sewage treatment plant. The effluent looks much like ordinary tap water; when it is chlorinated, it is safe for drinking. Discharging the effluent into a nearby stream does not create a health hazard. The effluent, however, is enriched with nutrients, and plant growth in streams and lakes may be increased by the addition of nutrients. Increased plant growth in water increases the consumption of oxygen, because more organic matter decomposition will eventually result. As a consequence, oxygen levels in water may become too low for fish to live. Weed growth may be stimulated by nutrients, and weeds interfere with boating and swimming. The problem is more acute today, because many cities have grown rapidly while the amount of water available for diluting and carrying away the effluent has remained about the same. At the same time, many cities are depleting their underground water supplies, which are a major source of water for use in sewage disposal.

To combat these problems, researchers at Pennsylvania State University set up experiments to use soil as a living filter for sewage effluent disposal. The researchers expected microorganisms to degrade detergents (similar to the problem of pesticide degradation) and, along with higher plants, to immobilize nutrients as the effluent water slowly percolated through the soil. By applying water in excess of the potential evapo-transpiration, they expected the filtered water to migrate downward and eventually recharge the aquifer for reuse. The water was applied with a sprinkling system (*see* Fig. 5-16).

The experiment has been a success. For a community of 10,000 people, producing 1,000,000 gallons of effluent a day, the researchers found that an application of 5 centimeters per week on 50 hectares of land was satisfactory. Under these conditions, over 80 percent of the effluent water migrated deep enough to be considered aquifer recharge water. After 3 years, the water, recovered at a depth of slightly over 1 meter, showed that over 90 percent of the hard detergents had been removed. There was almost a complete removal of nitrogen and phosphorus (two major nutrients that have been associated with eutrophication). Furthermore, tree and crop growth on the land were greatly stimulated.

Oil and Natural Gas Contamination of Soils

Recent construction of the Alaskan Pipeline has created an awareness and concern for oil spills on-land. Many soils along the Alaskan Pipeline have pergelic temperature regimes. There has been uncertainty about the effect of construction on the melting of permafrost. Oil spills and gas leaks near oil wells and pipelines, however, are as old as the petroleum industry. Many studies have been conducted on the effect of petroleum contamination on soils and plant growth; a summary of the major affects will be presented.

The first result of an oil spill or natural gas leak is the displacement of soil air and the creation of an anaerobic soil. Any vegetation is likely to be killed by a lack of soil

Figure 5-16
Application of sewage
effluent in a forest with a
sprinkler system in winter
at Pennsylvania State
University. (Photo
courtesy of USDA.)

oxygen. Over 100 species of microbes can decompose petroleum products. Crude oil and natural gas are a good energy source for microorganisms, and their growth is greatly stimulated. Reducing conditions accompanying the decomposition results in large increases in available iron and manganese. It is suspected that toxic levels of manganese for higher plants are produced in some cases.

In severe cases of oil spills, soil aggregates are broken down and dispersion results. There is usually an increase in micropores and an increase in bulk density. Cultivation and aeration of soil hasten the return of soil to an aerated state and the return of vegetation. In time, the oil or natural gas is decomposed and the soil returns to almost normal conditions. The low soil temperatures in Alaska would slow down dissipation of oil by microorganisms. A study of 12 soils exposed to contamination near Oklahoma City

showed that organic matter and nitrogen contents had been increased about 2.5 times. The increased nitrogen supply in soils that have been contaminated probably explains the increased crop yields observed on old oil spills.

Natural gas leaks are not uncommon in cities. The death of grass and trees is due to lack of soil air, which is displaced by natural gas or strongly reducing soil conditions. Sometimes, the problem can be diagnosed by smelling the loss of natural gas from a crack in a nearby sidewalk or driveway. (Methane is odorless; however, an odor has been added to public gas supplies.)

Use of Soil for Animal Waste Disposal

The concentration of large numbers of animals on specialized livestock farms has created serious animal waste disposal problems. Odors from stored or rotting manure are commonly offensive. Burning animal wastes can pollute the air. Laws exist that prohibit the disposal of animal wastes into water courses. Chemical processing of animal wastes is expensive. As a consequence, animal waste disposal is a major problem for many livestock producers. Soil is a natural medium for waste disposal. More attention is now being directed toward using soil for animal waste disposal.

Magnitude of the Animal Waste Disposal Problem

An interesting comparison between the production of human and animal wastes has been made for Minnesota. In Minnesota there are about 14,000,000 chickens and turkeys, 4,000,000 cattle, 1,500,000 dairy cows, 2,500,000 hogs, and 750,000 sheep. These animals produce waste equivalent to that produced by over 60,000,000 people. Since the population of Minnesota is about 4,000,000, the animal wastes produced are about 15 or more times greater than the human wastes.

Some feedlots in the United States carry as many as 10,000 head of livestock and produce waste equivalent to a city of 164,000 people (*see* Fig. 5-17). Some data on animal population and quantity of waste produced in the United States are given in Table 5-3. The data show that about 1,500,000,000 tons of waste are produced annually; about two-thirds is solid manure and about one-third is liquid. If the waste resulting from dead animal carcasses, paunch manure from abattoirs, and used bedding are included, the total amount of animal waste produced is about 2,000,000,000 tons per year.

Capacity of Soil for Waste Disposal

Under normal conditions, soil microorganisms exist in a near state of starvation. Readily available food or energy supplies are usually limited. The addition of decomposable wastes, such as manure, produce a rapid increase in the numbers and

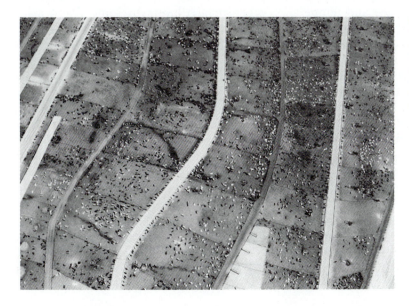

Figure 5-17
Aerial view at large feedlot at Coalinga, California. The waste disposal problem for each 10,000 cattle is equal to that of a city of 164,000. (Photo courtesy of EPA.)

Table 5-3
Animal Population and Waste Production
in the United States

Livestock	1965 Population, in Millions	Annual Waste Production, in Million Tons	
		Solid Waste	Liquid Waste
Cattle	107	1004.0	390.0
Horses	3	17.5	4.4
Hogs	53	57.3	33.9
Sheep	26	11.8	7.1
Chickens	375	27.4	—
Turkeys	104	19.0	—
Ducks	11	1.6	—
Total	—	1,138.6	435.4

From Wadleigh, 1968.

activity of microorganisms. As a result, soil can dispose of enormous quantities of animal waste. It seems reasonable to expect that a 2- to 3-centimeter layer of manure could be incorporated into the soil every year without difficulty, assuming favorable soil moisture and temperature. The nitrogen application would be about 800 to 1,000 kilograms per hectare (700 to 900 pounds per acre). On this basis, nitrate pollution of ground water would likely result in humid regions from the application of such a large quantity of manure. From a practical standpoint, it appears that considerations other than the decomposing capacity of soil limit the rate of waste application. Researchers in Canada analyzed manure on many kinds of farms and concluded that to prevent nitrate contamination of ground-water and not adversely affect corn yields, a 0.2-hectare minimum ($\frac{1}{2}$ acre) of land was needed for 1,000 broilers, 100 laying hens, 10 hogs (30 to 200 pounds), 2 feeder cattle (400–1,100 pounds), or 1 dairy cow (1,200 pounds).

Earth-Moving Activities of Soil Animals

All soil animals participate as consumers and aid in the cycling of nutrients and energy. Their harvesting and food storage activities result in transference of nutrients to their nests or burrows. Many of the larger soil animals move soil to such an extent that they affect soil formation. The emphasis in this section is on animals as earth movers.

Earthworm Activities

Earthworms are perhaps the best-known earth movers. Darwin made extensive studies of earthworms and found that they may deposit 4 to 6 metric tons of castings per hectare (10 to 15 tons per acre) per year on the soil surface, resulting in the buildup of a 2.5 centimeter surface layer every 12 years. This activity produces thicker than normal, dark-colored surface layers in some forest soils and buries stones and artifacts that are lying on top of the soil. The burying of artifacts is important to archeologists.

Ants and Their Activities

Although most persons appear to be more conscious of earthworms and their activities, the activities of ants are perhaps more important. Harvester ants are a pest in many places, including the southwestern part of the United States. Harvester ants denude the area surrounding their nests. Thorp estimated an average of 50 ant hills per hectare (20 per acre). Assuming an average denuded area of 4 meters in diameter, about 6 percent of the land surface would be denuded. Such harvesting of vegetation by ants can be of economic importance (*see* Fig. 5-18). On range lands, the forage for wildlife and cattle is reduced and the bare denuded areas are more subject to erosion. Harvester ants also gather seeds for food; this retards the reseeding of natural grasslands.

Ants transport large quantities of material from within the soil, depositing it on the surface. Some of the largest ant mounds are about 1 meter high and more than 3 meters in diameter. The effect of this transport is comparable to that of earthworms in creating thicker, dark-colored A horizons and burying objects lying on the surface. A study of ant activity on a prairie in southwestern Wisconsin showed that ants brought material to the surface from depths of about 2 meters and built mounds about 15 centimeters high and over 30 centimeters in diameter (*see* Fig. 5-19). Furthermore, it was estimated that 1.7 percent of the land was covered with mounds. Assuming that the average life of a mound is 12 years, the entire land surface would be reoccupied every 600 years. The researchers believe

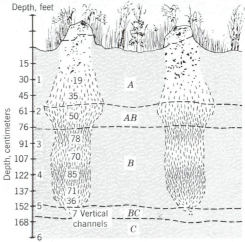

Figure 5-19
Ant *(Formica cinera)* in a Prairie soil (Mollisol) in southwestern Wisconsin. Upper photo shows ant mounds over 15 centimeters high and over 30 centimeters in diameter. The lower sketch shows soil horizons and location of ant channels; numbers refer to the number of channels observed at the depths indicated. (Photo courtesy of F. D. Hole, Soil Survey Division, Wisconsin Geological and Natural History Survey, University of Wisconsin.)

Figure 5-18
Denuded areas surrounding Red Harvester ant mounds in an alfalfa field. (Photo courtesy of USDA.)

that this evidence supports the view that the incorporation of subsoil material (Bt horizon) into the A horizon has helped to produce a thicker, dark-colored A horizon with a greater than normal clay content. The increase in clay content is supported by the fact that the clay contents of A horizons in a nearby forest are only 10 percent compared to 22 percent clay in the A horizons on the prairie.

Leaf-cutting ants march long distances to cut fragments of plant leaves and stems and to bring them to their nests and feed fungi. The fungus is used as food. Organic matter is incorporated into the soil depths; nutrients become concentrated in the nest sites.

Termite Activities

Termites inhabit tropical and subtropical areas. They exhibit great diversity in food and nesting habits. Some feed on wood, some feed on organic refuse, and others cultivate fungi. Protozoa in the digestive tract of many termites aid in the digestion of woody materials. Some species build huge nests up to 3 meters high and 10 to 15 meters in diameter. Most mounds are of a smaller scale. Some termites have nests in the soil and tunnels on the surface to permit foraging for food. Material is brought to the surface from depths as great as 3 meters. Nutrients and soil particles are transferred to the surface. Termites have had an enormous effect on many tropical soils where their earth-moving activities have been going on for hundreds of thousands of years. In summary, ants and termites create channels in soils and transport soil materials that tend to alter or obliterate soil horizons. A concentration of nutrients builds up where the mounds are located, because plant materials are stored and fecal material accumulated there. Some farmers in Southeast Asia recognize this and make use of the higher fertility of areas occupied by mounds.

Earth-Moving Activities of Rodents

Many rodents, including mice, ground squirrels, marmots, gophers, and prairie dogs, inhabit soil. A very characteristic microrelief called *mima mounds,* which consists of small mounds of earth, is the work of gophers in Washington and California. Mima mounds occur on shallow soils. They appear to be the response of gophers to building nests in dry soil near the tops of mounds. Successive generations of gophers at the same site create mounds that range from 0.5 to 1 meter high and 5 to 30 meters in diameter.

Extensive earth moving by prairie dogs has been documented. An average of 42 mounds per hectare (17 per acre) were observed near Akron, Colorado; they consisted of 39 tons of soil material. The upper 2 to 3 meters of soil was loess (wind-deposited silt), underlain by sand and gravel. All of the mounds observed, contained sand and gravel that had been brought up from depths over 2 meters. Prairie dog activity had changed the surface soil texture from silt loam to loam on one third of the area. Abandoned burrows filled with dark-colored surface soil are common in grassland soils and are called *crotovinas.*

Bibliography

Allison, F.E., *Soil Organic Matter and Its Role in Crop Production,* Elsevier, New York, 1973.

Andrews, S.A., Ed., *A Guide to the Study of Soil Ecology,* Prentice-Hall, Englewood Cliffs, N.J., 1975.

Arkley, R.J., and H.C. Brown, "The Origin of Mima Mound (Hogwallow) Microrelief in the Far Western States," *Soil Sci. Soc. Am. Proc., 18:*195–199, 1954.

Barley, K.P., "The Abundance of Earthworms in Agricultural Land and Their Possible Significance in Agriculture," in *Advances in Agronomy,* Vol. 13, Academic, New York, 1961, pp. 249–268.

Baxter, F.P., and F.D. Hole, "Ant (Formica cinerea) Pedoturbation in a Prairie Soil,"

*Soil Sci. Soc. Am. Proc., 31:*425–428, 1967.

Bormann, F.H., and G.E. Likens, "The Nutrient Cycles of an Ecosystem," *Sci. Am., 223:*92–101, 1970. Reprint 1202.

Bowen, E., *The High Sierra,* Time, New York, 1972.

Brady, N.,C., *The Nature and Properties of Soils,* 8th Ed., Macmillan, New York, 1974.

Buol, S.W., F.D. Hole, and R.J. McCracken, *Soil Genesis and Morphology,* Iowa State University, Ames, Iowa, 1973.

Burges, A., and F. Raw, *Soil Biology,* Academic, New York, 1967.

Clark, F.E., "Living Organisms in the Soil," in *Soil,* USDA Yearbook, Washington, D.C., 1957, pp. 147–165.

Cole, LaMont C., "The Ecosphere," *Sci. Am.,* April 1958, Reprint 144.

Dart, P.J., and S.P. Waniswam, "Non-symbiotic Nitrogen Fixation and Soil Fertility," *Trans. 12th Int. Cong. Soil Sci.,* Symposia Papers 1, Delhi, 1982, pp. 3–27.

Ellis, B.G., "Nutrient Cycling in Agricultural Systems," in *Envir. Quality: Now or Never,* C.L. San Clemente, Ed., Cont. Ed. Service, Michigan State University, 1972.

Ellis, R., Jr., and R.S. Adams, Jr., "Contamination of Soils by Petroleum Hydrocarbons," in *Advances in Agronomy,* Vol. 13, Academic, New York, 1961, pp. 197–216.

Kardos, L.T., "Waste Water Renovation by the Land—A Living Filter," in *Agriculture and the Quality of Our Environment,* AAAS Pub. 85, Washington, D.C., 1967, pp. 241–250.

Kevan, D. McE., *Soil Animals,* Philosophy Library, New York, 1962.

Martin, W.P., "Soil as an Animal Waste Disposal Medium," *Jour. Soil and Water Cons. 25:*43–45, 1970.

Marx, D.H., and W.C. Bryan, "The Significance of Mycorrhizae to Forest Trees," in *Forest Soils and Forest Land Man.,* Proc. Fourth Nor. Am. Forest Soils Conf., Laval University, Quebec, 1975, pp. 107–117.

Odum, E.P., *Ecology,* Holt, Rhinehart and Winston, New York, 1963.

Richards, B.N., *Introduction to the Soil Ecosystem,* Longmans, New York, 1974.

Schaller, F., *Soil Animals,* University of Michigan, Ann Arbor, Mich., 1968.

Stoeckeler, J.H., and H.F. Arneman, "Fertilizers in Forestry," *Adv. in Agronomy,* Vol. 12, Academic, New York, 1960, pp. 127–195.

Thorp, J., "Effects of Certain Animals That Live in Soils," *Sci. Month., 42:*180–191, 1949.

Wadleigh, C.H., "Wastes in Relation to Agriculture and Forestry," *USDA Misc. Pub.,* 1065, 1968.

Walters, E. M. P., *Animal Life in the Tropics,* George Allen and Unwin, London, 1960.

Webber, L.R., "Animal Wastes," *Dept. Soil Sci. Ann. Prog. Report,* University of Guelph, Ontario, 1967.

Wilson, P.W., *The Biochemistry of Symbiotic Nitrogen Fixation,* University of Wisconsin Press, Madison, Wis., 1940.

CHAPTER 6

SOIL ORGANIC MATTER

Almost all life found in soil is dependent on organic matter for energy and nutrients. For thousands of years, mankind has recognized the importance of organic matter in food production. Perhaps the most poetic expression of the effects of organic materials on plant growth was expressed by Omar Khayyam.

I sometimes think that never blows so red
The Rose as where some buried Caesar
bled.

Although organic matter in soils is very beneficial, Liebig, a famous German chemist, pointed out, over 100 years ago, that soils composed entirely of organic matter are naturally infertile. This chapter clarifies the role and importance of soil organic matter.

Humus Formation and Characteristics

In Chapter 5, we discussed the decomposition of plant residues and the synthesis of many compounds by soil organisms. As a result of these activities, soil contains an enormous number of organic compounds in various states of decomposition. *Humus* is the word used when referring to organic matter that has undergone extensive decomposition and is resistant to further alteration.

Humus Formation

The organic residues added to soils are not decomposed as a whole; however, the chemical constituents are decomposed independently of one another. In the formation of humus from plant residues, there is (1) a rapid reduction of the water-soluble constituents, of cellulose, and of hemicelluloses; (2) a relative increase in the percentage of lignin and lignin complexes, and (3) an increase in the protein content. The new protein is believed to be formed, for the most part, through the synthesizing activities of microorganisms. The lignin in humus originates mostly from plant residues with, perhaps, certain chemical modifications. Lignin has a 6-carbon ring structure that resists enzymatic decomposition. Reactions of lignin with amino acids and other substances form resistant compounds and enhance the accumulation of both lignin and protein materials in humus (Table 6-1). Fats and waxes have intermediate resistance to decomposition.

Normally, proteins are readily decomposed in soils. Additional mechanisms have been proposed to account for the accumulation of proteins in humus. Two will be mentioned. First, there is reason to believe that protein molecules can be adsorbed on the surface of clay minerals and rendered resistant to decomposition. Second, enzymes that decompose proteins may also be adsorbed by clay minerals so that the proteins are less susceptible to decomposition. That the clay plays an important role is supported by the fact that soils high in clay content tend to have a high organic matter content. This slow rate of decomposition of humus is obviously of considerable practical importance. It offers a means whereby nitrogen can be stored in soil and mineralized gradually.

A general schematic summary of the processes leading to humus formation is presented in Fig. 6-1. The material commonly referred to as humus includes the mass of plant residues undergoing decomposition, together with the synthesized cell substance and certain intermediary and end products. It is constantly changing in composition. It is better, therefore, to speak of humus not as a single group of substances but, indeed, as a state of matter, which is different under varying conditions of formation.

Characteristics and Properties of Humus

One of the most important and characteristic properties of humus is its nitrogen content, which usually varies from 3 to 6 percent; the nitrogen concentration may be frequently lower or higher than these figures. The carbon content is less variable and is commonly considered to be 58 percent. Assuming 58 percent carbon, the organic matter content can be estimated by multiplying the percentage of carbon by 1.724. The carbon-nitrogen ratio (C/N) is of the order of 10 to 12. This ratio varies with the nature of the humus, the stage of its decomposition, the nature and depth of the soil, and the climatic and other environmental conditions under which it is formed.

Table 6-1
Partial Composition of Mature Plant Tissue and Soil Organic Matter

Component	Percent	
	Plant Tissue	Soil Organic Matter
Cellulose	20–50	2–10
Hemicellulose	10–30	0–2
Lignin	10–30	35–50
Protein	1–15	28–35
Fats, waxes, etc.	1–8	1–8

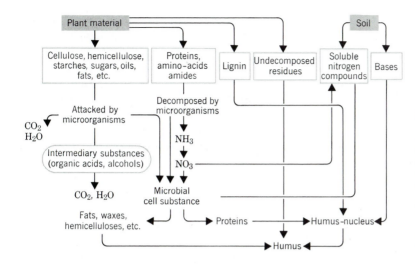

Figure 6-1
Schematic representation of the mechanism of the formation of humus in the decomposition of plant residues in the soil. (Reproduced from S. A. Wakesman, *Humus,* Williams and Wilkins Co. Copyright © 1938. By permission of the Williams and Wilkins Co.)

Humus is also an important reservoir of phosphorus (P) and sulfur (S). The ratio of C:N:P:S in humus is about 100 or 120:10:1:1. Another important property of humus is its high cation exchange capacity. Cation exchange is associated with several chemically active groups in both living and dead organic matter. During humification of organic matter, lignin is altered in such a way that there is an

increase in the carboxyl $-C\overset{\displaystyle O}{\underset{\displaystyle OH}{\diagup}}$ and

phenolic ⬡—OH groups. The hydrogen of the hydroxyls of both groups are replaceable by cations, giving rise to a cation exchange capacity. As a result, the cation exchange capacity of humus is many times greater than that of organic residues originally added to the soil. Cation exchange sites adsorb cations such as Ca, Mg, and K; in doing so, humus acts similarly to clay in retaining available nutrients against leaching and maintaining

the nutrients in a form available to higher plants and microorganisms. The cation exchange phenomenon of humus (and other kinds of soil organic matter) is illustrated by the following equation using a carboxyl group:

$$-R-C\overset{\displaystyle O}{\underset{\displaystyle OH}{\diagup}} + KCl \rightleftharpoons$$

$$-R-C\overset{\displaystyle O}{\underset{\displaystyle OK}{\diagup}} + HCl$$

The equation shows how water-soluble potassium chloride reacts with hydroxyls of carboxyl (and phenolic) groups. The potassium (K) is exchanged for the H of the hydroxyls. The K is adsorbed with enough energy to retard its loss from the soil by leaching; but the K is still readily available for plant use.

Humus absorbs large quantities of water and exhibits the properties of swelling and shrinking. It does not exhibit as pronounced properties of adhesion and cohesion as mineral colloids do. It is less stable, because it is subject to microbial decomposition. It has already been shown that soil humus is an important factor in structure formation. Humus possesses other physical and physiochemical properties that make it a highly valuable soil constituent.

The Dual Nature of Humus

Radiocarbon age determinations of soil organic fractions commonly show ages between 500 and 2,000 years. This means far less than 1 percent mineralization or decomposition per year. On the other hand, about 2 to 3 percent of the *N in humus* is mineralized each year in well-drained agricultural soils. These rather contradictory facts lead to the concept that soil humus is composed of at least two kinds of organic materials that differ greatly in resistance to decomposition and/or protection from decomposition.

When lignin reacts with various soil constituents, it forms compounds of great resistance. Inclusion of organic matter into the clay-mineral matrix of soil may render organic matter inaccessible for many years. The humus forms that comprise the bulk of humus contribute only a small amount of nutrients to plant growth in 1 year. Most of the nutrients mineralized in a given year come from a smaller, but more active, humus fraction that is very important in nutrient cycling. This fraction consists of dead plant and animal residues in various states of decomposition and short-lived organisms that soon become the substrate of other organisms.

Amount and Distribution of Organic Matter in Soils

As the rocks and minerals of the earth's crust decomposed, mineral elements were made available to plants. As supplies of nitrogen in usable chemical combinations were produced from the store of nitrogen in the air, plants grew, died, and contributed their remains to the soil. Thus, organic matter began to accumulate. As the supply of available plant nutrients in soil increased, the accumulation of soil organic matter increased accordingly. This condition continued until an equilibrium was reached, at which the rate of organic matter accumulation was equal to the rate of decomposition. This section will discuss factors that influence the amount of organic matter in soils, including climate, vegetation, drainage conditions, cultivation, and soil texture.

Influence of Climate and Vegetation on the Organic Matter Content of Soils

Generally, as the quantity of organic residues added annually to soils is increased, there is an increase in the total organic matter content. One would expect soils in deserts to contain very little organic matter, because the annual additions of organic matter from plant growth are very small. With increasing precipitation and an accompanying increase in the annual production of organic matter, there is an increase in the organic matter content of soils. On the plains of the United States from eastern Colorado to Indiana, the annual precipitation increases from about 40 to 90 centimeters. This is accompanied by a (1) shift from widely spaced bunch and short grasses to tall grass, and (2) an increase in the organic matter content of soils from about 180 to 360 metric tons per hectare (80 to 160 tons per acre) to a depth of 100 centimeters (*see* Fig. 6-2). Similar changes occur in Argentina, from the Andes Mountains to Buenos Aires, and in the southern part of the Soviet Union from a south to north direction.

The eastern United States is a forested

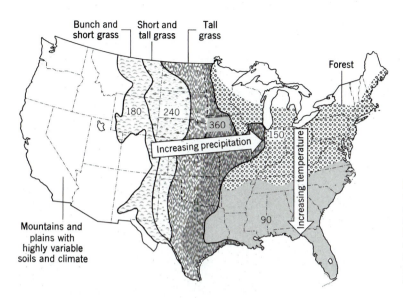

Figure 6-2
Generalized map showing organic matter content of soils, metric tons per hectare of 100 centimeters, as related to climate and vegetation. (Adapted from Schreiner and Brown, 1938.)

region, and the soils have considerably less organic matter than nearby soils developed under tall grass. Furthermore, with increasing average annual temperature in the forested area from north to south, the organic matter content of the soils decreases (*see* Fig. 6-2). Major causes are the increased rate of microbial activity and decomposition of organic matter with increasing temperature. In the tropics however, soils generally have more organic matter than soils in the southeastern United States. In the humid tropics, the production of organic matter per year is many times that of the temperate region. Many tropical soils have high clay content and more amorphous clays (with high specific surface), resulting in the effective protection of organic matter from decomposition.

Organic Matter in Forest versus Grassland Soils

The settlers that colonized America adapted to farming on soils that were developed under forest vegetation. By the early 1800s, settlers had spread westward to the tall, grass prairie lands of the central United States. The tough sod made it difficult to plow the land and new tools and techniques had to be developed. At first, the prairie lands were avoided; however, once they were broken, the superiority of grassland soils over forest soils was readily apparent. Now, the great productivity of the American prairies and the Argentine pampas are well known. One reason for the high productivity of these grassland soils is related to the amount and distribution of organic matter. Studies show that grassland soils, as compared to nearby forest soils, have: (1) about twice as much organic matter in the soil profile, and (2) a more gradual decrease of organic matter with increasing soil depth (*see* Fig. 6-3).

The explanation for the differences in the amount and distribution of organic matter in forest and grassland soils is related to differences in the growth of the plants and how the plant residues become incorporated into the soil. The roots of grasses are short lived; each year the de-

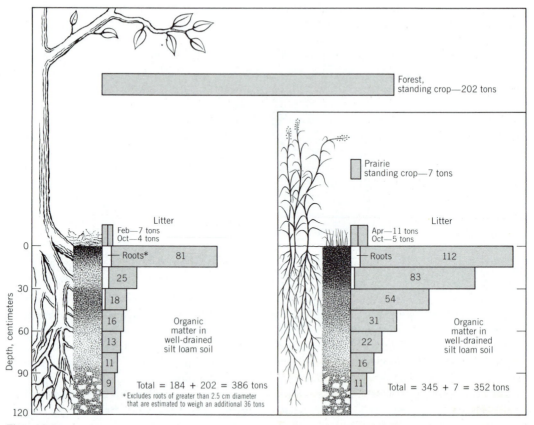

Figure 6-3
Metric tons per hectare distribution of organic matter in forest (white oak, black oak) and prairie (big bluestem, Indian grass) ecosystems in south central Wisconsin. (Adapted from Nielsen and Hole, 1963. Courtesy of F. D. Hole, Soil Survey Division, Wisconsin Geological and Natural History Survey, University Extension, University of Wisconsin.)

composition of dead roots contributes to the quantity of humified organic matter. Furthermore, the quanity of roots decreases gradually with increasing soil depth. In the forest, by contrast, the roots are long lived. The annual addition of plant residues is largely as leaves and dead wood that fall onto the surface. Some of the residues decompose on the surface, but small animals transport and mix some of the surface litter with a relatively thin layer of top soil. In the hardwood forest of southern Wisconsin (where earthworms are active), it was found that 81 tons of organic matter per hectare existed in the upper 15 centimeters of soil (A horizon), and only 25 tons per hectare in the next, deeper 15-centimeter layer of soil (E horizon).

Another interesting fact shown in Fig. 6-3 is that there is a similar amount of total organic matter in each ecosystem; however most of the organic matter in the forest exists in the *standing trees*, while over 90 percent of the organic matter on the prairie, exists within *soil*. When settlers cleared forests, they burned or harvested the trees and, in so doing, removed about one half of the organic matter. Breaking of the prairie land, by comparison, left virtually all organic matter in the soil, even if the grass was burned off before plowing. The differences in the amount and distribution of organic matter is one of several explanations for the larger crop yields on the grassland soils. Even today, with good soil management, the average yields of corn on well-drained grassland soils is about 10 to 20 percent greater than for well-drained forest soils in the central United States.

Organic Matter Content versus Soil Texture

Locally, there tends to be a correlation between the clay content of soil and the content of organic matter. The greater, combined supply of water and nutrients favors the production and accumulation of more organic matter in finer-textured soils. Clay also adsorbs decomposing en-

zymes that become inactivated. Organic molecules adsorbed on clays are partially protected from decomposition by microorganisms. As the content of organic matter in soil increases, the content of nitrogen and phosphorus increase, since they are important constituents of organic matter. The content of nitrogen and phosphorus in some New York soils is used to illustrate the general relationship between soil texture and organic matter content shown in Fig. 6-4.

Organic Soils

In the shallow water of lakes and ponds, plant residues accumulate instead of decomposing under anaerobic conditions in the water. Consequently, soils (Histosols) consisting almost entirely of organic matter develop. Plant tissue and pollen grains in peat can be easily identified. The remarkable preservation capacity of some swamp and bog waters can be illustrated by the fact that human bodies up to 3,000 years old have been found in peat bogs.

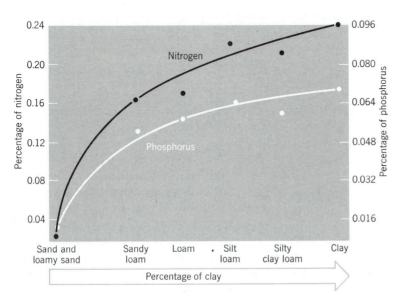

Figure 6-4
Nitrogen and phosphorus contents of several New York soil classes illustrate the geographically local trend of increasing organic matter content with increasing clay content of soil.

One of the best preserved is the 2,000-year-old Tollund man found in 1950 in Denmark. The facial expression at death and the bristles of the beard of the Tollund man were well preserved. Excellent fingerprints were made, and an autopsy revealed that his last meal consisted mainly of seeds, many of them weed seeds. When found, the Tollund man was buried under 2 meters of peat that had formed in the 2,000 years after his burial.

A characteristic feature of organic soils is a stratification or layering, which represents changes in the kinds of plants that produced the organic matter as a result of changes in climate or water level. Such a series of horizons that developed in Sweden is shown in Fig. 6-5. The woody peat layer is indicative of a dry period and was preceded and followed by wet periods, when the vegetation was sphagnum moss. The 5 meters of peat accumulated in 9,000 years or at the rate of 1 centimeter every 18 years.

Organic soils must be drained before they can be used for crop production. The upper soil layers then become aerobic and the peat begins to decompose. This converts undecomposed peat into well-decomposed muck. In time, the entire organic soil above the underlying mineral soil may disappear as a result of decomposition. This is a serious problem in warm climates, as in southern Florida, where the high annual temperature stimulates rapid decomposition.

Some Organic Matter Management Considerations

Organic matter plays many important roles in soils. Since soil organic matter originates from plant remains, it originally contained all nutrients needed for

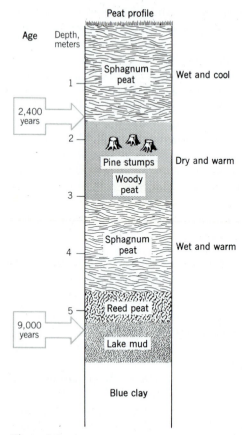

Figure 6-5
Stratification found in a peat soil in Sweden, where changes in climate produced changes in the kind of vegetation that formed the peat. (Data from Davis and Lucas, 1959.)

plant growth. Organic matter per se influences soil structure and tends to promote a desirable physical condition. Soil animals depend on organic matter for food and contribute to a desirable physical condition by mixing soil and creating channels. Naturally, there is much interest in managing organic matter to make soils more productive. A discussion of these kinds of considerations will help place the value of organic matter in soils in better perspective.

Decomposition and Mineralization of Nutrients

The pool of organic matter in a soil can be compared to a lake. Changes in the level of water in a lake depend on the difference between the amount of water entering and leaving the lake. This idea applied to soil organic matter is illustrated at the top of Fig. 6-6. Soil organic matter decomposes in mineral soils at a rate equal to about 1 to 4 percent per year. Assum-ing a 2-percent rate and 40,000 kilograms of organic matter per hectare furrow slice, 800 kilograms of soil organic matter would be lost or decomposed each year. On the other hand, if 800 kilograms of humus was formed from the residues added to the soil, the organic matter content of the soil would remain the same from one year to the next, and the soil would be at the equilibrium level (*see* Fig. 6-6).

Decomposition of 800 kilograms of hu-

Figure 6-6
Schematic illustration of the equilibrium concept of soil organic matter as applied to a hectare plow layer weighing 2,000,000 kilograms and containing 2 percent organic matter. (The values are similar on a pounds per acre basis.)

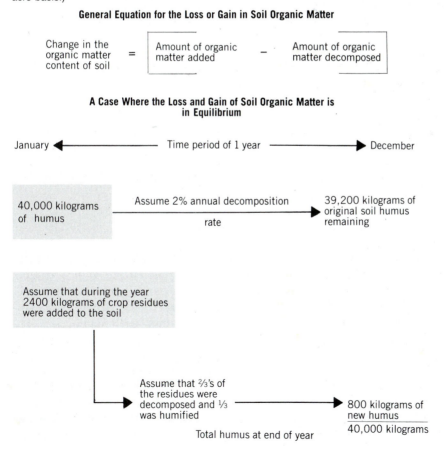

General Equation for the Loss or Gain in Soil Organic Matter

| Change in the organic matter content of soil | = | Amount of organic matter added | − | Amount of organic matter decomposed |

A Case Where the Loss and Gain of Soil Organic Matter is in Equilibrium

January ←——————— Time period of 1 year ——————→ December

40,000 kilograms of humus →(Assume 2% annual decomposition rate)→ 39,200 kilograms of original soil humus remaining

Assume that during the year 2400 kilograms of crop residues were added to the soil

Assume that ⅔'s of the residues were decomposed and ⅓ was humified ——→ 800 kilograms of new humus

Total humus at end of year

40,000 kilograms

mus would result in the mineralization of about 40 kilograms of nitrogen, assuming that soil organic matter is 5 percent nitrogen (800 × 0.05). Assuming that the ratio of N:P:S in soil organic matter is 10:1:1, there would be 4 kilograms of phosphorus and sulfur mineralized. Other nutrients are also mineralized, but the availability of most of the other nutrients is more related to mineral weathering, soil pH, and the like. The more organic matter that is added to soils each year, the more nutrients will be mineralized for plant growth.

Carbon-Nitrogen Ratio (C/N)

Soil microbes are the primary agents for organic matter decay and have certain dietary requirements. Of major concern, from a practical standpoint, is the amount of carbon relative to nitrogen in the decomposing organic matter. A problem arises when the nitrogen content of decomposing organic matter is small, because the microbes may become deprived of nitrogen and compete with the higher plants for whatever available nitrogen exists in the soil. Since the carbon content of organic materials is relatively constant, between about 40 to 50 percent, while the nitrogen content varies manyfold, the carbon-nitrogen (C/N) ratio is a convenient way to express the relative content of nitrogen. Thus, the carbon-nitrogen ratio of organic materials is an indication of the likelihood of a nitrogen shortage and of the competition between microbes and higher plants for whatever nitrogen is available in the soil.

The carbon-nitrogen ratios of some organic residues that are frequently added to soils are given in Table 6-2. They range from 10 to 12 for humus and immature sweet clover tissue to 400 for sawdust. Ma-

Table 6-2

The Carbon-Nitrogen (C/N) Ratio of Some Organic Materials

Material	C/N Ratio
Soil humus	10
Sweet clover (young)	12
Barnyard manure (rotted)	20
Clover residues	23
Green rye	36
Cane trash	50
Corn stover	60
Straw	80
Timothy	80
Sawdust	400

Data are taken from several sources. The values are approximate only, and the ratio in any particular material may vary considerably from the values given.

terials with small or narrow ratios are relatively rich in nitrogen, while those with higher or wider ratios are relatively low in nitrogen.

Mature plant residues that provide the raw material for microbial decomposition contain bout 50 percent carbon and 1 percent nitrogen (C/N = 50). The carbohydrates are quickly decomposed, and a large increase in microbial activity results. During decomposition, mineralization and immobilization of nutrients occur simultaneously. Of particular concern is whether the immobilization of nitrogen exceeds mineralization. If so, microorganisms will compete with higher plants for any available nitrogen, including that being mineralized from humus decomposition; higher plant growth will be reduced (see Fig. 6-7).

The *nitrogen factor* is a convenient term to express the extent to which a material is deficient in nitrogen for decomposition. It is defined as the number of units of in-

Figure 6-7
Paper mill sludge reduced the growth of young corn (maize) plants grown in a greenhouse when mixed with soil in quantities of one-quarter or one-half the total soil volume. The sludge contains about 40 percent paper fiber (cellulose) and little (if any) nitrogen, so it has a wide or high carbon-nitrogen ratio. Addition of the sludge reduced the supply of available nitrogen for the corn.

organic nitrogen that must be supplied to 100 units of organic material in order to prevent a net immobilization of nitrogen from the environment. A factor of approximately 0.9 represents straw. The nitrogen factor can be calculated for straw as follows. Assume that 100 kilograms of straw contains about 40 kilograms of carbon and 0.5 kilograms of nitrogen. Assuming that 35 percent of the carbon will be assimilated by the microbes and that one tenth as much nitrogen will be assimilated as carbon, the nitrogen factor will be 0.9

$$40 \times 0.35 = 14 \text{ kg of carbon assimilated}$$
$$\tfrac{14}{10} = 1.4 \text{ kg of nitrogen assimilated}$$
$$1.4-0.5 = 0.9 \text{ kg of nitrogen deficient}$$

The addition of 0.9 kilograms of inorganic nitrogen to the soil at the time 100 kilograms of straw are incorporated should prevent the immobilization of nitrogen from the soil environment and competition for nitrogen between a crop and microbes. The above calculations assume that 65 percent of the carbon in straw is converted to carbon dioxide during respiration in the decomposition process.

Immobilization exceeds nitrogen mineralization when carbon-nitrogen ratios are above 30 (*see* Fig. 6-8). In the range 15 to 30, immobiliztion and mineralization are about equal. Mineralization exceeds immobilization when the carbon-nitrogen ratio of the decomposing material is less than 15, as is the case with soil humus.

Several things may be done to prevent competition for nitrogen between microbes and higher plants. Straw and similar residues may be burned, instead of in-

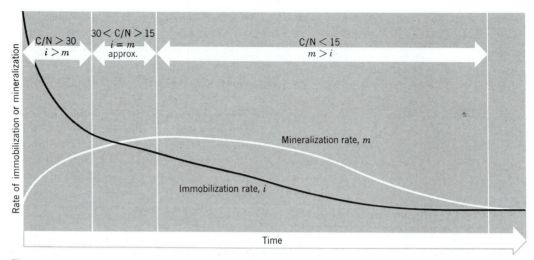

Figure 6-8
The relationship between carbon-nitrogen ratios and nitrogen availability during decomposition of plant residues. (Data from Broadbent, 1957.)

corporating them into the soil. Such a practice deprives soil of a source of organic matter, and the organic matter content of soil will be less than where straw is periodically returned to the soil. Second, nitrogen fertilizer can be added if crops are to be grown immediately after the turning under of wide carbon-nitrogen ratio materials. This nitrogen can be used early in the year for the decomposition of residues of a previous crop and, a month or so later, when the decomposition of residues is largely completed, the nitrogen can be used by the crop.

Many urban people have organic matter residues with wide carbon-nitrogen ratios such as tree leaves, grass clippings, or other plant wastes from a garden. The carbon-nitrogen ratio of these materials can be lowered by *composting*. Composting consists of storing organic materials in a pile while maintaining favorable moisture, aeration, and temperature relationships. As the organic matter is decomposed, much of the carbon, hydrogen, and oxygen are released as carbon dioxide and

water. Nutrients, like nitrogen, are continually reused or recycled by microbes and are conserved. Thus, while there is a loss of carbon, the amount of nitrogen remains about constant, resulting in a narrowing of the carbon-nitrogen ratio. There is also a general enrichment of all plant nutrients. The rotted material is easily incorporated into the soil, or it can be used as an organic matter mulch.

The low nitrogen content of the composting materials may greatly retard the rate of decomposition; for this reason, most composters add some nitrogen fertilizer. The quality of compost can be further improved by adding other materials. Some recommendations of the United States Department of Agriculture are given in Table 6-3.

Organic Matter Changes by Cultivation

Even on nonerosive land that is brought under cultivation, rapid losses of organic matter usually occur. It has been found at the Missouri Agricultural Experiment Sta-

Table 6-3

Materials Recommended for Making Compost

	Cups per Tightly Packed Bushel
For general purposes, including acid-loving plants:	
Ammonium sulfate	1
Superphosphate (20 percent)	$\frac{1}{2}$
Epsom salt	$\frac{1}{16}$
or:	
10-6-4 fertilizer	$1\frac{1}{2}$
For plants not needing acid soil:	
Ammonium sulfate	1
Superphosphate (20 percent)	$\frac{1}{2}$
Dolomitic limestone or wood ashes	$\frac{2}{3}$
or:	
10-6-4 fertilizer	$1\frac{1}{2}$
Dolomitic limestone or wood ashes	$\frac{2}{3}$

From Kellogg, 1957.

tion that, as a result of cultivation over a period of 60 years, soils in a noneroded condition lost over one third of their organic matter, the losses being much greater during the earler than later periods. The organic matter losses amounted to about 25 percent during the first 20 years, about 10 percent the second 20 years, and only about 7 percent the third 20 years. In other words, a new equilibrium level was almost attained after about 60 years (*see* Fig. 6-9).

At the Rothamsted Experiment Station north of London, England, a field was maintained free of vegetation beginning in 1870. The first 100 years, the organic matter content of the soil decreased from 2.4 to 1.4 percent. The rate of organic matter depletion decreased over time analogous to the results of the Missouri experiment. If the experiment is contin-

ued long enough, the organic matter content of the soil will approach zero.

Soils in arid regions naturally have very low organic matter contents. Irrigating arid-region land and producing crops result in large increases in the amount of organic matter returned to the soil each year. Consequently, irrigating arid lands and growing crops result in the establishment of a new equilibrium organic matter level higher than the original level.

Maintenance and Restoration

Although there is a rapid depletion of the organic matter in the soil immediately after virgin lands of humid regions are brought under cultivation, there is some consolation in the fact that this high rate does not continue indefinitely. It has been emphasized that after a period of rapid loss, a fairly constant level is attained during a long period of continued cultivation; the level is determined by the environment and soil management system associated with a particular soil. Once the organic content has reached a low level, restoring the organic matter to its original level would require that the original veg-

Figure 6-9

Decline of soil nitrogen (or organic matter) with length of cultivation period under average farming practices. (Data from Jenny, Missouri Agr. Exp. Sta. Bull., 324, 1933.)

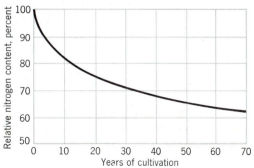

etation be reestablished. In time, a new equilibrium level of organic matter content would be achieved that would be the same or similar to the level that existed before the land was cultivated. While the land is being farmed, it is virtually impossible and much too expensive to maintain the organic matter content similar to that which existed in the virgin soil. It is, therefore, usually unwise and uneconomical to maintain the organic matter above a level consistent with good crop yields. Attention should be directed toward the frequent additions of small quantities of fresh organic materials, instead of toward practices of maintaining the organic matter content at any particularly high level.

In a consideration of the maintenance of soil organic matter, the amount of crop residues that must be returned to maintain a given organic matter content depends on soil and climatic conditions. It is interesting to note that 4,000 kilograms per hectare of crop residues were needed annually to maintain the organic matter content of Blackland soils near Dallas, Texas. Cropping systems that returned more than this amount showed an increase in the organic matter content after 12 years, whereas those returning less than this amount caused a decline in the organic matter content. Only the tops were measured; root residues in addition to top residues were returned to the soil. It would be reasonable to allow 1,000 kilograms for roots, making the total weight of residues added per year to maintain the organic matter content of these soils equal to about 5,000 kilograms per hectare.

Since World War II, there has been a large increase in crop yields in many countries. It may seem to some that the higher yields are more demanding or detrimental to soils than the production of lower yields. In general, this is not considered to be true. Larger crops yields, especially where only the grain is harvested, return more crop residues to the land, and there is more vegetative cover during the growing season. As a result, there is less erosion on sloping land and a higher level of organic matter is maintained in the soil. The relationship (interaction) between crop yields and the organic matter content of soil is illustrated in Fig. 6-10. The production and harvest, or removal, of 3,150 kilograms of corn (maize) grain per hectare (50 bushels per acre) returned enough residues (tops plus roots) to maintain a humus carbon content of 1.0 percent (1.7 percent organic matter). By comparison, a corn grain yield of 9,450 kilograms per hectare (150 bushels per acre) returned enough residues to maintain a humus carbon content of 2.3 percent (4.0 percent organic matter). Also, note that the 9,450-kilogram yield in a soil with 1 percent organic carbon would have resulted in an increase in humus carbon equal to about 1,000 kilograms per hectare (900 pounds per acre). The production of 3,150 kilograms in soil with 2.3 percent carbon would result in about a 1,000-kilogram-per hectare loss of humus carbon. Assuming that humus has a carbon-nitrogen ratio of 10, and that 3 percent of humus nitrogen is mineralized annually, the plow layer (weighing 2,000,000 kilograms per hectare with 2.3 percent carbon) would provide about 80 kilograms per hectare more of available nitrogen per year for crops than the plow layer with 1.0 percent humus carbon.

Effects of Green-Manuring

One of the oldest agricultural practices is the growing of legumes for soil improvement. The yields of nonleguminous crops

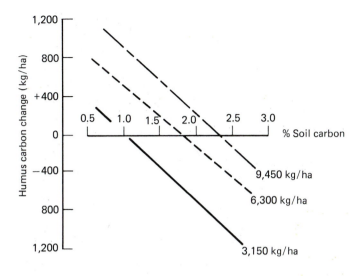

Figure 6-10
Estimated, annual soil-humus carbon changes as related to corn (maize) yields and carbon (humus) content of soil. The soil is a loamy Alfisol in southern Michigan with an estimated erosion loss of 8 metric tons per hectare (3.6 tons per acre) annually. (Adapted from Lucas and Vitosh, 1978.)

are usually greater when they are grown after legumes (such as alfalfa or clover) because of an increased nitrogen supply. In these cases, the legume crop is harvested and the benefit to crops grown later is a by-product. In green-manuring, a crop is planted just to be plowed under to add some organic matter to the soil. This is particularly beneficial for sandy soils that are very low in organic matter content. In these soils, little nitrogen is mineralized from soil organic matter, and added nitrogen fertilizer may be leached out of the soil before the crop has used it. In such cases, the green-manure crop is planted after harvest in the late summer or fall and plowed under just before planting the next crop in the spring. The gradual decomposition of the plowed-under crop provides plant nutrients, particularly nitrogen, for some weeks after planting. The major effect of the green-manure crop, in this case, is to increase the supply of nitrogen (and other nutrients), instead of providing a significant increase in the organic matter content of the soil. Other benefits commonly cited

are protection of soil from erosion and reduced loss of nutrients by leaching. However, it has been difficult to attribute economic benefits of green-manuring to any other effect than that of increased nitrogen supply. Where green-manure crops deplete soil moisture and contribute to a droughty situation, they may cause a reduction in yield of the next crop.

Use of Peats

Peats that are used for soil amendments are generally classified as moss peat, reed-sedge peat, and peat humus. Moss peat forms from moss vegetation, reed-sedge peat from reeds, sedges, cattails, and other associated plants, while peat humus is any peat that has undergone considerable decomposition. Peats are used largely for mulches and greenhouse soil mixes.

Some properties of common horticultural peats are given in Table 6-4. The nitrogen content (wide C/N ratio) of sphagnum peat indicates that its incorporation into soil may temporarily lower the available soil nitrogen supply. The low pH of

Table 6-4
Characteristics of Common Horticultural Peats

Type	Range in Nitrogen[a], %	Range of Water-Absorbing Capacity[a], %	Range in Ash Content[a], %	Range in Volume Weights[a], lb/ft^3	Range in pH
Sphagnum moss peat	0.6–1.4	1,500–3,000	1.0–5.0	4.5–7.0	3.0–4.0
Hypnum moss peat	2.0–3.5	1,200–1,800	4.0–10.0	5.0–10.0	5.0–7.0
Reed-sedge peat (low lime)	1.5–3.0	500–1,200	5.0–15.0	10.0–15.0	4.0–5.0
Reed-sedge peat (high lime)	2.0–3.5	400–1,200	5.0–18.0	10.0–18.0	5.1–7.5
Decomposed peat	2.0–3.5	150–500	10.0–50.0	10.0–40.0	5.0–7.5

From Lucus, et al., *Ext. Bull., 516,* Michigan State University, East Lansing, Michigan.
[a]Oven dry basis.

sphagnum peat, however, makes it desirable as a mulch for acid-requiring plants, such as azaleas and rhododendrons. Peat moss makes a neat-looking surface that sets off plants and protects the soil from the disruptive force of rain drips, thereby keeping the soil porous so that water rapidly enters the soil when irrigated.

Nutrient Accumulation Under Shifting Cultivation

When one flies over jungles of humid tropics, one can see small clearings where food crops are grown. These people have few animals that produce manure for the fields, chemical fertilizers are not available, and the highly weathered tropical soils are infertile. How can these people produce the subsistence food crops they need? They use a system of cultivation known as *shifting cultivation,* or slash and burn agriculture.

We have already noted the relatively efficient recycling of nutrients that can occur in forest (*see* Chapter 5) and the po-

tential for organic matter accumulation in trees. The shifting cultivator uses nutrients in the forest (trees, vines, leaves, etc.) through controlled cutting and burning, which kills most of the trees but does not destroy all of the organic matter. Crops are planted among the few, remaining living trees, stumps, and fallen trees. Nutrients are made available to the

Table 6-5
Nutrient Accumulation (or Immobilization) in Forest Fallow in the Congo

Age of Forest Fallow	Nutrients Immobilized in Vegetation, kg/ha.				
	N	P	S	K	Ca + Mg
2 years	188	22	37	185	160
5 years	566	33	103	455	420
8 years	578	35	101	839	667
18–19 years	701	108	196	600	820

Adapted from C.E., Kellogg, "Shifting Cultivation," *Soil Sci., 95:*221–230, 1963. Used by permission of Williams and Wilkins Co.

Figure 6-11
Landscape in Assam showing land use under shifting cultivation. Various stages of forest fallow can be observed, as well as a small burning in the right background. (Photo courtesy of C. E. Kellogg.)

crops from ashes and as organic matter decomposes. After about 1 to 5 years, nutrients are depleted; weed and diseases invade the cultivated land, resulting in extremely poor yields. The land is abandoned and the forest reestablishes itself. Perhaps 10 to 20 years are needed before enough nutrients have accumulated in the trees to permit another short period of cultivation. Nutrient accumulation by regenerating forest laying fallow in the Congo is given in Table 6-5. The 18- to 19-year-old forest vegetation contained about five times more nutrients than were contained in the 2-year-old forest fallow.

A common practice is to grow crops for 2 to 3 years and then have the forest lie fallow for about 15 years. A farmer would need 17 to 18 parcels. Each year, a new field would be brought into cultivation, and a field would be abandoned to forest fallow (*see* Fig. 6-11). A shifting cultivation system may require as much as 20 hectares or 50 acres per person.

Grasslands are not as effective nutrient accumulators as forests. The reason for this is that grasses are shallow rooted and

do not have as much potential for storage of biomass as standing vegetation. It is interesting to note that in humid and sub-humid tropics, forest soils have greater productivity than grassland or savannah soils under shifting cultivation; in humid temperate regions, the grassland soils are usually considered the most productive. One can not help but be impressed by the ingenuity that must have been required to perfect the shifting cultivation system. Today, over 200,000,000 people depend on the system for their livelihood.

Bibliography

Allison, F.E., *Soil Organic Matter and Its Role in Crop Production*, Elsevier, New York, 1973.

Broadbent, F.E., "Organic Matter," in *Soil*, USDA Yearbook, Washington, D.C., 1957, pp. 151–157.

Davis, J.F., and R.E. Lucas, "Organic Soils," *Michigan Agr. Exp. Sta. Spec. Bull., 425*, 1959.

Glob, P.V., "Lifelike Man Preserved 2,000 Years in Peat," *Nat. Geogr. Mag., 105:*419–430, 1954.

Jenny, H., "Causes of the High Nitrogen and

Organic Matter Contents of Certain Tropical Forest Soils," *Soil Sc., 69:*63–69, 1950.

Kellogg, C.E., "Home Gardens and Lawns," in *Soil,* USDA Yearbook, Washington, D.C., 1957, pp. 665–688.

Kellogg, C.E., "Shifting Cultivation," *Soil Sci., 95:*221–230, 1963.

Laws, W.D., "Farming Systems for Soil Improvement in the Blacklands," *Tex. Res. Found., Bull. 10,* 1961.

Liebig, Justus, *Chemistry in Its Application to Agriculture and Physiology,* L. Playfair, Ed., Wiley, New York, 1952.

Lucas, R.E., P.E. Rieke, and R.S. Farnham, "Peats for Soil Improvement and Soil Mixes," *Extension Bulletin 516,* Michigan State University, East Lansing, Mich.

Lucas R.E., and M.L. Vitosh, "Soil Organic Matter Dynamics," *Mich. Agr. Exp. Stat. Res. Report 358,* November, 1978.

Nielson, G.A., and F.D. Hole, "A Study of the Natural Processes of Incorporation of Organic Matter into Soil in the University of Wisconsin Arboretum," *Wis Acad. Sciences, 52:*213–227, 1963.

Schreiner, O., and B.E. Brown, "Soil Nitrogen," in *Soils and Men,* USDA Yearbook, Washington, D.C., 1938, pp. 361–376.

Smith, R.M., G. Samuels, and C.F. Cernuda, "Organic Matter and Nitrogen Buildups in Some Puerto Rican Soil Profiles," *Soil Sci., 72:*409–427, 1951.

Waksman, S.A., *Humus,* Williams and Wilkins, Baltimore, 1938.

CHAPTER 7

SOIL MINERALOGY

As recently as the eighteenth century, soil was viewed as little more than a mixture of rock and organic matter particles. It was believed that plant roots ingested soil particles and that these particles provided the sustenance of plants. Jethro Tull, an Englishman, believed that the swelling of plant roots caused a pressure that aided the entrance of soil particles into the roots. As a consequence, Tull invented the grain drill to plant grain in rows and a horse hoe (cultivator) to pulverize the soil when the grain was growing. Grain yields were increased. Tull thought this resulted from the effect of cultivation in loosening fine soil particles, which would make the particles more easily ingested. However, it was later learned that the grain grew better when cultivated because weed competition was reduced.

Shortly after 1800, de Saussure of Geneva used improved chemical techniques to discover the basic elements of photosynthesis and respiration. He showed that plants used carbon in the daytime and released carbon dioxide at night. The ash of plants was composed of nutrients obtained from the soil. In spite of the excellent experimental work of de Saussure, many continued to believe the *humus theory*, which held that the plant's source of carbon was humus ingested by the roots. It remained for Liebig to bring a death to the humus theory through the publication of his book *Chemistry in Agriculture and Physiology* in 1840. Liebig, one of the foremost chemists of the nineteenth century, correctly viewed mineral weathering as a source of nutrients that were absorbed as ions by roots.

Roentgen's discovery of X rays in 1895 led to the development of the use of X-ray diffraction to study the arrangement of atoms in minerals. The use of X-ray diffraction and more recently developed methods has resulted in a greater understanding of the role of mineral structure in the weathering and availability of plant nutrients.

Consideration of the characteristics of minerals found in soils, and their transformation from one form to another, is essential in understanding both the nature of soil's chemical properties and the origin of its fertility. Since soil develops from material composed of rocks and minerals from the earth's crust, we will direct our attention first to the chemical and mineralogical composition of the earth's crust.

Chemical and Mineralogical Composition of the Earth's Crust

About 90 chemical elements are known to exist in the earth's crust. When one considers the possible combinations in such a large number of elements, it is not surprising that some 2,000 minerals have been recognized. Relatively few elements and minerals, however, are of real importance in soils.

Chemical Composition of the Earth's Crust

Approximately 98 percent of the weight of the earth's crust is composed of eight chemical elements (*see* Fig. 7-1). In fact, two elements, oxygen and silicon, compose 75 percent of it. Many of the elements important in the growth of plants and animals occur in very small quantities.

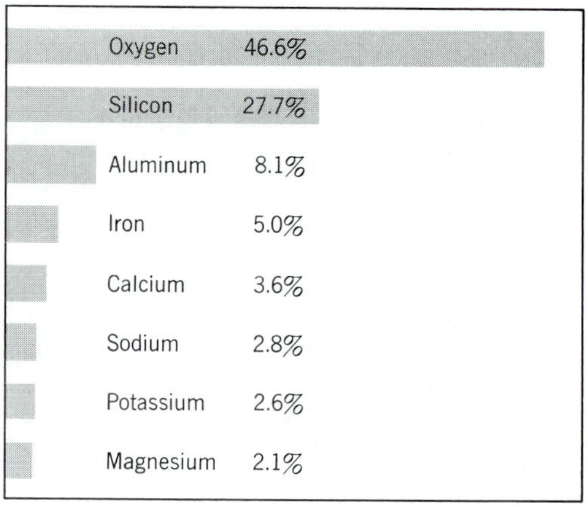

Oxygen	46.6%
Silicon	27.7%
Aluminum	8.1%
Iron	5.0%
Calcium	3.6%
Sodium	2.8%
Potassium	2.6%
Magnesium	2.1%

Figure 7-1
The eight elements in the earth's crust comprising over 1 percent by weight. The remainder of elements make up 1.5 percent. (Data from Clark, 1924.)

Needless to say, these elements and their compounds are not evenly distributed throughout the earth's surface. For example, in some places, phosphorus compounds are so concentrated that they are mined; in many other areas, there is a deficiency of phosphorus for plant growth.

Mineralogical Composition of Rocks

Most elements of the earth's crust have combined with one or more other elements to form compounds called *minerals*. The minerals generally exist in mixtures to form the *rocks* of the earth. The mineralogical composition of igneous rocks, shale, and sandstone are given in Table 7-1. Limestone is also an important sedimentary rock; it is composed largely of calcium and magnesium carbonates, with varying amounts of other minerals as impurities. The dominant minerals in these rocks are feldspar, amphibole, pyroxene, quartz, mica, clay, limonite (iron oxide), and carbonate minerals.

Table 7-1

Average Mineralogical Composition of Igneous and Sedimentary Rocks

Mineral Constituent	Origin	Igneous Rock, %	Shale, %	Sandstone, %
Feldspars	Primary	59.5	30.0	11.5
Amphiboles and Pyroxenes	Primary	16.8	—	[a]
Quartz	Primary	12.0	22.3	66.8
Micas	Primary	3.8	—	[a]
Titanium minerals	Primary	1.5	—	[a]
Apatite	Primary or secondary	0.6	—	[a]
Clays	Secondary	—	25.0	6.6
Limonite	Secondary	—	5.6	1.8
Carbonates	Secondary	—	5.7	11.1
Other minerals	—	5.8	11.4	2.2

Data from Clarke, 1924.
[a]Present in small amounts.

Weathering and Mineralogical Composition of Soils

Numerous examples of weathering abound and can be observed every day. Rusting of metal, cracking of sidewalks, and loss of mortar between bricks are a few examples. Weathering in soils results in the destruction of existing minerals and the synthesis of new minerals. Nutrients are made available for plants and clay minerals are formed. In a real sense, all life on earth is locked in the minerals and, through weathering, nutrients essential to life are made available. It is interesting to contemplate the amount of food or lumber that could be produced from the nutrients contained in the rocks of the Andes or the Rocky Mountains. Even life in the seas awaits nutrients that are released by weathering on the land then carried to the sea by rivers. For these reasons, we need to consider weathering and the mineralogical composition of soils.

Weathering—The Response of Rocks and Minerals to a New Environment

Rocks and minerals that are at or near equilibrium deep in the earth adjust to the greatly reduced pressure and temperature in the soil environment. The resulting adjustments or changes are called *weathering*. The changes are in the direction of a lower energy state and, to a large extent, are self-generating (exothermic). The response to reduced pressure is seen in the increased volume during *unloading*. Unloading is the removal of thick layers of sediment, overlying deeply buried rocks, by erosion or uplift. The release of pressure results in an accompanying bit of expansion that produces cracks and fissures. Strains from temperature changes and the pressures of freezing water, as well as the erosive action of water, wind, and ice, also cause a slow and unceasing breaking up of hard rocks.

Two of the most important weathering reactions are hydration and hydrolysis. Broken-edge bonds on mineral surfaces adsorb water molecules (hydration) and provide a source of H^+ in the immediate vicinity of the mineral surfaces. The small size of the hydrogen ion and its charge results in an effective replacement of basic cations of calcium, magnesium, potassium and sodium (Ca^{++}, Mg^{++}, K^+, and Na^+) in the mineral structures (hydrolysis) as follows:

$$KAlSi_3O_8 + H^+ \rightarrow HAlSi_3O_8 + K^+$$
(orthoclase)   (clay)

Other important weathering reactions include oxidation, dissolution, reduction and carbonation. An increase in volume accompanies these reactions and causes a peeling off of rock surfaces, thus producing exfoliation and spheroidal weathering. Hydration is considered to be effec-tive in producing spheroidal weathering (*see* Fig. 7-2). The net effect of weathering is: (1) the physical reduction in particle size, which results in the formation of gravel, sand, silt, and clay, and (2) the chemical decomposition of minerals, which results in the formation of altered minerals and decomposition products reacting to form new minerals and supply plants with es-sential nutrients.

Crystal Structure—A Clue to Weathering Rate

Particle size, through its effect on specific surface, is an important factor influencing the weathering rate of minerals. Chemical bonding within the mineral crystal, how-ever, is ultimately the major factor.

Sodium chloride exists as the natural mineral, halite. Sodium and chlorine exist in crystalline form as ions that attract each other. At the crystal faces, sodium and chloride ions, respectively, attract the neg-ative and positive poles of water mole-cules. Adsorption of water molecules dis-lodges sodium and chlorine from the

Figure 7-2
Hydration of minerals in the outer layer of a rock results in an increase in volume to cause a concentric spalling known as "spheroidal weathering." Weathered rock is basalt in Hawaii.

crystal and greatly increases their solubility. The mineral dissolves easily in water and is readily leached from soils. This accounts for the general absence of halite in the soils of humid regions.

We have noted that oxygen and silicon make up about 75 percent of the earth's crust on a weight basis, which explains the great abundance of silicate minerals in the crust. The large size of the oxygen atom, coupled with its abundance, causes oxygen to occupy over 90 percent of the volume of the earth's crust. When atoms such as oxygen gain electrons, the size increases and the valence becomes negative, depending on the number of electrons gained. When atoms such as silicon lose electrons, the size decreases and the valence becomes more positive, depending on the number of electrons lost (*see* Table 7-2). It is productive to think of minerals as being composed of ping-pong balls (represented by oxygen atoms), and that marbles (representing the smaller metallic cations) occupy the interstices between the ping-pong balls. The cations are arranged on the basis of their size and valence to produce electrically neutral crystals. The larger the cation, the greater the number of oxygens or other anions surrounding it (coordination number) (*see* Table 7-2).

The silicon atom fits in an interstice formed by four oxygens. The covalent bonding (shared electron pairs) between oxygen and silicon causes them to form a tetrahedron (*see* Fig. 7-3). Silicon-oxygen tetrahedra are the basic units in silicate minerals. The valence of silicon is +4; that of oxygen is −2. Therefore, each tetrahedron has a surplus charge of −4. Some, or all, of this negative charge can be dissipated by the common sharing of oxygen by adjacent tetrahedra. Different silicate minerals are formed, depending on the way the tetrahedra are linked together. In quartz, all oxygen atoms are shared by silicon, and the mineral has the formula SiO_2.

Olivine, $MgFeSiO_4$, is composed of individual silicon-oxygen tetrahedra bound together by magnesium and iron. Each magnesium and iron atom is surrounded by six oxygens—three from each of two faces of adjacent tetrahedrons (*see* Fig. 7-

Table 7-2

Size, Percent Volume in Earth's Crust, Coordination Number and Valence of the Most Abundant Elements in Soil Minerals

Element	Atomic Radius, A°	Volume, Percentage in Earth's Crust	Coordination Number with Oxygen	Valence
O	1.32	93.8	—	−2
Si	0.39	0.9	4	+4
Al	0.57	0.5	6 (and 4)	+3
Fe	0.83 (Fe^{+3} 0.67)	0.4	6	+2
Mg	0.78	0.3	6	+2
Na	0.98	1.3	8	+1
Ca	1.06	1.0	8	+2
K	1.33	1.8	8 (and 12)	+1

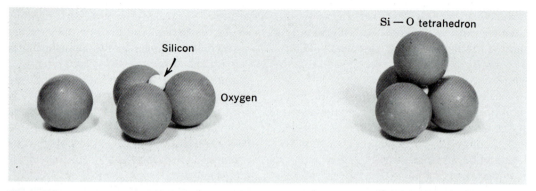

Figure 7-3
Models showing the tetrahedral (four-sided) arrangement of silicon and oxygen atoms in silicate minerals. The silicon-oxygen tetrahedron on the left has the apical oxygen set off to one side to show the position of the silicon atom.

4). The bonds between oxygen, iron, and magnesium are weak in comparison to the bonds in the tetrahedra between oxygen and silicon. During weathering (hydrolysis), exposed magnesium along the crystal face is dislodged by hydrogen ions; the magnesium then reacts with hydroxyl ions to form magnesium hydroxide. The ferrous iron along the crystal face readily looses an electron, and its size is reduced by about 20 percent (*see* Table 7-2). The smaller size of the ferric iron atom makes it easier for iron to leave the crystal and more likely to form a hydroxide. For these reasons, the weathering of olivine can be visualized, as shown in Fig. 7-4, as consisting of the separation of silicon-oxygen tetrahedra with the release of iron and magnesium. The magnesium and iron are then available for plant uptake. The magnesium compounds formed are sufficiently soluble to be leached from the

Figure 7-4
Models representing the weathering of olivine. Olivine is composed of silicon-oxygen tetrahedra held together by iron and magnesium. Every other tetrahedron is "inverted," as shown by the light-colored tetrahedra in the olivine model on the left. During weathering, the silicon-oxygen tetrahedra separate with the release of iron and magnesium.

soil; however, the iron compounds formed may be insoluble enough to accumulate. The silicon-oxygen tetrahedra often regroup, or polymerize, and form some other mineral by combining with other constituents in the soil solution. The tetrahedra can react with H^+ to form tetrahedral $Si(OH)_4$ which is silicic acid. It may be removed from the soil by leaching. As with halite, olivine weathers fairly easily, making it impossible for olivine to remain long in soils of humid regions.

In most silicate minerals, the silicon-oxygen tetrahedra are linked together by the common sharing of some oxygen atoms. The number of cations, other than silicon, needed to neutralize the crystal is reduced to the extent that oxygen sharing occurs. Increased oxygen sharing by silicon generally causes greater resistance to weathering, because a greater percentage of the crystal bonds are the strong covalent-like bonds that exist between silicon and oxygen. The ratio of oxygen to silicon in olivine is 4. It can be seen in Fig. 7-5 that increased oxygen sharing in the series olivine to pyroxene to amphibole to mica to quartz is associated with a decrease in the oxygen-silicon ratio from 4 to 2. The decreased ratio is associated with an increase in weathering resistance. In quartz (SiO_2), all oxygen atoms are neutralized by silicon, and quartz accumulates in many soils as the more weatherable minerals disappear.

Figure 7-5
Models showing the common arrangements of silicon-oxygen tetrahedra in silicate minerals and their relation to weathering resistance.

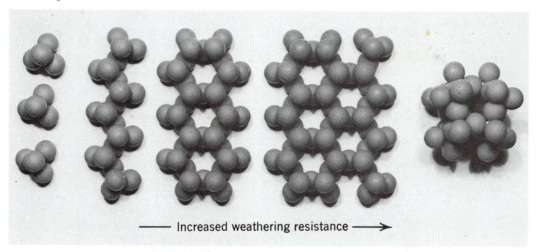

⟶ Increased weathering resistance ⟶

Arrangement of Si-O tetrahedra and representative minerals

Individual	Single chain	Double chain	Sheet	3-dimensional
Olivine	Pyroxene augite	Amphibole hornblende	Biotite (mica)	Quartz

Oxygen-silicon ratio				
4	3	2.7	2.5	2

Table 7-3
Representative Minerals and Soils Associated with Weathering Stages

Weathering Stage	Representative Minerals	Typical Soil Groups
	Early weathering stages	
1	Gypsum (also halite, sodium nitrate)	Soils dominated by these minerals in the fine silt and
2	Calcite (also dolomite, apatite)	clay fractions are the
3	Olivine-hornblende (also pyroxenes)	minimally weathered soils all over the world, but are
4	Biotite (also glauconite, non-tronite)	mainly soils of the desert regions where limited water
5	Albite (also anorthite, micro-cline, orthoclase)	keeps chemical weathering to a minimum. Many of these soils are Entisols, Inceptisols, and Aridisols.
	Intermediate weathering stages	
6	Quartz	Soils dominated by these minerals in the fine silt and
7	Muscovite (also hydrous mica)	clay fractions are mainly
8	2:1 layer silicates (including vermiculite, expanded hydrous mica)	those of temperate regions developed under grass or trees. Includes the major
9	Montmorillonite	soils of the wheat and corn belts of the world. Many of these soils are Alfisols, Mollisols, and Vertisols.
	Advanced weathering stages	
10	Kaolinite	Many intensely weathered soils
11	Gibbsite	of the warm and humid
12	Hematite (also goethite, limonite)	equatorial regions have clay fractions dominated by these
13	Anatase (also rutile, zircon)	minerals. The soils are frequently characterized by acidity and infertility. Many of these soils are Ultisols and Oxisols.

Based on Jackson and Sherman, 1953.

Mineralogical Composition of Soils versus Weathering Stage

The difference in the weathering resistance of minerals were used by Jackson and Sherman to establish 13 weathering stages that relate mineralogical composition of soils to weathering intensity. Representative minerals and typical soils associated with the weathering stages are given in Table 7-3.

Soils of Weathering Stage 1 may contain some gypsum and halite. Notice that soils containing significant amounts of olivine are representative of Stage 3, and biotite mica is representative of Stage 4. This sequence is in agreement with our discussion of the oxygen-silicon ratio and of weathering resistance. Soils with minerals representative of Stages 1 to 5 are considered to be in the early weathering stages. Such soils are frequently the minimally weathered soils throughout the world; however, they are primarily the soils of arid regions, where limited water restricts chemical weathering.

Soils of the intermediate weathering stages (stages 6 to 9) include most soils of humid temperate regions. Quartz is often abundant in these soils and is representative of Weathering Stage 6. Here, we notice minerals that we have not mentioned previously, such as hydrous mica, vermiculite, and montmorillonite (*see* Table 7-3). These latter two minerals are typically synthesized in soils, increase in abundance, and accumulate as fine-sized particles in the clay fraction.

Soils of the advanced weathering stages include the intensely weathered soils of humid tropics. Soils dominated by minerals of Weathering Stages 10 to 13 may have lost all or almost all original minerals of the parent material; they may consist mainly of stable minerals that have been synthesized during weathering. These minerals are kaolinite, gibbsite, hematite, and anatase (TiO_2). Thus, soils of the advanced weathering stages are usually characterized by extreme infertility and, where these soils exist, most of the nutrients in the ecosystem are circulating through the vegetation. Many intensely weathered soils of humid tropics are classified as *Ultisols;* soils that have undergone the "ultimate in weathering and leaching."

Structural Components of Soil Clays

Many soil clays are structurally related. Thus, learning the basic components in clays facilitates an understanding of the nature of soil clays and how one soil clay differs from another clay. Knowledge of clay structure is needed to understand how clay affects the physical and chemical properties of soils.

Silica, Gibbsite, and Brucite Sheets

Many soil clays are aluminosilicates. As already noted, silicon-oxygen tetrahedra share oxygen atoms to form sheets, or a planar structure, as illustrated by biotite in Fig. 7-5. The apical oxygen have one excess negative charge, and the silica sheet has the formula $Si_2O_5^{-2}$.

The *gibbsite* sheet is composed of octahedrons, with aluminum in six coordination with hydroxyl. An individual octahedron is $Al(OH)_6^{-3}$. The sharing of hydroxyls by adjacent octahedra forms an octahedral or gibbsite sheet with the composition, $Al_2(OH)_6$. Two of the three potential spaces for aluminum in the sheet are filled; the mineral is *dioctahedral* (*see* Fig. 7-6). The gibbsite sheet is an important part of many aluminosilicate clays; however, gibbsite also exists by itself in soils.

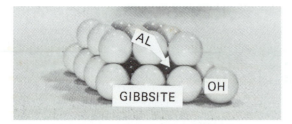

In a manner similar to aluminum and hydroxyl, magnesium and hydroxyl form an octahedral sheet called the *brucite* sheet. The magnesium is in six coordination with hydroxyl, and the sheet has the composition $Mg_3(OH)_6$. All three of the potential spaces for cations are filled with magnesium to produce a *trioctahedral* mineral. Both gibbsite and brucite are octahedral and have electrically neutral structures. They differ in that gibbsite is diocathedral and brucite is triocathedral.

The Kaolinite or 1:1 Layer

Crystalline silicate clays have plate-shaped particles composed of layers that are formed by the condensation of tetrahedral and octahedral sheets (*see* Fig. 7-7). The *kaolinite* layer is formed by the join-ing of one silica sheet and one gibbsite sheet. For simplicity, a pair of tetrahedra, instead of a sheet, are used to illustrate how the tetrahedral and octahedral sheets exist in the kaolinite layer in Fig. 7-8. Items 1 and 2 of Fig. 7-8 show the apical oxygens, which have one free bond, and the position of aluminum. Items 3 to 6 show the progessive addition of OH to complete the aluminum octahedron. Model 6 is representative of the structure of kaolinite, consisting of one tetrahedral and one octahedral sheet. For this reason, kaolinite is called a 1:1 clay (ratio of tetrahedral to octahedral sheets). Item 7 of Fig. 7-8 shows that apical oxygens are common to both the tetrahedral and octahedral sheets. In the octahedron, each Al neutralizes half a valence bond from each of the surrounding six anions, which

Figure 7-7
An electron micrograph of kaolinite particles magnified 35,821 times. Note the platy and hexagonal shape of the particles. (Photo courtesy of Mineral Industries Experiment Station, Pennsylvania State University.)

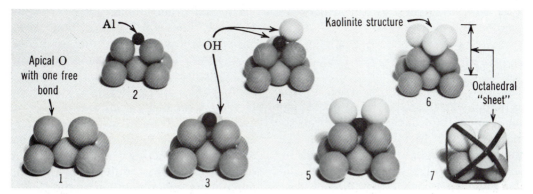

Figure 7-8
Models illustrating the structure of kaolinite. Item 1 shows that the apical oxygen of the tetrahedral sheet has a free bond that is neutralized by aluminum, as seen in item 2. Items 3 to 6 show the progressive addition of 4 OH to complete the octahedron or to put aluminum in six-coordination. Item 6 represents the kaolinite structure, and item 7 shows that the aluminum is surrounded by two oxygen and four hydroxyl. Item 7 is an example of an octahedron—four faces are seen and four faces are hidden.

results in a neutral structure. Each layer is 7.2 angstroms (7.2×10^{-8} centimeters) thick. The theoretical formula for the layer is $Si_4Al_4O_{10}(OH)_8$.

The Pyrophyllite Layer

The pyrophyllite layer has three sheets; a gibbsite sheet sandwiched between two silica sheets. All of the apical oxygens of the silica sheets are retained. Appropriate hydroxyls of the octahedra are replaced by oxygen of the silica sheets, leaving each aluminum surrounded with four oxygen and only two hydroxyls (*see* Fig. 7-9).

Pyrophyllite is a 2:1 mineral, because there are two tetrahedral sheets and one octahedral sheet per layer. The mineral is electrically neutral and has the theoretical formula $Si_8Al_4O_{20}(OH)_4$. Weak forces

Figure 7-9
Models illustrating a 2:1 structure consisting of two tetrahedral sheets and one octahedral sheet. On the left are two silicon-oxygen tetrahedral "sheets" facing each other. The center model shows the addition of aluminum that is shared by four apical oxygens. The right model shows the addition of two hydroxyls (one in rear is invisible) to place the aluminum in six-coordination in an octahedral arrangement. The aluminum shares one half of a valence bond from each of the six surrounding anions (four oxygen and two hydroxyl).

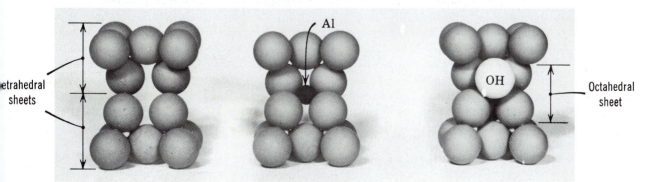

hold layers together to form particles composed of many layers.

The Mica Layer

The mica layer is a 2:1 structure similar to pyrophyllite; however, it differs in that about one in four of the Si^{+4} has been replaced by Al^{+3} in the silica sheets. The aluminum is slightly larger than silicon, but similar enough to proxy for the silicon. The replacement is called, *isomorphous substitution*. Each proxy-by-aluminum leaves one residual negative charge in the crystal lattice. This negative charge is balanced by a K^+ that fits between two adjacent layers. The potassium is about the size of oxygen (*see* Table 7-2); it fits in a hexagonal hole formed by the ring arrangement of the tetrahedra in the silica sheet. Note the ring arrangement of the tetrahedra in the sheet structure for biotite shown in Fig. 7-5. There are six tetrahedra per ring; thus, the potassium is in 12 coordination with six oxygen from each layer. This bonding forms a potassium bridge that firmly holds adjacent layers together. The mica structure is, therefore, a 2:1 nonexpanding mineral. The formula is $K_2(Si_6Al_2)Al_4O_{20}(OH)_4$, which indicates that two aluminums have proxied for two silicons in the tetrahedral sheet, and that the two potassiums maintain electrical neutrality. The potassium prevents adjacent layers from resting directly on each other. The top of one layer to the top of the next layer distance is 10 angstroms.

There are two micas, muscovite and biotite; they differ in their octahedral sheets. Muscovite has a gibbsite sheet and is dioctahedral, with its formula as previously given for the mica layer. Biotite has a brucite sheet with Fe^{+2} proxying for some of the Mg^{+2}. Biotite is tri-

octahedral and has the formula $K_2(Si_6Al_2)(Mg,Fe)_6O_{20}(OH)_4$. The presence of ferrous iron in the octahedral sheet results in a low resistance to weathering; biotite is in Weathering Stage 4 compared to stage 7 for muscovite (*see* Table 7-3). Soils with significant amounts of biotite tend to have an abundance of available potassium for plant growth. Due to the low weathering resistance of biotite (trioctahedral mica), soils tend to contain much more muscovite than biotite.

Origin, Structure, and Properties of Soil Clays

The clay separate contains the less than 2 micron (0.002 millimeters) mineral particles. This size limit would include a particle with its longest dimension equal to about 20,000 oxygen atoms side by side. There is no magic cutoff in mineral species at 2 microns, which results in the fact that some mineral species exist in both silt and clay. Quartz may be found in sand, silt, and clay of a given soil. For mineral particles to persist for thousands of years in particles as small as clay, they must have great resistance to weathering. This is the basis for referring to the dominant minerals in the clay separate as soil clays.

Soil clays may contain some resistant primary minerals, such as quartz and mica; however, the clay fraction is typically composed mainly of secondary minerals formed by the alteration of preexisting minerals and/or the synthesis of new minerals from the products of mineral degradation. Soil clays are placed in two major groups: (1) aluminosilicate, plate-shaped minerals (phyllosilicates) and (2) oxide clays. The origin, structure, and properties of silicate clays will be considered first.

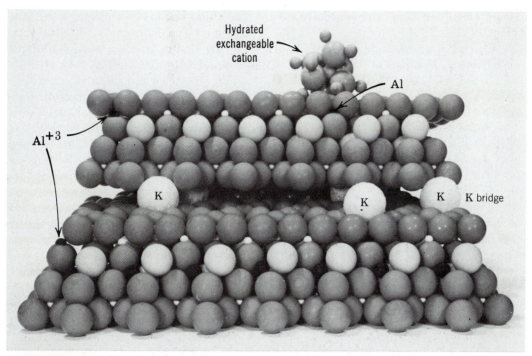

Figure 7-10
Model of the 2:1 clay mineral hydrous mica showing (1) some tetrahedral Al^{+3} substituted for Si^{+4}, which creates a negatively charged lattice, (2) "potassium bridges" that neutralize most of the lattice charge and hold the layers together to form a nonexpanding lattice, and (3) adsorption of a hydrated cation on an exchange site created by aluminum substitution. The K of the bridges fits in the cavities formed by rings of 6 oxygens of adjacent planar surfaces.

Hydrous Mica (Illite) and Vermiculite

Micas are 2:1 minerals with K^+ lodged between the silica sheets of adjacent layers. Leaching a soil that contains micas with dilute acid (such as carbonic acid), causes a slow removal of the interlayer potassium and a partial loss of the bonding between layers. The layers, however, do not expand or separate, and the layer thickness or vertical spacing remains 10 angstroms. This altered mineral is hydrous mica[1].

Hydrous mica has some negative charge on the planar surfaces where aluminum has been substituted for silicon and interlayer potassium has been lost. This creates a negatively charged lattice, that is related to the amount of potassium lost; these charged sites are satisfied by the adsorption of hydrated cations. Cation exchange sites also orginate on the edges of particles where exposed hydroxyls lose or dissociate a hydrogen ion that produces a negative charge. The adsorbed cations are in motion around the exchange sites and exchange places with each other. The extent of the adsorption of exchangeable cations is called the *cation exchange capacity* (CEC). The major features of hydrous mica are shown in Fig. 7-10.

---

[1]There is disagreement in the naming of this mineral. It has been called illite, hydromica and sericite.

Continued leaching and complete removal of interlayer potassium allows the layers to separate or expand because the silica sheets of adjacent layers acquire high negative charge. The complete loss of potassium results in an expanding lattice and a high cation exchange capacity. The new mineral formed is *vermiculite*. When the interlayer space is occupied by hydrated exchangeable cations, the vertical spacing is about 14 to 15 angstroms. Vermiculites are formed from both muscovite and biotite and can be either dioctahedral or trioctahedral. As with micas, trioctahedral vermiculite is less resistant to weathering. Vermiculite can be converted to hydrous mica and then to mica by leaching with a solution rich in potassium.

Smectite (Montmorillonite)

Smectite is replacing montmorillonite as a group name for 2:1-expanding clays with high exchange capacity, maximum expansion properties, and very small particle size. Smectites can orginate from the weathering of vermiculite or the synthesis obtained from the degradation products of other minerals. The dioctahedral members are most common in soils. One has cation exchange capacity from the proxying of magnesium for aluminum in the octahedral sheet. This mineral is called *montmorillonite*. (Montmorillonite has been used as both a group and specie name.) Its formation is favored by high magnesium in the weathering environment. Beidellite is dioctahedral and has cation exchange capacity orginating from the proxying of aluminum for silicon in the tetrahedral sheets. Generalized formulas indicating isomorphous substitution are:

$Si_8(AlMg)_4O_{20}(OH)_4$ montmorillonite
$(SiAl)_8Al_4O_{20}(OH)_4$ beidellite

The cation exchange capacity is intermediate between hydrous mica and vermiculite. The dominant features of montmorillonite are illustrated in Fig. 7-11.

The interlayer expansion of smectites is related to type of interlayer ions. In some cases, individual layers completely separate, producing extremely small-sized particles. Smectites give soils both swelling and shrinking properties with wetting and drying (*see* Fig. 7-12). The *shrink-swell potential* is one of the most important engineering properties of soils. It is related to the type and amount of clay. The shrink-swell potential is a measure of the volume change with wetting and drying; it is termed the *coefficient of linear extensibility* (COLE) and is defined as:

$$COLE = \frac{\text{length of moist sample}}{\text{length of dry sample}} - 1$$

If the COLE value exceeds 0.09, a significant *shrink-swell* can be expected. If it is greater than 0.03, considerable smectite is present. Soils with high COLE values cave-in basement walls, disrupt gas and water lines, cause heaving of highways and misalignment of telephone poles.

Chlorites and Interstratified Clays

Chlorites are complex clays with a mica (2:1) structure and gibbsite or brucite materials in the interlayer space. The interlayer material is considered to be a layer and the chlorites are 2:1:1 clays. Aluminum interlayering is favored in acid conditions: As mica-like minerals weather to form vermiculite and montmorillonite, hydroxy aluminum (gibbsite) is precipitated in the interlayer space. The interlayer material is positively charged due to an isomorphous replacement of cations, an excess of cations, or a vacancy of some anions. The positively charged interlayer

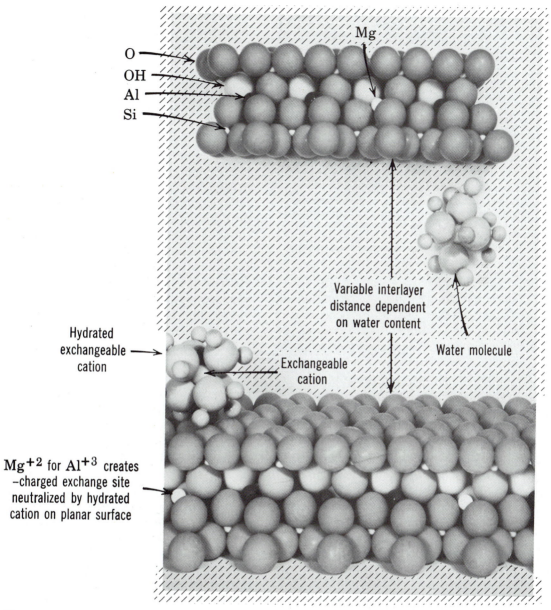

Figure 7-11
Models of 2:1 clay mineral montmorillonite showing: (1) expanding lattice and variable water content, (2) cation exchange site originating from isomorphous substitution of Mg^{+2} for Al^{+3}, and (3) adsorption of two hydrated cations on the planar surface cation exchange sites. In a moist soil, the interlayer water is believed to be all or primarily water associated with exchangeable cations.

Figure 7-12
Large cracks formed by drying of montmorillonite clay. The montmorillonite originated from weathering of basalt in the hills and was moved to the valley floor by erosion. The clay formed under an ustic moisture regime, where magnesium remains in the soil.

materials strongly attract the negatively charged, planar clay surfaces to reduce expandability and to block cation exchange sites. A wide variation exists in the extent of interlayer material from islands to complete sheets. The result is a considerable variation in the properties of chlorites.

Clay minerals are complex in their structure and properties; rather idealized forms have been presented. Many trace element ions proxy in the structures. These are potentially usable by plants. Furthermore, a clay particle may contain some layers of one kind of clay, along with some layers of a different clay. This is clay interstratification, and it is common in soils. The interstratification can be random or regular.

Kaolinite

Several mechanisms have been proposed for the genesis of 1:1 nonexpanding kaolinite. The stripping away of one of the silica sheets from 2:1 layers converts 2:1 clays into kaolinite. This type of formation is favored in acidic soils low in bases.

Kaolinite also forms from the direct weathering of orthoclase feldspar as follows:

$$2KAlSi_3O_8 + 11H_2O \rightarrow 2KOH$$
(orthoclase)
$$+ 4Si(OH)_4 + Al_2Si_2O_5(OH)_4$$
(kaolinite)

Feldspars have an open framework of silicon-oxygen (Si-O) tetrahedra similar to quartz. They differ from quartz in the proxying of some Al^{+3} for Si^{+4} and in the inclusion of cations such as potassium, calcium, and sodium to restore neutrality. The formation of kaolinite from feldspar requires a rather complete breakdown of the feldspar structure and a rearrangement of the silicon-oxygen tetrahedra into $Si_2O_5^{-2}$ sheets.

Breakdown of kaolinite can result in the formation of gibbsite, and its accumulation in the soil along with the loss of silica in the leaching water. Later, if the weathering environment is changed and the soil is invaded by silica-rich water, the gibbsite may be converted back into kaolinite and the kaolinite to montmorillonite. The presence of abundant kaolinite and gibbsite in a soil commonly indicates that most primary minerals (feldspars and micas) have disappeared due to weathering; the soil is very acidic and low in fertility. Ultisols ("ultimately weathered and leached" soils) and Oxisols (soils rich in iron and aluminum oxide clays) have kaolinite and gibbsite in abundance in the clay fraction.

Kaolinite is a 1:1 nonexpanding clay, and its properties are derived from its structure. The vertical layer to layer spacing is 7.2 angstroms. There is little cation exchange capacity, because there is little if any isomorphous substitution. The hydroxyls of the octahedral sheet rest on top of the oxygens of the silica sheet of adjacent layers. This causes hydrogen bonding between adjacent layers; hydrogen bonding firmly holds layers together. This bonding has two effects. First, it tends to favor the formation of large particles. Many kaolinite particles are silt sized, and soils with a high content of kaolinite may have a surprisingly high hydraulic conductivity when moist or wet. Second, water cannot penetrate between layers to cause expansion and contraction with wetting and drying. The major properties of kaolinite are shown in Fig. 7-13.

Allophane and Halloysite

Noncrystalline or amorphous minerals exist where volcanoes throw ash and cinders into the air; they solidify before the atoms become organized into crystals. In humid tropics, amorphous minerals weather rapidly into an amorphous clay called *allophane*. The soils formed lack discrete mineral particles and are like a gel; they feel like mucky soil when wet. There is an enormous amount of surface area and great protection against organic matter decomposition. Characteristic features of soils high in allophane include high organic matter content and water-holding capacity.

Slowly, allophane clays become more organized or crystalline. A mineral sequence from the weathering of volcanic ash is allophane-halloysite-kaolinite-oxides of aluminum and iron. Soils high in allophane are important in the Andes Mountains, Central America, Oregon, Washington, Alaska, and Hawaii.

Oxide Clays

All or most soils contain at least a small amount of colloidal-sized particles composed of oxides of iron and aluminum. Representative oxide clays include gibbsite ($Al(OH)_3$), of Weathering Stage 11, and hematite (Fe_2O_3) of Weathering Stage 12. Soils with properties dominated by oxide clays and kaolinite (Weathering Stage 10) are in the advanced weathering stages and are usually found in humid tropics.

Soils dominated by oxides and kaolinite clays are characterized by very stable soil aggregates; they exhibit a low degree of

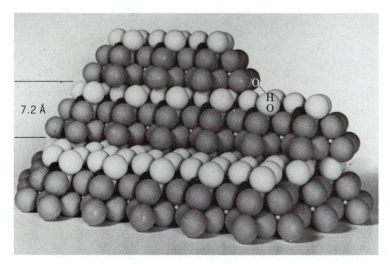

Figure 7-13
Model of the 1:1 clay mineral kaolinite showing three layers stacked one on top of the other and held together tightly by hydrogen bonding (attraction of hydrogen of the hydroxyl for the oxygen of the next adjacent unit). Each layer is 7.2 angstroms thick.

plasticity. Large amounts of oxide and kaolinite clays in a soil contribute to the formation of extremely stable soil aggregates, because the clays tend to neutralize each other. Kaolinite, as we have seen, has a net negative charge, while the oxide clays may have a net positive charge (in acidic soils). This promotes flocculation. Oxide gels also probably contribute to the hardness of aggregates that resist crushing when moistened and rubbed in the palm of the hand. Such soils act as sands, even though they contain 100 percent clay, as is common for soils in the Hawaiian Islands that have weathered from basalt. Infiltration of water is rapid and the soil resists erosion. Plasticity is low when wet and soils are easily tilled. Adsorption and release of water to plants occurs largely at high matric potential. Soil textures, by feel, are frequently silt loam and silty clay loam or, in some cases, sandy clay loam.

Summary Statement on Clays

1. There is a kinship among the crystalline silicate clays, due to the occurrence of tetrahedral and octahedral sheets in their structures. Hydrous mica can be weathered or converted into vermiculite, vermiculite to montmorillonite, and montmorillonite to kaolinite. The sequence hydrous mica to kaolinite represents the formation of clays with an increased weathering resistance.

2. Kaolinite can be converted to 2:1 clay, if the soil that contains kaolinite is invaded with silica-rich water. Vermiculite can be converted to hydrous mica by treatment with potassium-rich solutions.

3. A summary of the major silicate clays is given in Table 7-4.

4. Volcanic activity produces parent material with amorphous minerals

Table 7-4
Summary of Properties of Some Silicate Clays

Mineral	Type	Interlayer Condition	Cation Exchange Capacity, Meq per 100 grams[a]
Kaolinite	1:1 nonexpanding	strong H bonding	3–15
Hydrous mica (Illite)	2:1 nonexpanding	partial K loss, strong bonding	10–40
Vermiculite	2:1 expanding	complete loss of potassium, moderate bonding	100–150
Montmorillonite	2:1 expanding	very weak bonding, great expansion	80–150
Chlorite	2:1:1 partially or nonexpanding	brucite or gibbsite sheets	10–40

[a]Determined at pH 7.0.

that weather to form the amorphous clay, allophane. Over time, allophane may be altered to halloysite and kaolinite.

Ion Exchange Systems of Soil Clays

Some soils are dominated by layer silicate clays, some by oxide clays, and some soils have both kinds of clay in abundance. This results in three unique ion exchange systems in soils: (1) layer silicate, (2) oxide, and (3) oxide-coated, layer silicate systems. The type of ion exchange system controls the amount and nature of ion exchange; it has important consequences for plant growth and soil management.

Layer Silicate System Soils

Layer silicate system soils are overwhelmingly, negatively charged. The cation exchange capacity orginates mainly from isomorphous substitution. These negatively charged sites are not affected by pH and constitute the permanent charge. The charge characteristics may be modified, however, to a small extent by minor or thin coatings of oxide clays on particle surfaces. Some cation exchange capacity also orginates from the dissociation of hydrogen from edge Si—OH groups:

$$Si—OH + OH^- \longrightarrow Si—O^- + H_2O$$

The reaction is favored by a high hydroxyl concentration, which increases with increasing pH. As soil pH increases, the cation exchange capacity increases due to this pH-dependent (variable) charge. About 80 percent of the maximum cation exchange capacity of montmorillonite is due to a permanent charge, with 20 percent due to a pH-dependent charge (*see* Fig. 7-14). Layer silicate systems have much more cation exchange capacity than anion exchange capacity, as shown by montmorillonite, vermiculite and hydrous mica in Table 7-5.

Layer silicate system soils are minimally or moderately weathered and are represented by weathering stages 9 or lower. These soils are common in temperate regions and are not uncommon in the tropics, where relatively unweathered soils exist on alluvial plains, deltas, lake beds, and in desert and savannah regions.

Figure 7-14
Relation of cation exchange capacity to pH. Kaolinite and montmorillonite have some pH-dependent charge about pH 6. All the cation exchange capacity of humus is pH dependent and increases linearly with pH.

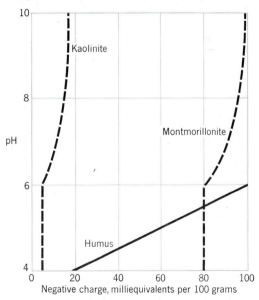

Oxide System Soils

Soils with oxide systems are characterized by weathering stage 10 or higher and are dominated by oxidic clays (*see* Fig. 7-15). Kaolinite particles (if present) are coated with thick, continuous layers of oxides, giving the particles the properties of oxidic clay. The clay particle surfaces have a large number of hydroxyls exposed and no isomorphous substitution. The cation exchange capacity is all pH dependent. Both cation and anion exchange capacity are low, as shown by gibbsite and goethite in Table 7-5.

Since all ion exchange capacity is pH dependent, the soil may have a negative or positive charge, or be neutral with a zero charge. This is shown by further elaboration of the mechanism that produces pH-dependent charge in layer silicate systems:

Figure 7-15
Nipe clay, an Oxisol in Puerto Rico, is about 60 percent iron oxide and has oxidic mineralogy.

$$
\begin{array}{ccc}
\text{H} & & \text{H} \\
\text{O} & & \text{O} \\
| & & | \\
\text{Al—OH}_2^+ \Longleftrightarrow \text{H}^+ + & \text{Al—OH} + \text{OH} \rightleftharpoons \\
| & & | \\
\text{O} & & \text{O} \\
\text{positive} & & \text{neutral}
\end{array}
$$

(zero point of charge)

$$
\begin{array}{c}
\text{H} \\
\text{O} \\
| \\
\text{Al—O}^- + \text{H}_2\text{O} \\
| \\
\text{O}
\end{array}
$$

negative

Oxide clays at the isoelectric point or zero point of charge (ZPC) are neutral. Increasing pH favors reaction with OH^- and creation of cation exchange capacity. By contrast, lowering pH favors reaction with H^+ and creation of anion exchange capacity. In theory, oxide system soils may be negatively or positively charged. In reality, they only very rarely have a net positive charge, as in the case of some highly oxidic subsoils where cation exchange capacity from organic matter is nil. Then, low pH may create a positively charged soil horizon. Soils high in anion exchange capacity preferentially adsorb chlorides, sulfates and nitrates, and preferentially repel and encourage leaching of calcium, magnesium, sodium, and potassium.

There may be a need to modify the concept of the nature of positive charge in oxidic clays, since a recent report indicates that titanium (Ti^{+4}) has been found to

Table 7-5

Charge Characteristics of Some Soil Clays

	Milliequivalents per 100 grams			
	Cation Exchange Capacity (CEC)			
Clay	Permanent	pH dependent	Total[a]	Anion Exchange Capacity (AEC)
Montmorillonite	112	6	118	1
Vermiculite	85	0	85	0
Hydrous mica	11	8	19	3
Kaolinite	1	3	4	2
Gibbsite	0	5	5	5
Goethite	0	4	4	4

Adapted from Sanchez, 1976. Represent clays removed from soils in Kenya.
[a]Determined at pH 8.2.

substitute for Fe^{+3} in iron oxides of some Oxisols, creating some permanent positive charge (*see* Tessen and Zauyah, 1982).

Oxide-Coated, Layer Silicate System Soils

The great bulk of the red and highly weathered tropical soils have oxide clay-coated silicate exchange systems. Both kaolinite and oxides are in abundance. The kaolinite particles, however, are not covered with thick and continuous coatings of oxides; exchange properties are due partly to the kaolinite and partly to the oxidic clay. In the field, soils have a net negative charge. Cation exchange capacity is low, because some of the negative charge of silicate clay is neutralized by positively charged oxide particles. Transferring temperate region soil technology to the tropics has been quite successful for soils with layer silicate or oxide-coated layer silicate systems. By contrast, the technology transfers have been more difficult for soils with oxidic systems.

Bibliography

Barnhisel, R. I., "Chlorites and Hydroxy Inter-layer Vermiculite and Smectite," in *Minerals in the Soil Environment,* Soil Sci. Soc. Am., Madison, 1977, pp. 331–356.

Barnhisel, R. I., and E. I. Rich, "Clay Mineral Formation in Different Rock Types of a Weathering Boulder of Conglomerate," *Soil Sci. Soc. Am. Proc., 31:*627–631, 1967.

Borchardt, G. A., "Montmorillonite and Other Smectite Minerals," in *Minerals in the Soil Environment,* Soil Sci. Soc. Am., Madison, 1977, pp. 293–330.

Clarke, F. W., "Data of Geochemistry," U.S. Geol. Sur. Bull. 770, Washington D.C., 1924.

Coleman, N. T., and A. Mehlich, "The Chemistry of Soil pH," in *Soil,* USDA Yearbook, Washington D.C., 1957, pp. 72–79.

Dixon, J. B., "Kaolinite and Serpentine Group Minerals," in *Minerals in the Soil Environment,* Soil Sci. Soc. Am., Proc., Madison, 1977, pp. 357–404.

Fanning, D. S., and V. Z. Keramidas, "Micas," in *Minerals in the Soil Environment,* Soil Sci. Soc. Am., Proc., Madison, 1977, pp. 195–292.

FitzPatrick, E. A., *Soils*, Longman, London, 1980.

Forsythe, W. M., "Soil-water Relations in Soils Derived from Volcanic Ash of Central America," in *Soil Management in Tropical America*, North Carolina State University, Raleigh, S. C., 1975, pp. 155–167.

Jackson, M. L., and G. D. Sherman, "Chemical Weathering of Minerals in Soils," *Adv. Agron., 5:*219–318, 1953.

Jenny, H., *The Soil Resource*, Springer-Verlag, New York, 1980.

Keng, J. C. W., and G. Uehara, "Chemistry, Mineralogy, and Taxonomy of Oxisols and Ultisols," *Soil and Crop Sci. Soc. of Florida, 33:*119–126, 1973.

Liebig, Justus, *Chemistry in its Application to Agriculture and Physiology*, L. Playfair, Ed., Wiley, New York, 1852.

Mackenzie, R. C., "The Classification of Soil Silicates and Oxides, in *Soil Components: Vol. 2, Inorganic Components*, Springer-Verlag, New York, 1975, pp. 1–28.

Pitty. A. F., *Geography and Soil Properties*, Methuen, London, 1979.

Sawnhey, B. L., "Interstratification in Layer Silicates," in *Minerals in the Soil Environment*, Soil Sci. Soc. Am., Proc., Madison, 1977, pp. 405–434.

Schofield, R. K., "The Effect of pH on Electric Charges Carried by Clay Particles," *Jour. Soil Sci., 1:*–8, 1949.

Thompson, L. M., and F. R. Troeh, *Soils and Soil Fertility*, 4th Ed., McGraw-Hill, New York, 1978.

Uehara, G., and G. Gillman, *The Mineralogy, Chemistry, and Physics of Tropical Soils with Variable Charge Clays*, Westview, Boulder, 1981.

CHAPTER 8

SOIL CHEMISTRY

The two essential things that plants absorb and remove from the soil are water and nutrients. Plants can be considered deficient in an essential nutrient element when: (1) there is virtually none in the soil, or (2) there is a large quantity in the soil, but too little of it is sufficiently soluble or available to supply plant needs. As a consequence, a total chemical analysis of the soil generally reveals little information that is important for plant nutrition. Instead, it is usually a relatively small part of the total element present in the soil that is available and important for plant growth. For these reasons, the major topics in soil chemistry are related to cation exchange capacity, ion exchange reactions, soil pH, solubilities and biochemical transformations.

Cation Exchange

Cation exchange is the interchange between a cation in a solution and another cation on the surface of any surface-active material. All soil components contribute, to some extent, to cation exchange sites; however, cation exchange in most soils is centered with clay and organic matter.

Nature of Cation Exchange

Cation exchange reactions in soils occur mainly near the surfaces of colloidal-sized clay and humus particles called *micelles*. Each micelle may have thousands of negative charges that are neutralized by the adsorbed cations. Assume for purposes of illustration that a micelle in a solution has 140 negatively charged sites that are neutralized by 40 calcium ions, 10 potassium ions, 10 hydrogen ions and 20 magnesium ions that react with some KCl and is represented as follows:

$$\begin{array}{cc} 40\ Ca^{+2} & 10\ K^{+} \\ \hline \multicolumn{2}{|c|}{\text{micelle}} \\ \hline 10\ H^{+} & 20\ Mg^{+2} \end{array} + 40\ KCl \rightleftharpoons$$

$$\begin{array}{cc} 36\ Ca^{+2} & 32\ K^{+} \\ \hline \multicolumn{2}{|c|}{\text{micelle}} \\ \hline 2\ H^{+} & 17\ Mg^{+2} \end{array} + \left\{ \begin{array}{l} 4\ CaCl_2 \\ 3\ MgCl_2 \\ 18\ KCl \\ 8\ HCL \end{array} \right.$$

The fertilizer salt KCl that is added will form K^+ and Cl^- ions. The increased number or concentration of K^+ in a solution will encourage an exchange with the exchangeable cations. In our illustration at equilibrium, there are an additional 22 K^+ adsorbed. This caused the exchange of 4 calcium ions, 3 magnesium ions, and 8 hydrogen ions that exist in the solution as chlorides. Of the "original" 40 KCl, 18, in effect, remain in the solution, and the system is in equilibrium.

The negatively charged micelle surfaces form a boundary along which the negative charge is localized, and where there is a strong attraction for cations. The cations neutralize the negatively charged surface. The exchangeable cations are hydrated and drag along hydration water molecules as they constantly move or oscillate around exchange sites and exchange places near the colloid surfaces. The concentration of cations is greatest near the micellar surfaces, where the negative charge is strongest and the concentration decreases rapidly with distance from the exchange surfaces. The negatively charged surfaces repel anions. At some distance from the micellar surfaces, the concentration of cations and anions is equal. The total number of cations involved is usually much greater than the number of anions. These relationships are shown in Fig. 8-1.

Root surfaces have cation exchange capacity (CEC). The negative charge is neutralized by hydrogen ions excreted by the root. During plant absorption of cations, hydrogen ions excreted by the root eventually effect an exchange with cations on micellar surfaces, such as potassium for example; the potassium ion is absorbed from the soil solution. In the process, the exchange sites of micelles become increasingly occupied by hydrogen ions, unless the nutrient cations are replaced by mineral weathering or fertilizers.

Cation Exchange Capacity of Soils

The *cation exchange capacity* (CEC) is defined as the sum total of exchangeable cations adsorbed, expressed in *milliequivalents per 100 grams of oven dry soil.* An equivalent weight is the quantity that is chemically equal to 1 gram of hydrogen. The number of hydrogen atoms in an equivalent weight is Avagadro's number (6.02×10^{23}). A milliequivalent weight is equal to 0.001 gram of hydrogen. If there was 1 milliequivalent weight of cation exchange capacity in 1 teaspoon of soil, it would contain 6.02×10^{20} per milliequivalent adsorbed monovalent cations. Such large numbers are difficult to comprehend, but will be better understood when these numbers are converted into weight of adsorbed cations later in this chapter. It suffices at this time to mention that this may amount to several thousands of pounds of exchangeable cations per acre furrow slice or thousands of kilograms in the plow layer of a hectare.

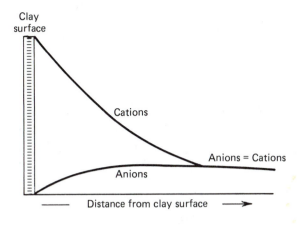

Figure 8-1
Distribution of cations and anions in the vicinity of the negatively charged surfaces of clay (and humus) particles.

The total cation exchange capacity of soil is the total number of exchange sites of both organic and mineral colloids. For most soils, organic matter is the component with the greatest cation exchange capacity. The cation exchange capacity of organic matter increases with humification. It is about 200 to 300 milliequivalents per 100 grams of well-humified organic matter or humus. The clays have a highly variable cation exchange capacity, ranging from less than 10 for oxide clays to over 100 for 2:1 clay (*see* Table 7-5).

The sand and silt fractions have exposed unsatisfied negative bonds, but due to a low specific surface, they contribute little to the cation exchange capacity of most soils. As a consequence, the cation exchange capacity of soils is affected mainly by the amount and kind of organic matter and clay.

The grassland soils of western Canada contain clay and organic matter that average 57 and 250 milliequivalents per 100 grams, respectively. Cation exchange capacity for these soils can be approximated with the following equation:

$$CEC = \text{percent organic matter}$$
$$\times\ 2.5 + \text{percent clay} \times 0.57$$

A soil containing 4 percent organic matter and 20 percent clay would have an estimated cation exchange capacity of 21.4 (10.0 + 11.4). The cation exchange capacity of soil horizons ranges from less than 5 to over 200, depending on the amount and kind of clay and organic matter. Organic matter usually decreases with increasing soil depth, and it contributes less to the cation exchange capacities of B and C horizons than A horizons. Clay accumulation in argillic horizons (Bt) may cause a maximum cation exchange capacity in the B horizon of some soils.

The cation exchange capacity of humus and oxide clay is all pH dependent; that of the silicate clays is partially pH dependent (*see* Fig. 7-14). Thus the cation exchange capacity is a function of soil pH. Generally, for silicate system soils, the cation exchange capacity is determined with buffer solutions at pH 7.0 or 8.2. For acidic-oxidic soils with considerable pH-dependent cation exchange capacity, the cation exchange capacity is commonly determined with unbuffered salt solutions, such as KCl at the soil's natural pH. This is called the *effective* cation exchange capacity, and it more nearly represents the soil conditions encountered by plant roots and soil organisms. The cation exchange capacity is determined routinely by saturating all exchange adsorption sites with a single cation, such as ammonium, and then replacing and measuring the ammonium adsorbed.

Kinds and Amounts of Exchangeable Cations

Mineral weathering is the natural source of cations that may be potentially adsorbed as exchangeable cations. The greater the supply of a given cation from weathering, the greater the likelihood that the cation will be adsorbed according to the law of mass action. The amounts and kinds of cations actually adsorbed, however, are significantly affected by cation valence and hydrated radius.

Cations with a greater valence are adsorbed more strongly or efficiently than cations of a lower valence. For a given valence, the cation with the smallest hydrated radius will move the closest to the micellar surface and be more strongly adsorbed. The energy of adsorption decreases as the square of the distance. The dehydrated and hydrated radius of four monovalent ions are given in Table 8-1. According to the dehydrated radius, it would appear that Li should have the greatest energy of adsorption and Rb the least. In fact, the reverse is true, because Li has the largest hydrated radius and Rb the least. The result is that the energy of adsorption, or the order of replacement, is reversed, with Rb first and Li fourth for the ions Li, Na, K, and Rb (see Table 8-1). Considering some of the most common exchangeable cations in soils, the replaceability series is usually Al > Ca > Mg > K > Na. Exchangeable hydrogen is difficult to put in the series because of the uncertainty of its hydration properties.

Table 8-1

Ionic Radii, Hydration, and Exchange Efficiency of Several Monovalent Ions

Ion	Radii of Ions Angstroms, (10^{-8} cm)		Order of Cation Exchange Efficiency
	Dehydrated	Hydrated	
Li	0.78	10.03	4th
Na	0.98	7.90	3rd
K	1.33	5.32	2nd
Rb	1.49	5.09	1st

The differences in the dehydrated and hydrated radius and valence between sodium and calcium is illustrated in Fig. 8-2. Calcium is adsorbed more strongly than sodium, because calcium has both a greater valence and a smaller hydrated radius. As a result, sodium is readily leached from soils of humid regions. Calcium is preferentially adsorbed; it is frequently the most abundant exchangeable cation. Note that the order of abundance for a Mollisol located in Iowa is similar to the energy of adsorption sequence: Ca > Mg > K > Na (see Table 8-2).

Figure 8-2

Calcium ions are more strongly adsorbed by clay than sodium ions, because calcium is divalent and has a smaller hydrated radius.

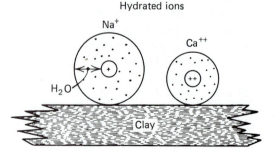

Table 8-2
Exchangeable Cations, Cation Exchange Capacity,
Percent Base Saturation, and pH of Horizons of Tama
Silty Clay Loam—a Mollisol

Depth, cm	Horizon	Exchangeable Cations, mEq/100 grams					Cation Exchange Capacity	Percent Base Saturation	pH
		Ca	Mg	K	Na	H			
0–5	Ap	14.0	3.4	0.5	0.1	9.3	27.3	66	5.7
15–28	A	13.8	4.2	0.4	0.1	11.4	29.9	62	5.8
28–50	AB	14.5	6.1	0.4	0.1	9.0	30.1	70	5.8
50–89	Bt	14.7	6.7	0.3	0.1	7.5	29.3	74	5.7
89–130	BC	14.8	5.7	0.3	0.1	5.6	26.5	79	6.0
130–155	C	16.1	5.6	0.4	0.2	4.1	26.4	84	6.5

Adapted from *Soil Survey Investigations Report,* No. 3, Iowa,
SCS, USDA, 1966.

For practical purposes, the sum of cations shown in Table 8-2 is considered synonymous with the cation exchange capacity, recognizing the fact that very small but agronomically important amounts of exchangeable iron, zinc, copper, manganese, and some other cations are present. The amount of exchangeable calcium in the Ap horizon, or plow layer, is 14 milliequivalents per 100 grams (*see* Table 8-2). An equivalent of calcium weighs 20 grams calculated as follows:

$$\frac{\text{atomic weight}}{\text{valence}} = \frac{40 \text{ grams}}{2} = 20 \text{ grams}$$

The milliequivalent weight of calcium is equal to 20 grams divided by 1,000 or 0.02 grams. Thus, 14 milliequivalents of calcium weigh 0.28 grams. It is easy to calculate the amount of exchangeable calcium in a typical plow layer that weighs 2,000,000 pounds by stating that since there are 0.28 grams of calcium in 100 grams of soil, there are 0.28 pounds of

calcium in 100 pounds of soil. For an acre furrow slice, the amount of exchangeable calcium is: $\frac{20}{1000} = .02 \times 14 = .28$

$$\frac{0.28 \text{ pounds Ca}}{100 \text{ pounds of soil}}$$

$$= \frac{x \text{ pounds Ca}}{2,000,000 \text{ pounds of soil}}$$

$$x = 5600 \text{ pounds of Ca}$$

The formula can also be used to calculate the kilograms of exchangeable calcium in the plow layer of a hectare by substituting kilograms for pounds in the equation. A soil with 14 milliequivalents of exchangeable calcium per 100 grams has 0.28 kilograms of calcium per 100 kilograms of soil. A 20-centimeter-thick plow layer, for a hectare that has a bulk density of 1 gram per cubic centimeter, weighs 2,000,000 kilograms. Thus, there would be 5,600 kilograms of exchangeable calcium in the plow layer of a hectare. Considering that plants obtain calcium from soil below the Ap horizons or plow layers, it is easy to see why many soils can have enough available calcium to last the plants growing in fields or natural ecosystems for hundreds or thousands of years, and why calcium deficiency symptoms are not commonly observed in the field. Taking into consideration the differences in the milliequivalent weights of various cations, there would be 816 kilograms of exchangeable magnesium and 390 kilograms of exchangeable potassium per 2,000,000 kilograms of soil in the Ap horizon. This is useful in predicting whether the addition of magnesium and potassium fertilizers would increase plant growth.

Percent Base and Hydrogen Saturation

The exchangeable bases are generally considered to include calcium, magnesium, potassium, and sodium[1]. There are 18 millequivalents of exchangeable bases and 9.3 milliequivalents of exchangeable hydrogen for the Ap horizon of the soil cited in Table 8-2; this results in a cation exchange capacity of 27.3. The 18 milliequivalents of bases represent 66 percent of the cation exchange capacity, which means that the soil is 66 percent base saturated. Conversely, the 9.3 milliequivalents of exchangeable hydrogen means the soil is 34 percent hydrogen saturated. High base saturation, in general, reflects an appreciable supply of bases from weathering and/or a limited removal of bases by leaching.

Anion Exchange

In general, plants absorb as many anions as cations. The availability to plants of three important anions (nitrate, phosphate, and sulfate) is importantly related to mineralization of organic matter. Phosphate and sulfate, however, are significantly involved in anion exchange in soils.

Origin of Anion Exchange Capacity

Anion exchange sites arise from the protonation of hydroxyls on the surfaces of clays. Gibbsite is an oxide clay that consists of aluminum in six coordination with hydroxyls. In the normally acidic environment of highly weathered tropical soils, hydroxyls take on hydrogen atoms (protons), or are protonated as follows.

$$
\begin{array}{ccc}
\mid & & \mid \\
Al & & Al \\
\mid\ \diagdown & \xrightarrow{\text{protonated}} & \mid\ \diagdown \\
O\quad OH + H^+ & \longrightarrow & O\quad OH_2^+ \\
\mid\ \diagup & & \mid\ \diagup \\
Al & & Al \\
\mid & & \mid
\end{array}
$$

neutral   proton   protonated and
oxide   positively charged
surface   oxide surface

Some information on anion exchange capacity of clays is given in Table 7-5.

Importance of Anion Exchange

Cation exchange capacity increases as soil pH increases, and anion exchange capacity increases as soil pH decreases. Few soil horizons are positively charged; however,

---

[1]Technically, a base is a proton acceptor like the OH ion, while an acid is a proton donor like the H ion. However, the exchangeable cations Ca, Mg, K, and Na are all associated with compounds in the soil such as $CaCO_3$, $MgCO_3$, K_2CO_3, and Na_2CO_3, which are more basic than acidic in reaction. For this reason, Ca, Mg, Na, and K are commonly referred to as *exchangeable bases*, while H is commonly called an *exchangeable acid*.

the importance of the phenomenon can hardly be overestimated, since most important soil-plant nutrition relationships are reversed. In summary, soils with net, positively charged colloids:

1. Adsorb anions like nitrate and chloride ions.
2. Cations like calcium, magnesium, and potassium are repelled and remain very susceptible to leaching in the soil solution.

The high negative charge at the clay surfaces of soils with layer silicate systems, or high cation exchangeable clays, repels anions to such an extent that anion exchange is of little importance in such soils.

Factors Affecting Soil pH

The pH of soil is one of its most important properties, since there are many pH and nutrient availability relationships; there also are many relationships that exist between pH and the overall genesis and properties of soils. Soil pH is commonly determined by (1) mixing one part of soil with two parts of distilled water (or other suitable material such as a neutral salt solution), (2) occasionally mixing it to allow the soil and water to approach equilibrium, and then (3) measuring the pH of the soil-water suspension. There are many components in soils that influence the H^+ concentration of a soil solution. The situation is complicated by the great diversity of soil materials and their interactions. This section begins with a consideration of pH and the factors that control the pH in most soils, which is generally in the range of 4 to 10. Soil pH less than 4 is usually associated with the presence of a strong acid, such as sulfuric acid.

pH Defined

Water is neutral, because the concentration of H^+ and OH^- are equal. At neutrality, the pH is 7. For our purposes, it is sufficient to consider water as composed of water molecules, hydrogen ions, and hydroxyl ions. Water dissociates or ionizes as follows:

$$HOH \rightleftharpoons H^+ + OH^-$$

At equilibrium, the reaction is strongly to the left, producing the following composition of water expressed as moles per liter: (at 25°C, a liter of water weighs 997 grams and 1 mole of water weighs 18 grams, resulting in 55.4 moles of water per liter).

HOH	55.399,999,8 moles per liter
H^+	0.000,000,1 moles per liter
OH^-	0.000,000,1 moles per liter

Two points should be noted. First, only one water molecule in 554,000,000 is ionized. Second, the number, concentration, or moles per liter of H^+ is equal to OH^-.

The pH scale has been devised for conveniently expressing the extremely small concentrations of H^+ found in water and many important biological systems. The pH is defined as:

$$pH = \log \frac{1}{[H^+]}$$

(where $[H^+]$ equals moles of H^+ per liter)

The pH of pure water is calculated as follows:

$$pH = \log \frac{1}{0.000,000,1}$$
$$= \log 10,000,000 = 7$$

Each unit change in pH is associated with

a 10-fold change in the concentration of H^+ and OH^- (*see* Fig. 8-3).

Role of Carbonates in Slightly to Strongly Alkaline Soils

Parent materials have a wide range in pH; young soils inherit the pH of these materials. Throughout the world, however, there are many parent materials and soils that contain carbonates and are calcareous. When calcareous soil is treated with a dilute hydrochloric acid, carbon dioxide is produced and escapes as a gas; the soil is said to effervesce. The reaction is:

$$CaCO_3 + 2HCl \rightarrow CaCl_2 + H_2O + CO_2 \nearrow$$

Hydrolysis of carbonates produces an excess of OH^- relative to H^+ to produce alkaline soil:

Figure 8-3
Changes in H+ and OH^- concentrations with changes in pH of solutions.

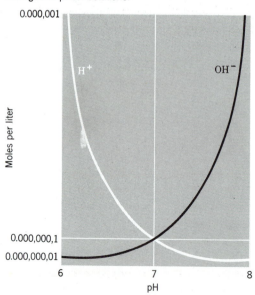

$$CaCO_3 + H_2O \rightarrow Ca^{+2} + HCO_3^- + OH^-$$

The hydrolysis of calcium carbonate is sufficient to account for a pH as high as 8.3. If the soil contains Na_2CO_3 that is more soluble than $CaCO_3$, there is a greater hydrolysis and production of hydroxyl; the pH may go as high as about 10.

Calcareous soils are 100 percent base saturated. These exchangeable bases also react with water and, by hydrolysis, produce OH^-:

$$\boxed{\text{micelle}}\,Ca^{+2} + 2H_2O \rightleftharpoons$$

$$\boxed{\text{micelle}}\begin{array}{l}H^+\\H^+\end{array} + Ca^{+2} + 2OH^-$$

The strongly adsorbed calcium that tends to dominate the exchange surfaces hydrolyzes very little and is not very important. When exchangeable sodium saturation is 15 percent or more, there is sufficient hydrolysis to produce a pH as high as 10. Thus, soils in the pH range of about 8, and up to 10, have a pH that is controlled mainly by carbonate hydrolysis. To a lesser extent, it is controlled by hydrolysis of exchangeable bases, except in soils with 15 percent or more of exchangeable sodium.

Role of Exchangeable Hydrogen and Exchangeable Bases in Slightly Acidic and Alkaline Soils

Mineral weathering in soils may contribute bases that accumulate as carbonates for the maintenance of the carbonate system. In the absence of leaching in soils with an aridic soil moisture regime and an inheritance of carbonates from the parent material, the soil may remain calcareous and have a pH dominated by carbonate hydrolysis. Soils developing where soil moisture regimes are udic may become leached of carbonates; in time, a point may be reached when the soil is noncalcareous and is 100 percent base saturated. The cation exchange capacity of these soils is commonly dominated by 2:1 clays and organic matter, and pH of about 8 is common. If the release of bases from mineral weathering equals the loss of bases from soil by leaching, cropping, and so on, the soil may remain 100 percent base saturated and the pH remains above 7. If, however, the amount of water moving through the soil increases and/or the release of bases from weathering decreases as primary minerals are depleted, exchangeable bases will be lost from the soil; exchangeable H^+ will begin to saturate the cation exchange capacity. Many slightly or moderately weathered soils have a pH of 7 when the hydrogen saturation is 20 percent and the base saturation is 80 percent (*see* Fig. 8-4). Such moderately weathered soils are generally dominated by silicate clay and humus ion exchange systems.

Exchangeable hydrogen contributes H^+ to the soil solution via dissociation from the exchange as follows:

$$\boxed{\text{micelle}}\ H^+ \rightleftharpoons H^+$$

Soil pH is caused by the competing effects of exchangeable hydrogen dissociation and the production of OH^- from exchangeable base hydrolysis in soils ranging from slightly alkaline to slightly acidic. As leaching continues and soil pH declines, the cation exchange capacity declines, in large part due to a reduction in a pH-dependent charge of organic matter. The exchangeable hydrogen on the pH-dependent sites of organic matter shows less tendency to ionize as pH declines. Thus, as soil pH declines below about 6.0, the hydrogen associated with the pH-dependent sites of organic matter becomes increasingly less exchangeable and increasingly more bound and exchangeable hydrogen becomes less important as a source of soil acidity.

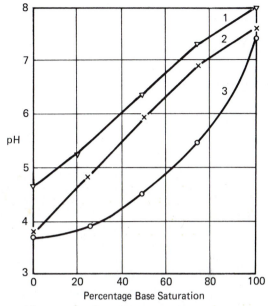

Figure 8-4
The percentage base saturation relationships for soils having different sources of cation exchange capacity. Curve 1—kaolinite; curve 2—organic matter; and curve 3—montmorillonite clay.

Role of Aluminum in Moderately and Strongly Acidic Soils

When a montmorillonite clay is leached with an acid, becoming 100 percent hydrogen saturated, it will be converted into a clay that is essentially 100 percent aluminum saturated. Aluminum comes out of the clay structure and occupies the cation exchange sites. This exchangeable aluminum (Al^{+3}) is strongly adsorbed, but it maintains an equilibrium with the soil solution. Al^{+3} in solution hydrolyzes to produce H^+ as follows:

$$Al^{+3} + H_2O \rightarrow AlOH^{+2} + H^+$$
$$(\text{hydroxy-Al})$$

The hydroxy aluminum ions may be readsorbed to the exchange and, in turn, may also hydrolyze:

$$Al(OH)^{+2} + H_2O \rightarrow Al(OH)_2^{+1} + H^+$$
$$(\text{hydroxy-Al})$$

Thus, the major source of H^+ in moderately and strongly acidic soils is aluminum hydrolysis; pH ranges about 5.5 to 4.0. If the soil is leached with a neutral salt solution, such as KCl, the tightly bound, pH-dependent charge hydrogen associated with organic matter is not removed to any appreciable extent. Instead, the principal cations removed are aluminum, calcium, magnesium, and some potassium. When the plow layers of three Ultisols from the coastal plains of the southeastern United States were leached with a KCl solution, exchangeable aluminum was the dominant exchangeable cation extracted. (see Table 8-3).

The three soils are very acidic (their pH ranging from 4.6 to 4.9), more so than most soils; they have a very high aluminum saturation. The data, however, illustrate the dominant role that aluminum can play in contributing to soil acidity. The sum of the extractable cations for these soils can be considered equal to the effective cation exchange capacity, since there would be a very small amount of potassium. The three soils averaged only a 12 percent base saturation and an 88 percent acidic saturation (aluminum plus hydrogen).

These Ultisols are highly weathered, and the clay fraction is dominated by kaolinite and oxidic clays. Thus, as these soils evolve, there are important changes in mineralogy that are associated with important changes in soil pH. In Ultisols (and Oxisols), there is generally a lack of primary minerals in the sand and silt that can weather to release bases, even though the high acidity favors rapid mineral weathering. These soils, compared to layer silicate system soils, have a more pH-dependent charge that more weakly adsorbs bases (such as calcium). Thus, even though some of these soils have few milliequivalents of calcium per 100 grams of soil, they may provide sufficient calcium

Table 8-3

The pH, KCl- Extractable Cations, Effective Cation Exchange Capacity and Percentage Base and Acidic Saturation of Three Ultisols

Soil	pH	Ca	Mg	Al	H	Cation Exchange Capacity	Acidic (Al + H)	Base
		(mEq/100 grams)						
Norfolk	4.9	0.06	0.02	1.43	0	1.51	95	5
Rains	4.7	0.11	0.02	1.62	0	1.75	93	7
Dunbar	4.6	1.00	0.11	3.67	0	4.78	77	23
Average		0.39	0.05	2.24	0	2.68	88	12

Adapted from Evans and Kamprath, 1970.

for plants. The high, exchangeable aluminum saturation makes these soils prone to having so much soluble aluminum as to be toxic to many plants. When aluminum saturation is more than 60 percent, appreciable aluminum appears in the soil solution; for many plants, aluminum toxicity will occur when the solution aluminum exceeds one part per million. An example of an intensely weathered and very acidic Oxisol dominated by kaolinitic and oxidic clay, and with high aluminum saturation, is shown in Fig. 8-5.

Role of Pyrite and Sulfuric Acid in Very Strongly Acidic Soils

In coastal areas where salt marshes occur and soils are flooded and water saturated most of the time, sulfur-reducing bacteria produce sulfide (H_2S). Most of these soils contain available iron and pyrite (FeS_2) forms. As long as soil remains flooded and reducing conditions exist, marsh plants grow well and pH may be almost neutral. However, due to the lowering of the sea level or elevation of the land, the soils become aerated; sulfur-oxidizing bacteria may convert the sulfur within the

Figure 8-5
Oxisol in Brazil that is typical of deeply and intensely weathered soils dominated by oxidic and kaolinitic clays and having high aluminum saturation.

pyrite to sulfuric acid. The pH may drop below 4, even as low as 2.0, and conditions can become too acidic for plants to survive; iron and aluminum toxicity for plants occurs. Many of these soils are used in Asia for rice production; water management is the most important soil management practice. Soil pH can change drastically and quickly depending on whether soil is subjected to alternate periods of flooding and drainage. These soils are called acid sulfate soils and *cat-clays*. A typical environment for their development is shown in Fig. 8-6.

Summary of the Soil pH Continuum

It has been emphasized that a given soil pH tends to be associated with a particular set of soil conditions. Soils with a pH of 8 and above are usually dominated by carbonate hydrolysis, and they have developed mainly from calcareous parent material. Weathering and leaching have been minimal. Carbonate hydrolysis and, to a lesser extent, exchangeable base hydrolysis, control pH in many young Entisols and Inceptisols, soils with aridic, soil moisture regimes, the Aridisols, and many Vertisols, where the high content of swelling clay inhibits base and carbonate removal via leaching.

Calcareous soils that have been leached of carbonate are 100 percent base saturated. Assuming that the parent materials were rich in feldspars and micas, the clays will be 2:1 and pH will be about 8. Hydrous mica, vermiculite, and montmorillonite clays with a high permanent charge tend to dominate the cation exchange capacity. As soil becomes more leached, exchangeable hydrogen will appear; soil acidity is affected by the dissociation of exchangeable hydrogen. These properties are characteristic of Mollisols, which are

Figure 8-6.
Mangrove swamps are one of the typical environments where acid sulfate soils develop.

required to have a 50 percent or more base saturation at a depth of 1.8 meters below the soil surface or 1.25 meters below the upper part of any argillic horizon that may be present (unless the soil is shallow and underlain by bedrock, and so on). These properties that are characteristic of many Mollisols are summarized in Table 8-4.

Slightly more advanced weathering is typical for Alfisols than is typical for Mollisols. Acidity may be sufficient to cause the appearance of hydroxy-Al. Some of the hydrogen ions in the soil solution are caused by hydroxy-aluminum hydrolysis. Clays are mainly 2:1, but chloritization (2:1:1 clay) is evidenced by interlayering with gibbsite. Base saturation is lower than that required for Mollisols. It must be 35 percent or more at depths 1.25 me-

Table 8-4
Some Generalized Properties of Mollisols, Alfisols, Ultisols and Oxisols

Property	Mollisols	Alfisols	Ultisols	Oxisols
Clay types	2:1	2:1 with chloritization	1:1 and oxidic	1:1 and oxidic
Cation exchange capacity	High	High	Low	Low
Sources of acidity	H^+	H^+ and hydroxy aluminum	Hydroxy aluminum and Al^{+3}	Hydroxy aluminum and Al^{+3}
Base saturation, percent	50 or more[a]	35 or more[b]	Less than 35[b]	—
Base status	High	High	Low	Low

[a]Determined at pH 8.2.
[b]Determined at pH 7.0.

ters below the upper part of the argillic horizon or 1.8 meters below the soil surface (unless it is shallow for bedrock, and so on). Both Mollisols and Alfisols typically contain a significant amount of weatherable minerals and are considered to be high base-status soils, as shown in Table 8-4.

Base saturation requirements, at depths of 1 to 2 meters below the soil surface, recognize that a given soil horizon becomes more weathered over time; at the same time, the depth of weathering in that soil increases. Many Mollisols have acidic A horizons, almost neutral B horizons, and calcareous C horizons. Ultisols have less than 35 percent base saturation, with the depth requirements the same as for Alfisols. This means that in most Ultisols, there are few primary minerals left to supply bases in the root zones of most crops.

Oxisols appear to be more weathered by one stage than Ultisols. Both soils are low base-status soils dominated by 1:1 and oxidic clays. Their acidity comes mainly from exchangeable aluminum hydrolysis. There is no base saturation requirement for Oxisols; however, it is generally very low and similar to that of Ultisols. Both Oxisols and Ultisols are generally considered low base-status soils (*see* Table 8-4).

These generalizations have assumed that clays in soil are formed by weathering within that soil, and the clays have not necessarily been determined by any inheritance from the parent material. There are some Alfisols that are dominated by inherited kaolinite. Furthermore, some Oxisols appear to have developed before the climate became arid. Subsequently, they have experienced an increase in base saturation. One must expect to find many exceptions to these generalizations.

High base-status soils are dominant in arid or desert regions (Aridisols) and in temperate region grasslands (Mollisols) and forests (Alfisols). Low base-status soils (Ultisols) are common on the coastal plains and piedmont of the southeastern United States. Both Ultisols and Oxisols are dominant in warm and humid forest regions throughout the world. Generalized locations are given in Figs. 10-1 and 10-2.

Other Factors Affecting Soil pH

A few other factors that influence soil pH are worthy of mention. Sulfur is a by-product in industrial gases. It is sometimes responsible for soil acidity in nearby soils as a result of the formation of sulfuric acid. A small amount of nitric acid is a natural component of rain.

About 30 years ago, a marked decrease occurred in the pH of rainfall in the northeastern United States. Rain with pH as low as 2.1 was collected. There has been an increase in the amount of fossil fuel burned; the taller stacks distribute the SO_2 over a wider area. The resulting increased acidity of rain is affecting lakes, vegetation, soil acidity, and weathering of buildings. The effect of "acid rain" on plants and soil acidity appears to have reduced the growth of forests in the northeastern United States and Sweden. Much of the SO_2 causing acid rain in Sweden probably comes from industrial areas of England and the Ruhr. The effect of acid rain on soils will depend on the nature of the soil. Calcareous soils will be little affected and plants may be benefited from the nitrogen and sulfur. By contrast the additional acidity could seriously impair low-base status soils that are very acidic.

A survey of rain acidity was conducted by 16,000 high school students in March, 1973. Many low readings of 3.5 were reported for Chicago, New York, Cleveland, Boston, and Los Angeles. Normal rainfall has a pH of about 5.6. A map based on this survey, showing the distribution of rain acidity in the United States, is given in Fig. 8-7.

Significance of Soil pH

Some organisms have a rather small tolerance to variations in pH, but other organisms can tolerate a wide pH range. Studies have shown that the actual concentrations of H^+ or OH^- are not very important, except under the most ex-

treme circumstances. It is the associated conditions of a certain pH value that is most important.

Nutrient Availability and pH Relationships

Perhaps the greatest general influence of pH on plant growth is its effect on the availability of nutrients (see Fig. 8-8). Soil pH is related to the percentage of base saturation. When the base saturation is less than 100 percent, an increase in pH is associated with an increase in the amount of calcium and magnesium in

Figure 8-7
The pH of rain in the United States. (From A. E. Klein, "Acid Rain in the U.S.," *Science Teacher,* pp. 36–38, May 1974.)

pH less than 5.0

pH 5.0 to 5.5

pH over 5.5

lime Calculation

80 - 57.6 = 22.4

33 x .224 = 7.39 x .05 = .37 x 20,000 =
7,400 lbs /acre

3. .21   CEC = 10.08

$\dfrac{2.78}{10.08}$ = .278 = 28%

4. K) $\dfrac{39}{1}$ x $\dfrac{1}{1000}$ = .039 : $\dfrac{.21}{100 \, g \, soil}$ x $\dfrac{.039}{MEQ}$ x $\dfrac{2,000,000 \, lbs}{Acre}$ =
163.8 lbs/acre

Mg) $\dfrac{24}{2}$ x $\dfrac{1}{1000}$ = .012 : $\dfrac{1.07}{100g}$ x $\dfrac{.012}{MEQ}$ x $\dfrac{2,000,000 \, lbs}{Acre}$ =
256.8 lb/ac

Ca) $\dfrac{40}{2}$ x $\dfrac{1}{1,000}$ = .020 : $\dfrac{1.5}{100g}$ x $\dfrac{.02}{MEQ}$ x $\dfrac{2,000,000 \, lbs}{Acre}$ =
600 lb/ac

5) .70 - .28 = .62
10 x .62 = 6.20 Cmol / lt

1) CEC = 29.32   Base Sat $\dfrac{11.92}{29.32}$ = .41

90 - 41 = 49
29. x .49 = 14.21 Cmoles x $\dfrac{1000g}{MEQH}$ x $\dfrac{.05 CaCO_3}{MEQ}$ = .71
.71 x 20000 = 74,200 lbs/acre

4. $\dfrac{24}{2}$ x $\dfrac{1}{1000}$ = .012 $\dfrac{3.47}{100g}$ x $\dfrac{.012}{MEQ}$ = .0416
MEQ

100 - 70 = .30 x 22.2 = 6.66 x .05

.5

90 − 73 = 17

$26.75 \times .17 = 4.55 \times .05 = \dfrac{.228}{100} \ MEQ \times 20,000 = 4560\#$

$.012 + 8.58 = .103 = \dfrac{t}{1,600,000} = 1,030 \ ppm$
$\div 100$

$\dfrac{11.53}{21.33} = .54 \ Base \ Sat.$

CEC 21.33

90 − 54 = 36

$21.33 \times .36 = 7.68 \times .05 = \dfrac{7.68}{1000} \ MEQ \times 2,000,000 \ lbs = 7,680 \ lbs/acre$

$.012 \times 2.67 = .032 = \dfrac{x}{1,000,000} = 320.4 \ ppm$
$\div 100$

$.012 \times \dfrac{5}{100} + 20,000$

$25 \times .06 = 1.5 \ cal \times .05 = \dfrac{.075}{100} \ MEQ \times 20,000 \ lb = 1500 \ lbs \quad \dfrac{1500 \ lbs}{acre} \div .75 \ ton$

$\dfrac{264}{300} = 88$

$x = 20 + 20 = 88$

$\dfrac{x}{1,000,000} \times 20,000$

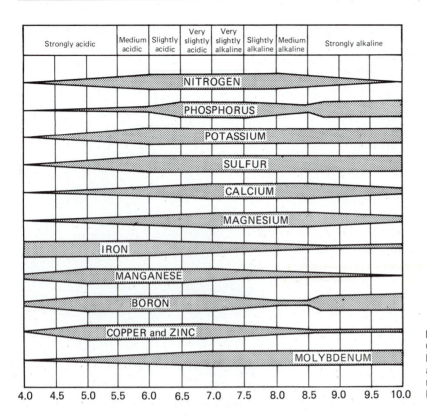

Figure 8-8
General relationship between soil pH and availability of plant nutrients; the wider the bar, the more availability.

the soil solution, since they are usually the dominant exchangeable bases. Many studies have been conducted that relate increases in plant growth with increases in the percentage of calcium in plants, and with increasing pH or percentage base saturation. The general relationship between pH and the availability of calcium and magnesium are shown in Fig. 8-8.

Another nutrient whose availability is increased as the pH is increased on the low end of the pH range in soils is molybdenum. At low pH, molybdenum forms insoluble compounds with iron, and it is rendered unavailable. Under these conditions, plants susceptible to molybdenum deficiency, such as cauliflower, clover, and citrus, will show a growth response to an increase in pH. The fact that increased availability of molybdenum occurs as the pH increases is responsible for molybdenum toxicity for plants in some soils.

Potassium availability is usually good in neutral and alkaline soils that reflects the limited leaching of exchangeable potassium (*see* Fig. 8-8).

The availability or solubility of some plant nutrients decreases with an increase in pH. Iron and manganese are two good examples (*see* Fig. 8-9). Iron and manganese are commonly deficient in calcareous soils. Phosphorus and boron also tend to be unavailable in calcareous soils that result from reactions with calcium. Phosphorus and boron also tend to be unavailable in very acidic soils. Copper and zinc have reduced availability in both highly acidic and alkaline soils. For the plant nutrients as a whole, good overall nutrient availability is found near pH 6.5 in high base-status soils. Soil pH in low base-status soils generally never exceeds 6.0 or 6.2.

Effect of pH on Soil Organisms

The greater capacity of fungi over bacteria to thrive in highly acidic soils was mentioned in Chapter 5. The pH requirements of some disease organisms can be used by soil management practices as a means to control diseases. One of the best-known cases is that of the maintenance of acidic soils to control potato scab. Damping-off disease in nurseries is controlled by maintaining the pH at 5.5 or less. Nitrifying organisms also become inhibited when the pH is less than 5.5. The availability of nitrogen in soils is related mainly to the effect of pH on decomposition of organic matter (*see* Fig. 8.8). High soil acidity has also been shown to inhibit earthworms in soils. Peter Farb relates an interesting case where soil pH influenced both earthworm and mole activity.

On a certain tennis court in England, each spring after the rains and snow had washed away the court marking, the marking lines could still be located because moles had unerringly followed them. This intriguing mystery was investigated, and it produced the following results. The tennis court had been built on acidic soil and chalk (lime) was used to mark the lines. As a result, the pH in the soil below the marking lines was increased. Earthworms then inhabited the soil under the markings, and the remainder of the soil on the tennis court remained uninhabited. Since earthworms are a primary food of moles, the moles re-

Figure 8-9
Pin oak is a popular ornamental tree that is susceptible to iron deficiency when grown on neutral or alkaline soil. Iron-deficient leaves have dark-green veins and yellow intervein areas. This tree is seriously deficient in iron, as indicated by the death and defoliation of many leaves on the upper branches.

stricted their activity to the soil only under the marking lines.

Aluminum, Iron, and Manganese Toxicity in Acidic Soils

The data in Fig. 8-8 show that iron and manganese are most available in highly acidic soils. Manganese toxicity may occur when the pH is about 4.5 or less. Evidence indicates that the high exchangeable aluminum in many acidic soils of the southeastern United States restricts root growth in subsoils. Plants (even varieties of the same species) exhibit differences in tolerance to high levels of aluminum, iron, or manganese, as well as to other soil conditions associated with soil pH. This gives rise to pH preferences of plants.

pH Preferences of Plants

Blueberries are well known for their highly acidic soil requirement. In a study, it was shown that poor blueberry growth occured when calcium saturation exceeded 10 percent. Normally, this would mean that the soil had a very low base saturation and pH. In Table 8-5, the optimum range for highbush blueberries is given as 4.0 to 5.0. Other plants preferring a pH of 5 or less include azaleas, orchids, sphagnum moss, jack pine, black spruce, and cranberry. The plants listed in Table 8-5 that prefer a pH of 6 or less are italicized. Most field and vegetable crops prefer a pH of about 6 or higher.

The fact that plants have specific soil pH requirements gives rise to the need to alter soil pH for successful growth of many plants. We examine this topic next.

Alteration of Soil pH

There are two approaches to assure that plants will grow without serious inhibition from unfavorable soil pH:(1) plants can be selected that will grow well with the existing soil pH, or (2) the pH of the soil can be altered to suit the preference of the plants. Considerations for altering soil pH will be examined first.

Soil pH can be lowered and soil acidity increased by the addition of sulfur or sulfur-containing compounds. The sulfur is converted into sulfuric acid. Most changes in soil pH are directed toward increased pH and reduced soil acidity. Lime, ($CaCO_3$) is commonly used; it hydrolyzes to produce OH^-, and the calcium increases base saturation.

Benefits of Liming

The benefits of lime depend on the soil and crop conditions. In Mollisols, there is little or no aluminum saturation and no danger of aluminum toxicity. Therefore, liming will not produce a benefit by reducing the amount of aluminum in the solution. However, some crops (such as alfalfa) may benefit from the increased nitrogen fixation that results from increased pH and calcium availability. By contrast, the major benefit of liming Oxisols is the inactivation of aluminum and manganese in the soil solution, which reduces or prevents aluminum and manganese toxicity. In addition, lime acts as a fertilizer to supply calcium when it is in very short supply. Overliming soils can either produce adverse problems associated with deficiencies of zinc, manganese, iron, boron and copper or increase molybdenum sufficiently to produce molybdenum toxicity. The major benefits of liming and the adverse effects of overliming, as well as the desirable pH range for high base- (and permanently charged) and low base-status soils (with high pH-dependent charges) is summarized in Table 8-6.

Lime Requirement of Low Base-Status Soils with a High pH-Dependent Charge

When exchangeable aluminum saturation exceeds 60 percent of the effective cation-exchange capacity, the amount of aluminum in the solution increases significantly. The amount of lime recommended is related to the amount of exchangeable aluminum, since inactivation of aluminum

Table 8-5
Optimum pH Ranges of Selected Plants Growing on High Base-Status Soils

Field Crops		**Forest Plants / Flowers**		**Forest Plants**	
Alfalfa	6.2–7.8	Alyssum	6.0–7.5	Oak, Pine	5.0–6.5
Barley	6.5–7.8	*Azalea*	*4.5–5.0*	Oak, White	5.0–6.5
Bean, field	6.0–7.5	Barberry, Japanese	6.0–7.5	*Pine, Jack*	*4.5–5.0*
Beets, sugar	6.5–8.0	Begonia	5.5–7.0	*Pine, Loblolly*	*5.0–6.0*
Bluegrass, Ky.	5.5–7.5	Burning bush	5.5–7.5	*Pine, Red*	*5.0–6.0*
Clover, red	6.0–7.5	Calendula	5.5–7.0	*Pine, White*	*4.5–6.0*
Clover, sweet	6.5–7.5	Carnation	6.0–7.5	*Spruce, Black*	*4.0–5.0*
Clover, white	5.6–7.0	Chrysanthemum	6.0–7.5	Spruce, Colorado	6.0–7.0
Corn	5.5–7.5	*Gardenia*	*5.0–6.0*	*Spruce, White*	*5.0–6.0*
Flax	5.0–7.0	Geranium	6.0–8.0	*Sycamore*	*6.0–7.5*
Oats	5.0–7.5	*Holly, American*	*5.0–6.0*	Tamarack	6.0–7.5
Pea, field	6.0–7.5	Ivy, Boston	6.0–8.0	Walnut, Black	6.0–8.0
Peanut	5.3–6.6	Lilac	6.0–7.5	Yew, Japanese	6.0–7.0
Rice	5.0–6.5	Lily, Easter	6.0–7.0		
Rye	5.0–7.0	*Magnolia*	*5.0–6.0*	**Weeds**	
Sorghum	5.5–7.5	*Orchid*	*4.0–5.0*	Dandelion	5.5–7.0
Soybean	6.0–7.0	*Phlox*	*5.0–6.0*	Dodder	5.5–7.0
Sugar Cane	6.0–8.0	Poinsettia	6.0–7.0	Foxtail	6.0–7.5
Tobacco	5.5–7.5	Quince, flowering	6.0–7.0	Goldenrod	5.0–7.5
Wheat	5.5–7.5	*Rhododendron*	*4.5–6.0*	Grass, Crab	6.0–7.0
		Rose, hybrid tea	5.5–7.0	Grass, Quack	5.5–6.5
Vegetable Crops		Snapdragon	6.0–7.5	*Horse Tail*	*4.5–6.0*
Asparagus	6.0–8.0	Snowball	6.5–7.5	*Milkweed*	*4.0–5.0*
Beets, table	6.0–7.5	Sweet William	6.0–7.5	Mustard, Wild	6.0–8.0
Broccoli	6.0–7.0	Zinnia	5.5–7.5	Thistle, Canada	5.0–7.5
Cabbage	6.0–7.5				
Carrot	5.5–7.0	**Forest Plants**		**Fruits**	
Cauliflower	5.5–7.5	Ash, White	6.0–7.5	Apple	5.0–6.5
Celery	5.8–7.0	*Aspen, American*	*3.8–5.5*	Apricot	6.0–7.0
Cucumber	5.5–7.0	Beech	5.0–6.7	Arbor Vitae	6.0–7.5
Lettuce	6.0–7.0	*Birch, European*		Blueberry, High	
Muskmelon	6.0–7.0	*(white)*	*4.5–6.0*	Bush	4.0–5.0
Onion	5.8–7.0	*Cedar, White*	*4.5–5.0*	Cherry, sour	6.0–7.0
Potato	4.8–6.5	Club Moss	4.5–5.0	Cherry, sweet	6.0–7.5
Rhubarb	5.5–7.0	*Fir, Balsam*	*5.0–6.0*	Crab apple	6.0–7.5
Spinach	6.0–7.5	Fir, Douglas	6.0–7.0	*Cranberry, large*	*4.2–5.0*
Tomato	5.5–7.5	*Heather*	*4.5–6.0*	Peach	6.0–7.5
		Hemlock	*5.0–6.0*	*Pineapple*	*5.0–6.0*
Flowers and Shrubs		Larch, European	5.0–6.5	Raspberry, Red	5.5–7.0
African violet	6.0–7.0	Maple, Sugar	6.0–7.5	Strawberry	5.0–6.5
Almond, flowering	6.0–7.0	*Moss, Sphagnum*	*3.5–5.0*		
		Oak, Black	6.0–7.0		

Data from Spurway, 1941. Plants with an optimum pH range of 6.0 and below are italicized.

Table 8-6

Benefits of Liming, the Adverse Effects of Overliming, and the Desirable pH
Ranges for High and Low Base-Status Soils

Soil	Benefits of Lime	Detrimental Effects of Overliming	Desirable pH Range
High base-status and permanent charge	Increase: soil pH calcium availability nitrogen fixation	Decrease: zinc availability manganese availability Increase molybdenum to toxic levels	6.0–6.8
Low base-status and high pH-dependent charge	Inactivate aluminum and manganese Add calcium (and magnesium)	Decrease availability of: zinc manganese copper boron Increase molybdenum to toxic levels	5.0–6.2

appears to be the major benefit of liming. Some workers have suggested that the lime requirement is the amount of lime equal to 1.5 times the milliequivalents of exchangeable aluminum. Thus, the lime requirement of the Rains soil noted in Table 8-3 is equal to:

1.5×1.62 meq
= 2.43 meq lime per 100 grams of soil

Assume that the lime is $CaCO_3$. The milliequivalent weight of 1 milliequivalent of $CaCO_3$ equals:

$$\frac{\text{molecular weight}}{\text{valence}} = \frac{100 \text{ grams}}{2}$$

$$= 50 \text{ grams}$$

$$= \text{equivalent weight}$$

$$\frac{50 \text{ grams}}{1,000} = 0.05 \text{ grams}$$

per meq of $CaCO_3$

The 2.43 milliequivalents of lime weighs

0.12 grams (2.43 milliequivalents × 0.05 gram). These values can be substituted into the equation used earlier to calculate pounds of exchangeable cations per plow layer for an acre or hectare. Thus:

$$\frac{0.12 \text{ pounds of } CaCO_3}{100 \text{ pounds of soil}}$$
$$= \frac{x \text{ pounds of } CaCO_3}{2,000,000 \text{ pounds of soil}}$$

$x = 2,400$ pounds $CaCO_3$

Assuming a 2,000,000 kilogram plow layer for a hectare, the lime requirement is 2,400 kilograms. The rate must be adjusted to account for different thickness or weight other than 2,000,000 pounds or kilograms of soil. The effect of liming on the reduction of exchangeable aluminum of a low base-status soil is shown in Fig. 8-10.

In many soils, there is also aluminum toxicity in the subsoil that restricts the root growth and the uptake of water. Soils are droughty, which is partly due to a lack

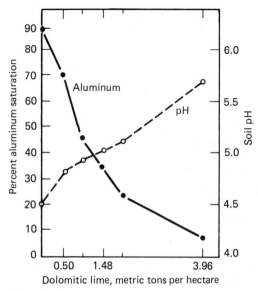

Figure 8-10

Effect of liming on soil pH and percentage aluminum saturation of a low base-status soil with a high pH-dependent charge. (Data from an annual report on tropical soils research of North Carolina State University, 1976.)

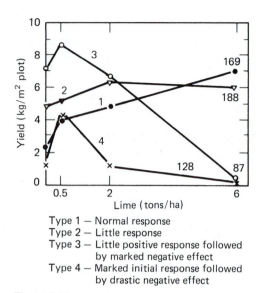

Type 1 — Normal response
Type 2 — Little response
Type 3 — Little positive response followed
by marked negative effect
Type 4 — Marked initial response followed
by drastic negative effect

Figure 8-11

Fresh root yields of cassava cultivars after growing 6 months on a strongly acidic Oxisol. The different cultivar responses also characterize the kinds of liming responses typically found between different crops. Note the large plant responses from small amounts of lime. (Data from Spain, et al., 1975.)

of roots in the subsoil. This problem is difficult to solve, since overliming the surface soil layer, to get some lime to move downward, may create micronutrient toxicities or deficiencies and reduce plant growth. Calcium carbonate is also quite insoluble and moves downward very slowly. Experiments have been conducted in North Carolina to increase rooting depth by incorporating lime deeply.

A characteristic feature of liming very acidic low base-status soils is that great variation exists in the plant response between cultivars or varieties of a given crop and the generally great variation in response between crops. Frequently, a small amount of lime can greatly increase plant growth; a small amount of additional lime can greatly decrease plant growth (*see* Fig. 8-11).

Lime Requirement of High Base-Status Soils with a High Permanent Charge

Two major benefits of liming high base-status soils with a high permanent charge are pH adjustment and the increase in base saturation (*see* Fig. 8-12). Basically, this means a significant decrease in exchangeable hydrogen, which is a major source of acidity, and an increased calcium availability. Within these soils, there is a relationship between pH and base saturation or hydrogen saturation. For broad geographic areas where soils have similar mineralogy, a general relationship exists. In southern Michigan, the data from thousands of soil samples were analyzed and, for mineral soils, the relationship is as follows:

$$pH \times 24 = 187 - 0.3 \text{ (CEC)}$$
$$- \text{ percent H saturation}$$

Percent base saturation

Figure 8-12
Growth of soybeans versus base saturation,
increasing from 25% on the left to 85% on the right,
on a high-base status Alfisol.

Using this equation, Fig. 8-13 was obtained. From the figure, it can be seen that a pH of 7 was associated with a 15 percent H saturation and 85 percent base saturation (assuming a cation exchange capacity of 13). Furthermore, for the soils studied, the maximum pH that could be developed by the dissociation of cations from the exchange was 7.6.

At 50 percent H and base saturation, the pH was 5.5. This is due to the exchangeable H being less tightly adsorbed to the micelles than the exchangeable bases, which are predominately divalent calcium and magnesium. In kaolinitic soils, where the exchange sites are mainly from dissociation of H from OH on the exposed edges of clay particles, a pH between 6 and 7 may be associated with 50 percent base saturation, because OH-bound H is much more tightly adsorbed than is H adsorbed on an exchange site that results from isomorphous substitution. At pH 6.0, kaolinite is 40 percent base saturated, montmorillonite is over 80 percent base saturated, and humus is intermediate base saturated (*see* Fig. 8-4).

In theory, increasing the pH of the soils typified in Fig. 8-13 from pH 5.5 and 50

percent base saturation to pH 7.0 and 85 percent base saturation would require a 35 percent increase in base saturation. Assuming a cation exchange capacity of 13, it would require 4.55 milliequivalents (13 milliequivalents $\times$ 0.35) of lime per 100 grams of soil. Using $CaCO_3$, which has a milliequivalent weight of 0.05 grams, it would require 0.2275 grams of $CaCO_3$ for 100 grams of soil. The lime requirement for a 2,000,000 pound acre furrow slice is:

$$\frac{0.2275 \text{ pounds of } CaCO_3}{100 \text{ pounds of soil}} =$$

$$\frac{x \text{ pounds of } CaCO_3}{2,000,000 \text{ pounds of soil}}$$

$x = 4,500$ pounds

Similarly, for a 2,000,000 kilogram furrow slice for a hectare, the lime or $CaCO_3$ requirement is 4,500 kilograms.

Role of Cation-Exchange Capacity in Altering Soil pH

Suppose the soil just used to illustrate the theoretical lime requirement had a cation exchange capacity twice as large: 26 instead of 13. Using the equation relating pH and hydrogen saturation, one finds that at pH 5.5 the soil with twice as much cation exchange capacity has a 53 percent base saturation compared to 50 for the soil with a cation exchange capacity of 13. Notice this in Table 8-7. This means that the same concentration of H^+ in the soil solution (same pH) is produced with only a slight difference in the percentage of H or base saturation. The soil with a cation exchange capacity of 26 has nearly twice the exchangeable H as the soil with a cation exchange capacity of 13, when both have a pH of 5.5. The data in Table 8-7 also show that similar percentage base saturation increases are needed to in-

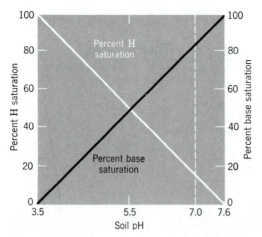

Figure 8-13
Relationship between soil pH and percentage base and hydrogen saturation for loamy-textured soils of southern Michigan assuming a cation exchange capacity of 13.

Soil acidity has two components: (1) the active or solution H^+, and (2) the exchangeable or reserve acidity. These two forms tend toward equilibrium so that a change in one produces a change in the other. When a base is added to an acid soil, the solution H is neutralized and some exchangeable H ionizes to reestablish the equilibrium. The amount of exchangeable H is slowly decreased, the solution H is decreased, and the pH slowly increases. The resistance of the soil pH to change gives rise to the buffering phenomenon in soils. Soils with the largest cation exchange capacities offer the greatest resistance to change in pH and are the most strongly buffered.

Liming produces less pH change if the cation exchange capacity is mainly pH dependent. This results from increased cation exchange capacity as lime neutralizes soil acidity. There is an increase in milliequivalents of exchangeable bases, but little change in the percentage of base saturation. For this reason, liming of many tropical soils rich in oxide clays (Oxisols) is based on a reduction of toxic effects of low pH, such as Al toxicity, instead of in-

crease pH from 5.5 to 7. This results in the need for almost two times more lime for the soil with twice the cation exchange capacity (*see* Table 8-7), (4.55) versus 9.36 milliequivalents per 100 grams of soil. The greater capacity of soils with greater exchange capacities to adsorb bases for a given pH change is illustrated in Fig. 8-14.

Table 8-7
Comparisons of Base Saturation Relationships of Soils with Varying Cation Exchange Capacity

	Cation Exchange Capacity of Soil, meq/100 grams	
	13	26
Base saturation at pH 5.5	50%	53%
Exchangeable H at pH 5.5	6.5 meq	12.2 meq
Base saturation at pH 7	85%	89%
Base saturation change needed to increase pH from 5.5 to 7	35%	36%
Base needed per 100 grams of soil	4.55 meq	9.36 meq

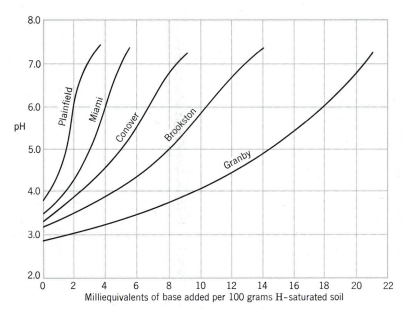

Figure 8-14
Titration curves showing the different amounts of base needed to increase the pH of soils having increasing cation exchange capacities from left to right (different degrees of buffering). All soils had zero percent base saturation when titration was started.

creasing pH to some predetermined goal as 6.5.

Soil Tests to Determine Lime Requirement

It is common practice in soil-testing laboratories to make a direct measurement of the lime requirement by using a buffer solution. One such test was developed at Ohio State University and is widely used. A buffer solution with a pH of 7.5 is mixed with a known quantity of soil. The exchange acidity is replaced from the exchange sites. The depression in pH from 7.5 (buffer solution pH) to the pH of the mixture of buffer solution and soil is a measure of the total acidity. The lime requirement is obtained from tables that have been developed, which relate depression of buffer pH to tons of limestone required to raise the pH of the soil to a desired value.

A less accurate but useful method, especially for the homeowner, is to determine the pH of the soil with inexpensive indicator dyes (*see* Fig. 8-15), and to obtain the lime requirement from a table that relates soil pH and texture (used as an indication of cation exchange capacity) to lime requirement. Such a table used for liming a garden is presented in Table 8-8. The lower lime requirement values for the soils of the southern states is a reflection of the lower cation exchange capacity resulting from kaolinitic clays. Caution must be observed to prevent overliming, because some micronutrients may become unavailable, resulting in nutrient deficiencies.

Forms of Lime

Chemically, lime is CaO, but an extension of the meaning of the word lime is used to include all limestone products used to neutralize soil acidity. Limestone deposits are widely distributed and constitute the most important source of lime (*see* Fig. 8-16).

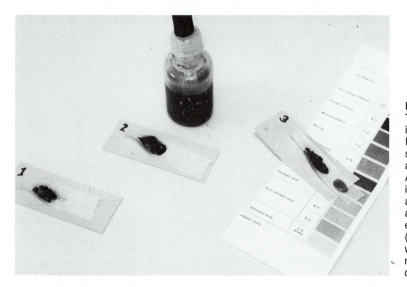

Figure 8-15
To determine soil pH with indicator dye solution: (1) Place a small amount of soil in the folded crease of a strip of wax paper, (2) Add 1 or 2 drops more indicator than the soil can absorb, tilt up and down to allow indicator to equilibrate with soil, and (3) Separate excess dye with the tip of a pencil and match dye color with the color chart.

Limestone is a carbonate form of lime with $CaCO_3$ and $MgCO_3$ as the major components. The oxide form, CaO, is produced by heating calcium carbonate and driving off carbon dioxide. Hydroxide forms of lime are produced by "slaking" or adding water to the oxide forms. The reaction of $CaCO_3$ in the neutralization of soil acidity is as follows:

$$\boxed{\text{micelle}}\begin{matrix} H^+ \\ Al^{+3} \end{matrix} + 2CaCO_3 + 3H_2O \rightarrow$$

$$\boxed{\text{micelle}}\begin{matrix} Ca^{+2} \\ Ca^{+2} \end{matrix} + 2H_2CO_3 + Al(OH)_3$$

$$H_2O \quad CO_2$$

Calcium occupies the exchange. The exchangeable H is converted into water when carbonic acid breaks down into water and carbon dioxide. The exchangeable aluminum is converted to insoluble aluminum hydroxide (gibbsite), which precipitates from the solution.

On a weight basis, the three forms of lime have different neutralization capacities. The neutralizing value of liming ma-

terials is calculated on the basis of pure calciuim carbonate as 100 percent. When pure calcium carbonate is burned in a kiln, 100 grams of the dry material give off 44 grams of carbon dioxide gas, leaving 56 grams of calcium oxide. When the 56 grams of calcium oxide are moistened, they react with 18 grams of water to form 74 grams of hydrated lime. It is obvious that 100 grams of pure calcium carbonate, 74 grams of pure calcium hydroxide, and 56 grams of pure calcium oxide all contain the same amount of calcium; all have the same power to neutralize soil acidity.

The relative ability of different liming materials to neutralize acidity is frequently expressed on a percentage basis. Thus, the neutralizing power of the different forms of lime in the pure state is determined by their molecular weights. The molecular weight of calcium carbonate is 100, of calcium hydroxide is 74, and of calcium oxide is 56. By dividing 100 by 74, the figure 1.35 is obtained, which means that 1 gram of calcium hydroxide supplies the same amount of calcium as

Table 8-8

Suggested Applications of Finely Ground Limestone to Raise the pH of a 7-inch Layer of Several Textural Classes of Acidic Soils, in Pounds per 1,000 ft^2

Texture Class	pH 4.5 to 5.5		pH 5.5 to 6.5	
	Northern and Central States	Southern Coastal States	Northern and Central States	Southern Coastal States
Sands and loamy sands	25	15	30	20
Sandy loams	45	25	55	35
Loams	60	40	85	50
Slit loams	80	60	105	75
Clay loams	100	80	120	100
Muck	200	175	225	200

From Kellogg, 1957.

1.35 of calcium carbonate. In other words, if expressed on a percentage basis (1.35 × 100), pure calcium hydroxide has a neutralizing value of 135 percent, relative to calcium carbonate. Likewise, calcium oxide (100/56 × 100 = 178) has a neutralizing power of 178 percent. Pure magnesium carbonate with a molecular weight of 84 has a neutralizing value of 119 percent (100/84 × 100 = 119). A

Figure 8-16
Many areas have limestone near the surface that can readily be mined, ground, and applied to acidic soils as an inexpensive corrective for excess soil acidity.
(Monroe County, Wisc.)

limestone containing 80 percent calcium carbonate and 20 percent magnesium carbonate would have a neutralizing value of 103.8 percent (80 + 20 × 1.19 = 103.8). Frequently magnesium limestones have a neutralizing power of 107 or 108 percent. The molecular weights, neutralizing values, and calcium carbonate equivalents for the common chemical forms of liming materials in a pure state are given in Table 8-9. Locally, many other materials are used for liming. These include marl, by-product lime from manufacturing, and oyster shells.

Importance of Limestone Particle Size

With all factors being equal, the finer a limestone is ground, the more rapidly it will dissolve and the more thoroughly it can be mixed with the soil. The effectiveness of a ground substance of a given neutralizing power is determined not only by its rate of solubility, but also by its contact with the colloidal particles. However, the finer the stone is ground, the greater its cost will be and the lesser its lasting qualities. Furthermore, a very finely ground limestone is difficult to handle

and unpleasant to distribute. Therefore, it is generally recommended that a ground limestone of mediuim fineness be purchased. Such a grade can be ground rather cheaply. It contains a sufficient quantity of fine material to give immediate effects and a sufficient amount of coarse material to give it lasting qualities. Such a ground limestone would be one that would pass an 8-mesh screen; between 25 and 50 percent would pass a 100-mesh screen. The fractions 8 to 20 mesh and larger are not very effective in increasing the pH of acid soils (*see* Fig 8-17).

Method and Time of Applying Lime

The principal requirement of any method of applying lime is that it should be distributed evenly and, except when applied to grasslands, it should be thoroughly mixed with the soil. Lime, even as it dissolves, moves to no appreciable extent horizontally and only to a limited extent vertically. Movement is not sufficient to distribute the lime evenly over the field, or to mix it thoroughly with the soil. Since soil acidity is due largely to colloidal acids,

Table 8-9
Relative Neutralizing Power of Different Forms of Lime

Form of Lime	Molecular Weight	Neutralizing Value, %	Pounds Equivalent to 1 Ton of Pure CaCO$_3$	Kilograms Equivalent to 1 Metric Ton of Pure CaCO$_3$
Calcium carbonate	100	100	2,000	1,000
Magnesium carbonate	84	119	1,680	840
Calcium hydroxide	74	135	1,480	740
Magnesium hydroxide	58	172	1,160	580
Calcium oxide	56	178	1,120	560
Magnesium oxide	40	250	800	400

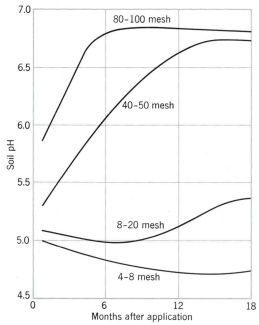

Figure 8-17
Effect of dolomitic lime of different particle sizes on the pH of soil at various times after application. (Adapted from T. A. Meyer and G. W. Volk, "Effect of Particle Size of Limestone on Soil Reaction," *Soil Science, 73:*37–52, 1952. Used by permission of author and the Williams and Wilkins Co., copyright © 1952.)

it is essential that lime come in contact with all soil particles so far as possible. This requires a thorough and even mixing of lime with soil. An even distribution may be accomplished by the using of most of the standard lime spreaders on the market. The only way of mixing the lime thoroughly with the soil is by tillage operations.

Generally, lime may be applied any time during the year when it is most convenient. Often, however, the type of the rotation, the system of farming, and the form of lime used will be the deciding factors.

It is advisable, of course, to apply lime where it can be used to the best advantage

in the rotation; for example, preceding the legume crop or in connection with a green-manure crop. It is usually best to apply lime considerably in advance of seeding legumes, so that the lime will have time to correct the acid condition of the soil. To insure best results, lime should be in the soil at least 6 months prior to seeding legumes, although successful legume seedings sometimes are obtained by applying lime immediately preceding or with the legume seeding.

The data from lysimeter studies conducted in Ohio, cited in Table 3-2, showed that the annual loss of calcium in the drainage water could be equal to or greater than the calcium removed in some harvested crops. This shows that in humid regions leaching may be as important as crop production in the removal of exchangeable bases and the development of soil acidity. Factors that affect leaching, such as soil permeability and slope, will influence the frequency of liming. In general, it can be stated that several hundred pounds of lime per acre will be needed annually in humid regions. Weathering can be expected to supply part of the need, but it is obvious that the maintenance of a favorable pH will require liming once every 5 to 10 years.

Acidulation of Soils

The addition of acid sphagnum peat to soil may have some acidifying effect. However, significant and dependable increases in soil acidity are perhaps best achieved through the use of sulfur. Sulfur is slowly converted to sulfuric acid so the change in soil pH is gradually decreased over several months or a year. The milliequivalent weight of sulfur is 0.16 grams, and 320 pounds of sulfur per acre furrow slice or 320 kilograms for a 2,000,000 kilogram plow layer of a hectare will theoretically increase exchangeable acidity or exchangeable hydrogen equal to 1 milliequivalent per 100 grams of soil. In a forest nursery near Orono, Canada, the use of sulfur resulted in significant increases in soil acidity and the growth of red pine seedlings. About 900 pounds of sulfur for plow layers per acre or 900 kilograms per hectare lowered soil pH one unit in the experiment.

As with lime, the amount of sulfur required varies with the cation exchange capacity of the soil. Recommendations for using sulfur on small areas are given in Table 8-10. The effect of nitrogen fertilizer on soil acidity is discussed in Chapter 12.

Managing the pH of Calcareous Soils

Many millions of acres of soil are calcareous in arid regions, on flood plains, and on recently drained lake plains. Some of these soils are excellent for agriculture, such as those located in the Palouse Country in Washington, the flood plains of the Red, Platte, and Mississippi rivers, and on the lake plains around the Great Lakes.

Plants growing on calcareous soils are sometimes deficient in iron, manganese, zinc, copper, or boron. To lower the pH of calcareous soils appreciably, the calcium carbonate must be leached out; this is impractical. As a result, crops are fertilized with appropriate nutrients that are deficient, or crops are selected that are adapted to the alkaline soils.

Plants show considerable differences in their ability to use certain nutrients in calcareous soils. Sorghum may be iron deficient, but alfalfa will not be iron deficient when grown on the same calcareous soil. Varieties of the same species may exhibit remarkable differences in tolerating toxicity concentrations of some nutrients, or in removing certain nutrients from soil. Since the pH of calcareous soils cannot be practically altered, crop selection and fertilization with deficient nutrients are commonly the only practical solutions for growing many crops on calcareous soils.

Gypsum Requirement of Sodic Soils

Sodic soils are nonsaline soils characterized by about 15 percent or more exchangeable sodium. Hydrolysis of ex-

Table 8-10

Suggested Applications of Ordinary Powdered Sulfur to Reduce the pH of 100 ft^2 of a 20 Centimeter Layer of Sand or Loam Soil

| Original pH | Pints of Powdered Sulfur, for Desired pH | | | | | | | | | |
| | 4.5 | | 5.0 | | 5.5 | | 6.0 | | 6.0 | |
	Sand	Loam	Sand	Loam	Sand	Loam	Sand	Loam	Sand	Loam
5.0	$\frac{2}{3}$	2								
5.5	$1\frac{1}{3}$	4	$\frac{2}{3}$	2						
6.0	2	$5\frac{1}{2}$	$1\frac{1}{3}$	4	$\frac{2}{3}$	2				
6.5	$2\frac{1}{2}$	8	2	$5\frac{1}{2}$	1	4	$\frac{2}{3}$	2		
7.0	3	10	$2\frac{1}{2}$	8	2	$5\frac{1}{2}$	1	4	$\frac{2}{3}$	2

Taken from Kellogg, 1957.

changeable sodium tends to cause structure deterioration and pH in the range of 8.5 to 10. The gypsum requirement is the amount of gypsum required to reduce the sodium saturation to some acceptable level for a given quantity of soil.

For each milliequivalent of exchangeable sodium per 100 grams of soil, it requires 0.086 grams of gypsum ($CaSO_4 - 2H_2O$) calculated as follows:

$$\frac{\text{molecular weight gypsum}}{\text{valence} \times 1,000} = \frac{172 \text{ grams}}{2,000}$$
$$= 0.086 \text{ grams gypsum}$$

Therefore, each milliequivalent of exchangeable sodium that must be replaced by calcium and converted into sodium sulfate, which can be leached from the soil, requires gypsum equal to 0.086 grams per 100 grams, or 0.086 pounds per 100 pounds, or 0.086 kilograms of gypsum per 100 kilograms of soil. For plow layers weighing 2,000,000 pounds, each milliequivalent of exchangeable sodium requires:

$$\frac{0.086 \text{ pounds gypsum}}{100 \text{ pounds soil}}$$
$$= \frac{x \text{ pounds gypsum}}{2,000,000 \text{ pounds soil}}$$

$x = 1,720$ pounds of gypsum per acre furrow slice

It requires 1,720 kilograms of gypsum for a 2,000,000-kilogram furrow slice for a hectare for each milliequivalent of exchangeable sodium per 100 grams of soil.

Effects of Flooding on Soil Chemical Properties

In well-aerated soils, the soil atmosphere contains an abundant supply of oxygen for microbial respiration and organic matter decomposition. Oxygen serves as the dominant electron acceptor in the transformation of organic substrates to carbon dioxide and water. When soils are flooded and puddled for paddy production, a unique oxidation-reduction profile is created. A thin, oxidized surface soil layer results from the diffusion of oxygen from the water layer into the soil surface. This layer is underlain by a much thicker layer that is reduced (*see* Fig. 2-21). Oxygen is deficient in the reduced zone; other electron acceptors are used, which initiates a series of reactions that drastically alter the soil's chemical environment.

Dominant Oxidation and Reduction Reactions

Flooded soils with a good supply of readily decomposable organic matter may be depleted of oxygen within 1 day. Then, anaerobic and facultative organisms multiply and continue the decomposition process. In the absence of adequate oxygen, other electron acceptors begin to function, depending on their tendency to accept electrons. Nitrate is reduced first, followed by manganic compounds, ferric compounds, sulfate, and lastly, sulfide. The reactions are as follows; reduction to right and oxidation to the left:

(1) $2NO_3^- + 12 H^+ + 10e^- \rightleftarrows$
$$N_2 + 6H_2O$$
(2) $MnO_2 + 4H^+ + 2e^- \rightleftarrows$
$$Mn^{+2} + 2H_2O$$
(3) $Fe(OH)_3 + e^- \rightleftarrows Fe(OH)_2$
$$+ OH^-$$
(4) $SO_4^{-2} + H_2O + 2e^- \rightleftarrows$
$$SO_3^{-2} + 2OH^-$$
(5) $SO_3^{-2} + 3H_2O + 6e^- \rightleftarrows$
$$S^{-2} + 6OH^-$$

Soils may have various levels of reducing conditions, depending on the quantity of decomposable organic matter and the amount of oxygen brought into the paddy by irrigation water and temperature. The formation of S^{-2} is produced in a strongly reduced environment, resulting in the formation of FeS and H_2S. In some cases, toxic amounts of H_2S are produced for rice production. About 2 weeks are needed to create rather stable conditions.

Changes in Soil pH

From the reactions just cited, there is an increase in hydroxls from the reduction of ferric iron, sulfate, and sulfide. In most paddy soils, iron is the most abundant electron acceptor; iron reduction tends to cause an increase in soil pH. The production of carbon dioxide and the accompanying formation of carbonic acid has an acidifying effect. When soils are alkaline, they tend to be low in iron. The production of carbon dioxide results in a reduction in pH. In acid soils, the abundance of iron results in a net increase in pH from a reduction of iron. The net effect of these reactions is to make the pH of paddy soils move toward neutrality (*see* Fig. 8-18). As a consequence, most paddy soils tend to have favorable nutrient availability and do not need lime. The increase in pH of highly acidic soils is sufficient to reduce aluminum toxicity.

Figure 8-18
Flooding soil increases pH of acidic soils and decreases pH of alkaline soils. The pH change is importantly related to the soil's content of iron and organic matter. (Data from Ponnamperuma, 1976.)

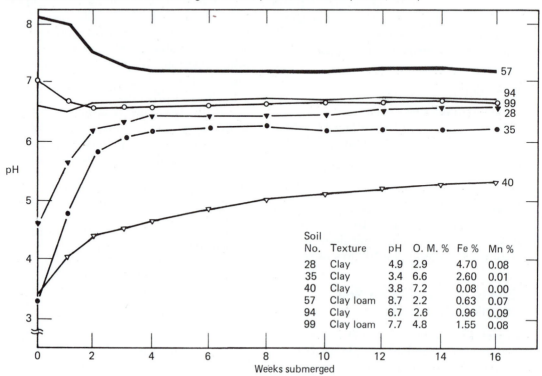

Soil No.	Texture	pH	O. M. %	Fe %	Mn %
28	Clay	4.9	2.9	4.70	0.08
35	Clay	3.4	6.6	2.60	0.01
40	Clay	3.8	7.2	0.08	0.00
57	Clay loam	8.7	2.2	0.63	0.07
94	Clay	6.7	2.6	0.96	0.09
99	Clay loam	7.7	4.8	1.55	0.08

Bibliography

Allaway, W. H., "pH, Soil Acidity, and Plant Growth," in *Soil*, USDA Yearbook, Washington, D.C., 1957, pp. 67–71.

Coleman, N. T., and G. W. Thomas, "Basic Chemistry of Soil Acidity," in *Soil Acidity and Liming*, R. W. Pearson and F. Adams, Eds., Agronomy 12, Madison, 1967, pp. 1–41.

Evans, C. E., and E. J. Kamprath, "Lime Response as Related to Percent Al Saturation, Solution Al, and Organic Matter Content," *SSSA Proc., 34:*893–896, 1970.

Farb, P., *The Living Earth*, Harper, New York, 1959.

Foy, C. D., and G. B. Burns, "Toxic Factors in Acid Soils," *Plant Food Review, 10,* 1964.

Jenny, H., *The Soil Resource*, Springer-Verlag, New York, 1980.

Kamprath, E. J., "Potential Detrimental Effects From Liming Highly Weathered Soils to Neutrality," *Soil and Crop Soc. Florida, 31:*200–203, 1971.

Kellogg, C. E., "Home Gardens and Lawns," in *Soil*, USDA Yearbook, Washington, D.C., 1957, pp. 665–688.

Klein, A. E., "Acid Rain in the United States," *The Science Teacher, 41:*36–38, May 1974.

Likens, G. E., and F. H. Bormann, "Acid Rain: A Serious Regional Environmental Problem," *Science, 184:*1176–1197, 1974.

McLean, E.O., "Potentially Beneficial Effects From Liming: Chemical and Physical," *Florida Soil Sci. Proc., 31:*189–196, 1971.

Meyer, T. A., and G. W. Volk, "Effect of Particle Size of Limestone on Soil Reaction," *Soil Sci., 73:*37–52, 1952.

Mohr, E. C. J., and F. A. van Baren, *Tropical Soils*, Interscience, New York, 1954.

Mullin, R. E., "Soil Acidulation with Sulfur in a Forest Tree Nursery," *Sulfur Jour., 5:*2–3, 1969.

Patrick, R., V. P. Binetti, and S. G. Halterman, "Acid Lakes from Natural and Anthropogenic Causes," *Science, 211:*446–448, January 30, 1981.

Pearson, R. W., "Soil Acidity and Liming in the Humid Tropics," *Cornell Int. Agr. Bull., 30,* Ithaca, 1975.

Ponnamperuma, F. N., "Specific Soil Chemical Characteristics for Rice Production in Asia," *IRRI Res. Paper Series No. 2,* Manila, 1976.

Sanchez, P. A., *Properties and Management of Soils in the Tropics*, Wiley, New York, 1976.

Smith, G. D., W. H. Allaway, and F. F. Riecken, "Prairie Soils of the Upper Mississippi Valley," *Adv. Agron., 2:*157–205, 1950.

Smith, R. J., "Administration Views on Acid Rain Assailed," *Science, 214:*38, October 2, 1981.

Soil Survey Staff, *Soil Survey Mannual*, USDA Agr. Handbook, *18,* Washington, D.C., 1951.

Spain, J. M., C. A. Francis, R. H. Howeler, and F. Calvo, "Differential Species and Varietal Tolerances to Soil Acidity in Tropical Crops and Pastures," in *Soil Management in Tropical America*, Proc. of Seminar at CIAT, Cali, Columbia, 1975.

Spurway, C. H., "Soil Reaction (pH) Preferences of Plants," *Mich. Agr. Exp. Sta. Spec. Bull., 306,* 1941.

St. Arnaud, R. J., and G. A. Septon, "Contribution of Clay and Organic Matter to Cation-Exchange Capacity of Chernozemic Soils," *Can. Journ. Soil Sci., 52:*124–126, 1972.

Thompson, L. M., and F. R. Troeh, *Soils and Soil Fertility*, McGraw-Hill, New York, 1978.

Turner, F., and W. W. McCall, "Studies on Crop Response to Molybdenum and Lime in Michigan," *Mich. Agr. Exp. Sta. Quart. Bull., 40:*268–281, 1957.

CHAPTER 9

SOIL GENESIS AND THE SOIL SURVEY

Soil is the product of the interaction of five factors: (1) parent material, (2) climate, (3) organisms, (4) topography, and (5) time. Because there are many degrees or variations of soil-forming factors, the potential for creating different kinds of soil is enormous. A study of soil genesis leads to an understanding of how soils develop, why soils differ in their properties and productivity, and how soils can be managed for various uses. This information is also needed to understand the geographic distribution of soils and the construction of soil maps.

Time as a Soil-Forming Factor

Soils, as products of evolution, are constantly changing as the landscape changes. They have a life cycle in the same sense that landscapes evolve through a cycle. The life cycle of soils typically includes the stages of parent material, immature soil, mature soil, and old age. Consideration of time as a soil-forming factor assumes that the other soil-forming factors remain constant or nearly so.

Time Required for Soil Development

Soil development is a two-step process. First, there is the formation of the parent material, which is considered generally to be unconsolidated. Second, there is soil development or horizon development. If the time required to weather granite or quartzite is included in the soil-forming time, the time required can be enormous (*see* Fig. 9-1). Once parent material has been formed, however, soil formation can proceed rapidly in permeable, unconsolidated material in a warm and humid climate. Plant growth can occur on freshly exposed parent material, so soil development need not precede plant growth.

Figure 9-1
A case where soil development is limited by the slow rate of granite weathering or parent material formation.

This is readily seen in areas where a plant cover is established on freshly exposed road cuts alongside highways. On spoil banks resulting from railroad construction in the central United States, a 13-centimeter-thick A horizon containing 3.1 percent organic matter developed within 75 years. A summary of soil formation rates by Buol, et al. (1980), indicates a range of 1.3 years per centimeter for young soil (Entisol) developed from volcanic ash to 750 years per centimeter for a 1-meter-thick Oxisol in Africa.

Some soil properties and horizons develop much faster than others. A quasi-equilibrium organic matter content occurs in many soils within about 100 years. Weakly developed B horizons may develop within a few hundred years under favorable conditions. Evidence of clay migration from the A horizon to the B horizon was observed in 450-year-old soils developed from alluvium in Pennsylvania. Enough clay migration to produce an argillic horizon took at least 2,000 years in Oregon and about 10,000 years in California (*see* Fig. 2-6). Soil genesis studies have shown that considerable soil formation can occur within a few thousand years; however, some of the world's oldest soils, found in northern Australia, are about 4,000,000 years old.[1]

Aridity and rapid removal of soil via erosion on steep slopes can delay or prevent the development of mature soils. It becomes clear then that the rate of development varies greatly from one soil to another. A given period of time may produce much change in one soil and little in another. For this reason, the maturity of soil is expressed in the degree of horizon development instead of the number of years. Conditions that hasten the rate of

[1]Personal communication from E. G. Hallsworth.

soil development are: (1) a warm, humid climate, (2) forest vegetation, (3) permeable, unconsolidated material low in lime content, and (4) flat or depressional topography with good drainage. Factors that tend to retard development are: (1) a cold, dry climate, (2) grass vegetation, (3) impermeable, consolidated material high in lime, and (4) a steeply sloping topography.

Rate of Soil Develoment

As is typical of many processes in nature, the rate of soil development as a whole varies over time, as do many of its individual processes. First, it can be mentioned that the characteristics of a soil change most rapidly when the soil is young, and that detectable changes occur more slowly with age.

Second, the individual processes vary in intensity over time. Changes in the organic matter content of a soil can be separated into three phases. In young soils, the organic matter content is increasing rapidly, because the rate of addition exceeds the rate of decomposition. Maturity is characterized by a constant organic matter content as additions are counterbalanced by losses. Old age is characterized by a lower and declining organic matter content, indicating that the rate of addition is waning as the soil becomes more weathered. The fertility declines and the reduced rate of organic matter production allows decomposition to exceed the rate of addition during the period when the soil changes from mature to old age.

Another illustration of the rate of soil development is silicate clay formation. A youthful soil that has a low clay content and a high content of primary minerals might be characterized by a high rate of clay formation. In a mature or old soil, in which most of the primary minerals have already been weathered, silicate clay formation will necessarily be low. The high clay content, however, encourages a relatively high rate of clay decomposition. Thus, it is seen that some of the processes are more operative in youthful soils, whereas others are more operative in old soils.

Stages in Soil Development

The major processes in the development of grassland soils from calcareous and permeable, glacial parent materials are:

1. Addition and transformation of organic matter by soil organisms.
2. Removal of carbonates and then exchangeable bases by leaching.
3. Uptake and recycling of bases by plants.
4. Weathering of primary minerals.
5. Synthesis of clays.
6. Clay translocation.

The parent material may be transformed into an *immature* or young soil in a relatively short period of time, if conditions are favorable. This stage is characterized by organic matter accumulation in the surface soil, and by little weathering, leaching, or translocation of colloids. Only the A and C horizons are present. Soil properties, to a large extent, have been inherited from the parent material. The mature stage is attained with the translocation of clay and the formation of an argillic horizon or Bt horizon. The mature stage has the greatest capacity to support vegetation and the greatest organic matter content. Weathering and leaching have been moderate, the soil is only moderately acidic, and many bases are still being released by weathering. This represents the condition of many soils in the Corn Belt of the United States. It is the stage of maximum productivity for many agricultural crops (*see* Item c, Fig. 9-2).

Eventually, if sufficient time has elapsed, the mature soil may become highly differentiated and the B horizon becomes an impermeable clay-pan. The clay-pan creates a "perched" water table that, in the wet season, restricts root growth. With continued eluviation, an E horizon develops immediately above the Bt horizon and below an A horizon that is now thinner. The soil is very acidic and infertile and has progressed to the old age stage. The soil is less productive for most agricultural crops than are the mature stage soils. Under these conditions, soil development proceeds from parent material to immature soil (Entisol) to mature soil (Mollisol) to old age (Alfisol). A summary of developmental stages is presented in Fig. 9-2.

Mohr and van Baren have recognized five stages in the development of tropical soils:

1. Initial stage—the unweathered parent material.
2. Juvenile stage—weathering has started, but much of the original material is still unweathered.
3. Virile stage—easily weatherable minerals have largely decomposed; clay content has increased and a certain mellowness is discernible.
4. Senile stage—decomposition arrives at a final stage, and only the most resistant minerals have survived.
5. Final stage—soil development has been completed and the soil is weathered out under the prevailing conditions.

The names used to refer to the stages are very descriptive; for instance, virile, refer-

ring to the stage at which the capacity of the soil to support vegetation is at a maximum.

Soil Development in Relation to Climate

Important climatic influences that affect soil development are precipitation and temperature. The climate also influences soil development indirectly in determining natural vegetation. It is not surprising that there are many parallels in the distribution of climate, vegetation, and soil on the earth's surface.

Role of Precipitation in Soil Genesis

The presence of water in soil enhances soil genesis through its effect on biomass production, mineral weathering and clay formation, translocation of colloids within the soil, and removal of soluble products from the soil. The water that enters soil in arid and subhumid regions is mainly lost by evaporation and transpiration. Some mineral weathering may occur, but the soil is dry much of the time. The limited precipitation prevents the formation of surplus water and the removal of soluble products from soil. The result is that soils tend to be rich in many soluble nutrients, near neutral or alkaline, 100 percent base saturated, and may contain some soluble salt. The depth of leaching is a function of the depth of water penetration and Ck layers are common (*see* Fig. 9-3). Low precipitation also results in low biomass production and in soils with low contents of organic matter.

Soils of humid regions may have water available for weathering and plant growth all the time except when the soil is frozen. In general, an increase in precipitation is associated with an increase in weathering and clay formation. The amount of clay

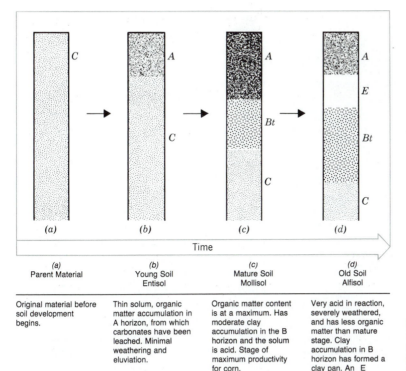

(a) Parent Material	(b) Young Soil Entisol	(c) Mature Soil Mollisol	(d) Old Soil Alfisol
Original material before soil development begins.	Thin solum, organic matter accumulation in A horizon, from which carbonates have been leached. Minimal weathering and eluviation.	Organic matter content is at a maximum. Has moderate clay accumulation in the B horizon and the solum is acid. Stage of maximum productivity for corn.	Very acid in reaction, severely weathered, and has less organic matter than mature stage. Clay accumulation in B horizon has formed a clay pan. An E horizon exists.

Figure 9-2
A summary of the stages of development of soils in the central United States under tall-grass vegetation.

formed in some soils of Illinois, however, has been found to be more related to the frequency of wetting and drying cycles in summer than to the total amount of precipitation. The author and some other investigators have found that the most rapid rate of clay formation occurs at the 5- to 25-centimeter depth in some soils of the grasslands which is the zone of greatest root growth. Small rains may moisten only the upper portion of the soil in summer, leaving dry soil underneath and a thin surface layer that is quickly dried by evaporation.

Surplus water is effective in the translocation of colloids and the removal of soluble-weathering products from the solum. Humid region soils, compared to arid and subhumid region soils, tend to have a greater clay content, a greater acidity (pH less than 7), a lower percentage of base saturation (less than 100), and a lower fertility level. Greater biomass production in humid regions results in greater organic matter content, as shown in Fig. 9-4, for a transect across the central United States.

An increase in both clay and organic matter content contributes to an increased cation exchange capacity. In the central United States, an increase in cation exchange capacity of up to about 65 centimeters of annual precipitation is associated with 100 percent base saturation for loamy upland soils. As the annual precipitation increases beyond 65 centimeters, surplus water appears; leaching results in base saturation less than 100 percent and in the development of soil acidity (*see* Fig. 9-5).

Indirectly, precipitation has another influence if leaching causes the pH of the

Figure 9-3
Grassland soil developed under ustic soil moisture regime, where there has been sufficient water for clay formation and movement and development of a Bt horizon. Limited leaching is expressed by the shallow depth (68 centimeters) of a k layer (accumulation of carbonates). (Photo courtesy of USDA.)

clay content occur with an average increase in soil temperature. Jenny summarized the results of many studies of soils formed from igneous rocks (rocks without clay content) in the eastern United States. Soils that ranged in latitude from 35 to 41 degrees north contained 26 percent clay compared to 38 percent clay for soils 30 to 35 degrees north. The comparison was made in soils of nonglacial origin to reduce the likelihood that the southern soils had weathered for a longer period of time.

The quantity of organic matter in a soil represents the balance between addition and decomposition. Accordingly, climatic factors that affect the quantity of organic material developed in, or returned to, the soil and the activity of decay organisms, have a bearing on the amount accumulated. Studies have shown that when average annual temperatures increase and the moisture and other relations remain constant, the quantity of organic matter decreases in temperate region soils of similar characteristics and covered by the same type of vegetation. The decrease is somewhat greater in grassland soils than in forest soils (*see* Fig. 9-6). Although this relationship exists in the continental United States, it cannot be validly extrapolated to the equatorial regions.

Many humid tropical soils have a greater organic matter content than humid temperate region soils. Several likely explanations exist. Although organic matter decomposes more rapidly in the warm tropics, the type of vegetation produced is quite different; annual production may be 90 to 180 metric tons per hectare compared to 2 to 6 tons in temperate regions. Many tropical soils are dominated by oxide and amorphous clays that effectively protect organic matter from microbial decomposition. Many red-colored soils in the tropics have a greater content of or-

surface layers to become about 4.5 or less. Under these conditions, microbial decomposition of organic matter may become so restricted that the litter added to the surface of the soil accumulates instead of decomposes. This accounts for the existence of prominent O horizons in some forest soils.

Effect of Temperature on Soil Genesis

Every increase in temperature of 10°C increases the rate of chemical reactions two to three times. Increased weathering and

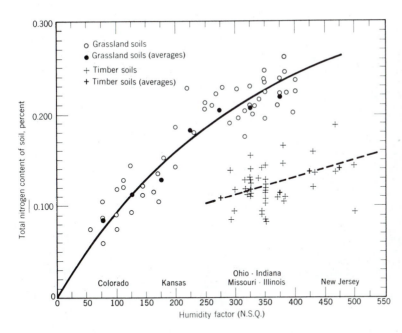

Figure 9-4
Soil nitrogen-humidity factor relationship along the annual isotherm of 11°C. (From *Missouri Res. Bull.,* 152.) (N.S.Q. is the ratio of precipitation to the absolute saturation deficit of the air. These values are used instead of precipitation values as such, since rainfall alone is not a satisfactory index of soil-moisture conditions because of the great variations in evaporation. The N.S.Q.s include, therefore, the effect not only of temperature but also of air humidity on evaporation.)

Figure 9-5
The relationships between annual precipitation and cation exchange capacity, exchangeable bases, and hydrogen in soils of the central United States. (From "Functional Relationships between Soil Properties and Rainfall," H. Jenny and C. D. Leonard, *Soil Sci.* 38:363–381, 1934. Used by permission of the author and the Williams and Wilkins Co., copyright © 1934.)

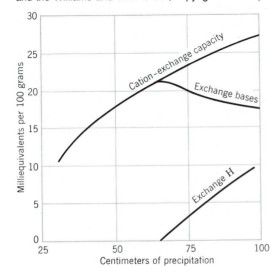

ganic matter than nearby black soils. It is interesting that a very high organic matter content is characteristic of cold soils of tundra regions. Both precipitation and temperature are low, but with meager biomass production, mineralization is much inhibited and the accumulation rate barely exceeds the mineralization rate; over time, however, much organic matter accumulates.

Climate and Kinds of Clay Minerals

Several generalizations concerning the type of clay found in soils and climate can be made. In the soils of the northern part of the United States, the clay fraction is generally dominated by 2:1 and 2:1:1 silicate clay minerals (hydrous mica, montmorillonite, chlorite). Kaolinite or 1:1 silicate clays and oxides of iron or aluminum are more common in the soils of the southeastern United States as well as in many tropical areas. In humid tropics, intense weathering can result in an al-

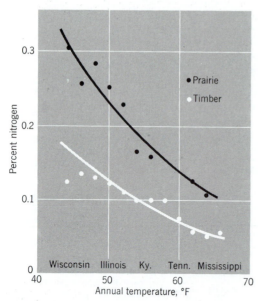

Figure 9-6
Nitrogen-temperature relation in humid prairie (upper curve) and humid forest soils (lower curve) for silt loams. (From *Missouri Res. Bull.*, 152, 1930.)

most complete loss of silica, and the clay fraction of the soil will then be high in iron and aluminum oxides. There are many exceptions to these generalizations, since they only apply to the clay formed in soil.

Climate Change and Soil Properties

It seems that only the youngest soils have had a fairly constant climatic influence during soil genesis. Many soils throughout the world have been influenced by more than one climate because of climate shifts; the soils are polygenetic in character. The properties of buried soils, *paleosols*, have been studied for clues to past climates. In the southwestern United States, the uplift of the Rocky Mountains cut off moisture-laden winds and created a drier climate. Today, some soils have a strongly developed agrillic horizon, which is indicative

of a more humid climate when much clay was formed and translocated. Today, however, many of these soils have high base saturation and Bk and Ck horizons. The explanation is that after developing in a humid climate, recalcification of the soil has occurred, with drifting calcareous dust probably playing an important role.

Biotic Factor in Soil Genesis

Plants affect soil genesis through the addition of organic matter, the cycling of ions, and the movement of water through the hydrologic cycle. Soil animals affect soil genesis as consumers and decomposers of organic matter and, perhaps, most importantly as earth movers (*see* Chap. 5).

Natural vegetation may be divided, very broadly, into two general classes of trees and grass. The soils supporting them are termed forest soils and grassland soils, respectively. There are several characteristics in soils developed in association with grass that are of considerable agricultural significance. The different effect of each kind of vegetation on the soil supporting it is brought out in the following discussion.

Amount and Distribution of Organic Matter in Soil

In Chapter 6, the differences in the amount and distribution of organic matter in grassland and forest ecosystems were compared (*see* Fig. 6-3). Considering the discussion in Chapter 6, it is sufficient now to restate that *under comparable environmental conditions* grassland soil profiles contain more organic matter that is more uniformly distributed with depth than do forest soil profiles (*see* Fig. 9-7).

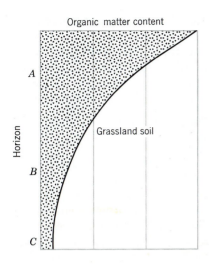
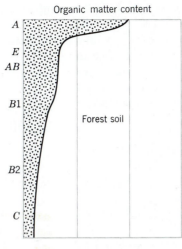

Figure 9-7
Grassland soil profiles contain about twice as much organic matter that is more uniformly distributed through the profile than forest soils under similar environmental conditions.

Differences in Nutrient Cycle

Plants absorb nutrients from the soil and transport these nutrients to the tops of plants. When the tops die and fall onto the soil surface, decomposition of organic matter releases the nutrients in a self-fertilizing "do-it-yourself manner." Bases returned to the soil surface in this manner retard the loss of exchangeable bases, by leaching, and the development of soil acidity. Wide differences in the uptake of ions and, consequently, in the chemical composition of plant tissues have been well substantiated. There are large differences even between tree species, and this plays a role in soil development. Species that normally absorb large quantities of alkaline earths and alkali metals will delay the development of soil acidity because of the large amount of bases returned to the surface of the soil in vegetative residues. The data in Table 9-1 support the fact that hardwoods maintain a higher pH and percentage base saturation than spruce, when grown on parent material with the same mineralogical composition.

Rate of Eluviation and Leaching

Under the same climatic conditions, where both forest and grasslands exist side by side and have a comparable parent material and slope, the forest soils will show evidence of greater eluviation and

Table 9-1
Effect of Tree Species on Soil pH and Base Saturation

Forest Type	Horizons	pH	Percentage Base Saturation
Spruce	Oe	3.45	13
	E	4.60	20
	Bs1	4.75	27
	Bs2	4.95	27
	C	5.05	23
Hardwood	Oe	5.56	72
	A	5.05	47
	Bw1	5.14	36
	Bw2	5.24	34
	C	5.32	34

Adapted from R. Muckenhirn, 1949.

leaching. Three possible causes for this difference have been offered. First, the forest vegetation returns fewer alkaline earths and alkali metals to the surface in vegetation each year. Second, water is intercepted for transpiration at a greater depth by trees, so the water is more effective in leaching before it is absorbed by roots. Third, water entering the soil is more acidic. Organic acids produced during the decomposition of organic residues in the O horizon, which is more prominently developed under trees, cause greater replacement and leaching of exchangeable bases.

Closely associated with the leaching of bases is the translocation of clay. That clay eluviates downward is evident from the higher clay content of the B horizon and from the occurrence of more pronounced clay coatings on the peds of the B horizon. Greater movement of clay in forested soil is based on the higher clay content of the B horizon and the lower clay content of the A horizon of forested soil, as compared to grassland soil. Thus, the permeability and other physical properties of subsoil also exhibit a degree of difference.

Two important points stand out in summarizing the differences between forest and grassland soils of temperate regions. Forest soil has about half as much organic matter in its solum as grassland soil does, and it is less uniformly distributed vertically. Forest soil shows evidence of greater age or development. The horizons of the solum are more acidic and have a lower percentage base saturation. Relatively more clay has been translocated from the A to B horizon. It can be seen that the differences are of degree and not of kind. This supports the view that the same basic processes have been operative in both. Eventually, both kinds of soils can evolve into clay-pan soils, and these modest but agriculturally important differences become less important. Soils may become strikingly similar in old age regardless of the vegetative cover under which they evolved.

Role of Parent Material in Soil Genesis

The nature of the parent material will have a decisive effect on the properties of young soils; it may exert an influence on even the oldest soils. Properties of the parent material that exert a profound influence on soil development include texture, mineralogical composition, and degree of stratification.

Consolidated Rock as a Source of Parent Material

Consolidated rock is not strictly parent material, but serves as a source of parent material. Soil formation may begin immediately after the deposition of volcanic ash, but must await the physical disintegration of hard rock, where granite is exposed. During the early stages of soil formation, rock disintegration may limit the rate and depth of soil development. Where the rate of rock disintegration exceeds the rate of removal of material by erosion, productive soils with thick solums may develop from bedrock. This is the case in the Blue Grass region of Kentucky where soils developed from limestone.

Perhaps, most of the world's soils have developed from sediments that were originally derived from rocks such as glacial debris, alluvium, colluvium, loess, volcanic ash, and so on. Even in mountainous areas, it is common to find many thick sediments, along with steep slopes where bare rock is exposed as shown in Fig. 9-8.

Figure 9-8
Sediments many meters thick exist in the alluvial valley in the foreground and on the forest-covered, lateral moraine on the left. Glacial till, outwash, and colluvium are important sediments at the intermediate altitudes, with bare rocks exposed at the highest elevations (Rocky Mountain National Park).

Effect of Rock Composition on Parent Materials and Soils

Granite and rhyolite are igneous rocks that have the same chemical composition. Rhyolite has the finer texture, or smaller mineral-grain size, because during formation it was subjected to more rapid cooling. This causes rhyolite to weather more slowly and results in a finer-textured soil than that developed from granite. A similar comparison can be made between basalt and gabbro in young soils; however, the textures of the old soils developed from these two materials are similar, because all the minerals are weatherable. Since the minerals in basalt weather more easily than those in granite, the finer-textured soil will develop from basalt. Large areas of deep, clay-textured soils have developed from basalt in India and Australia. The complete weathering of minerals in basalt in humid tropical regions produces soils composed only of clay-sized mineral particles.

For some sedimentary rocks, a few generalizations can be drawn. Sandstones high in quartz weather to produce sandy soils. Soils developed from limestone and shale are usually fine textured. Some cherty limestones, however, result in the formation of stony soils. Parent materials and soils derived from limestone are composed mainly of impurities in the limestone (*see* Fig. 9-9).

It has been pointed out that soil acidity encourages mineral decomposition, translocation of colloids, and the overall development of the soil profile. Therefore, where parent materials are rich in lime, development of soil is delayed, and it remains in the immature stage for a longer period of time.

Figure 9-9
Parent material derived from limestone is the residual accumulation of impurities in the limestone after water has leached away carbonate materials. Since clay is a major impurity, parent materials and soils derived from limestone tend to be fine textured.

Water-Deposited Sediments

Alluvial deposits are scattered in narrow, irregular strips that border streams and rivers. A common characteristic of this material is its stratification; its layers of different-sized particles overlying each other. Mineralogically, alluvium is related to soils that served as a source of the material.

Most alluvium is carried and deposited during floods, because it is at this period that erosion is most active and the carrying capacity of streams is at a maximum. When a flooding stream overflows its banks, its carrying power is suddenly reduced as the flow area increases and velocity decreases. This causes coarse sands and gravels to settle along the bank, where they sometimes form conspicuous ridges called *natural levees.* As the water reaches the *floodplains* of the valley, the rate of flow is slow enough to permit silt to settle. Finally, water is left in quiet pools, from which it seeps away or evaporates, leaving the fine clay. Levees are characterized by good internal drainage

during periods of low water, whereas floodplains exhibit poor internal drainage.

Terraces are developed from floodplains as streams cut deeper channels because of lowered outlets. Several terraces may be found along a stream or a lake that has undergone repeated changes in level. In glaciated areas, extensive terraces were formed as the glaciers receded and their outwash plains were no longer covered with water. Terraces usually are quite well drained and may be droughty. They exhibit stratification.

Streams flowing from hills or mountains into dry valleys or basins drop their sediments in a fan-like deposit as the water spreads out. These *alluvial fans* are usually well drained and coarsely textured, being composed of sands and gravels.

Sediments not deposited as floodplains are carried to the lake, gulf, or other bodies of water into which a stream empties. The decrease in velocity at the stream's mouth, together with the coagulating effect of the salt content of the receiving water body, results in the deposition of much of the suspended material, thus producing a *delta.* These deposits are poorly drained; however, where drainage is provided, they constitute important crop-producing areas as evidenced by the deltas of the Nile, Po, Tigris, Euphrates, and the Mississippi rivers.

Floodplains as well as deltas are generally rich in plant nutrients and comparatively high in organic matter content. Terraces and alluvial fans, on the other hand, are more likely to be less fertile. Special crops, such as vegetables and fruits, frequently are grown on the latter formations, because the soil warms up quickly and their good drainage and coarse texture permit free root development.

It is common to find *marine* deposits along coastlines. This material was derived from sediments carried by streams and deposited in the ocean and gulf through decreased current velocity and chemical coagulation. Much of it is sandy, but it is interspersed with beds of silt and clay that were deposited in estuaries, other sheltered bodies of water, or farther out in the ocean. When raised above water level, these deposits were subjected to soil-forming processes. Water-deposited sediments are important in Areas 7, 13 and 14 of Fig. 9-10.

Glacial Materials

Ice was the transporting agent for much of the mantle of northern Europe, Asia, and North America. In the United States, the Ohio and Missouri rivers form a general southern boundary for ice-carried material. As the great continental ice sheets moved southward from their accumulation centers, they first followed and filled the great drainage valleys, and then gradually spread out over the intervening upland and divides. As the ponderous ice mass moved forward, it pushed before it

Figure 9-10
A generalized physiographic map of the United States. (Drawn from a map prepared by the Division of Soil Survey and presented in *USDA Bull.,* 96.) Legend of areas: 1. Pacific Coast region. 2. Northwest intermountain region. 3. Great Basin region. 4. Southwest arid region. 5. Rocky Mountain region. 6. Great Plains region. 7. Glacial lake and river terraces. 8. Glacial region. 9. Loessial deposits. 10. Limestone valleys and uplands. 11. Appalachian Mountains and plateaus. 12. Piedmont plateaus. 13. River floodplains. 14. Atlantic and Gulf coastal plains. 15. Loessial deposits over glacial material.

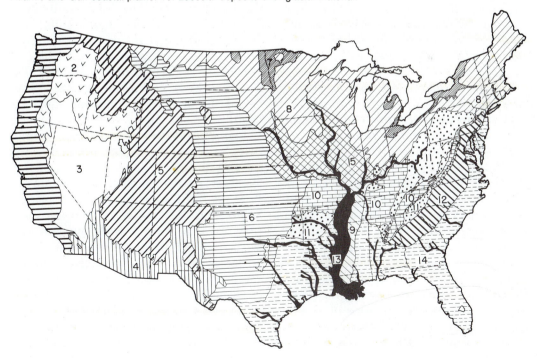

and gathered within itself a large part of the unconsolidated surface layer. It also scooped up great rock fragments, which scraped the rock floor over which they passed. Sharp corners and edges of even the hardest rocks were ground smooth by this abrasive action to form the rounded rocks and boulders that are characteristic of glaciated landscapes. Large quantities of weathered and unweathered rocks, varying in size from fine rock powder to massive boulders, were thus incorporated into the ice and carried along in the glacier.

The movements of this continental ice sheet depended on the changes in climatic conditions that took place during the glacial age. During mild periods, the ice melted rapidly. In cold seasons, melting ceased and the ice front would creep southward. Sometimes during extremely mild periods, the ice would melt faster than it was pushed forward. This would lead to a rapid recession of the ice front; all debris carried in the ice was, of course, dropped. Generally, after this type of recession, the land surface appeared as a rolling plain, called a *till plain* or *ground moraine*.

At certain times, climatic conditions allowed the glacier to melt back just as fast as its rate of advance. This process resulted in the front of the ice remaining at one place for some time. All debris carried by the ice was brought to the line of the stationary ice front and dumped there as melting proceeded. This process resulted in the formation of ridges or a series of hills, called *terminal* and *recessional moraines*. Lateral moraines formed along the sides of the ice sheets (*see* Fig. 9-8). Moraines are usually composed of an unassorted, heterogeneous mass of boulders, rocks, sand, silt, and clay, called *till*. However, in places, a water sorting also occurred. As would be expected, the proportions of these materials vary greatly; hence, some moraines are relatively high in sand content, whereas others contain a large proportion of fine particles. Not only does the material vary in texture, but the shape of the surface is also variable.

As the ice melted, giving rise to moraines, great volumes of water rushed away. These waters carried quantities of sediment, the coarser of which was deposited as the current diminished. These coarse-textured, comparatively level deposits are known as *outwash plains* (*see* Fig. 9-11). Most of the finer silt and clay were carried into slowly moving water or lake basins, where they settled out to form lake bed or *lacustrine* plains. In Michigan, there is evidence of as many lakes that have become extinct as there are lakes in existence, whereas along the Great Lakes extensive areas have been exposed through the disappearance of ice barriers, the lowering of the outlet at Niagara Falls, and the tilting of the land surface. The disappearance of Lake Agassiz has laid bare a great land surface in Minnesota and the Dakotas, and the basin of old Lake Bonneville occupies an immense area in Utah.

The glaciers also produced *kames* and *eskers*, but these formations are only of local significance. Minor readvances of the ice sheet resulted in the modificaiton of these glacial formations in many places. Sometimes, till was relaid over a lake plain, or sandy and gravelly outwash was pushed up to form gravelly moraines. Major readvances streamlined glacial material in some parts of the country to form *drumlins*. It is customary to designate all the material deposited by glaciers and their melt waters as *glacial drift*. In general, the effect of glaciation was to scrape off and smooth the tops of high land

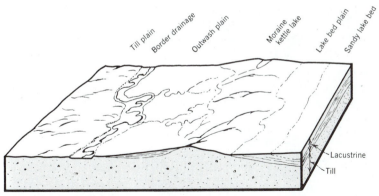

Figure 9-11
A generalized view of physiographic features in a glaciated area. On the left is a till plain, separated from the moraine by an outwash plain, and a border-drainage way developed while the ice front was at the moraine line and melting as rapidly as it advanced to deposit the morainic material. To the right, a glacial lake has receded to leave a plain of fine-textured sediment partially covered with lacustrine sands (deposit nearer the water's edge).

forms while filling valleys and depressions. Thus, continental glaciation decreased the local relief intervals and flattened the topography.

Wind-Transported or Aeolian Material

There are three classes of wind-moved soil material: (1) sand of variable fineness, which may be collected into low swells or steep ridges, such as dunes, (2) volcanic ash, and (3) silt-like material, called *loess*, which occupies large areas in the United States, Europe, and Asia.

Dune sand is of little agricultural value, although crops are produced on it to a limited extent in humid regions. At times, dunes are a hazard to agriculture; in their movement, they sometimes cover good land.

Loess was deposited in the central United States after the recession of the ice sheet. This material was derived in part from sediments deposited by huge rivers that were fed by the melting continental glaciers in a broad belt, even beyond the

southern limits of glaciation. A period of aridity after the recession of the glaciers, with strong westerly winds, set the stage for the transportation of this wind-blown material to its present resting place (*see* Fig. 9-12). The great thickness of the deposits of loess on the east and northeast banks of the Mississippi and Missouri rivers is one of the facts that have led to this explanation of the accumulation of the material (see Areas 9 and 15 of Fig. 9-10). Glaciers from the Rocky Mountains probably supplied the sediments making up the western part of the loessial deposits. Deserts are also sources of loess. Extensive deposits of loess are found along the Rhine River and its tributaries, and over a large part of the immense valley of the Hwang Ho in China. Other deposits occur in southern Russia, several Balkan countries, and northern France, Belgium, Poland, and the Argentine pampa.

Loess is composed largely of silt, and it is commonly grayish-yellow or buff. This wind-blown material stands in almost vertical walls, so that gullies and streams cut-

Figure 9-12
A loess deposit about 15 meters thick along the eastern side of the Mississippi River floodplain in western Tennessee.

ting through it have very steep banks. Its content of mineral plant nutrients was originally high, as was the quantity of calcium compounds. Loess is one of the most uniform of soil parent materials, but even it varies considerably in texture and mineralogical composition. Soils developed from these deposits are frequently referred to as fertile; however, no single material always gives rise to a productive soil, as the parent material is only one of the factors involved in soil development.

Volcanic ash is amorphous, fine, dust-like particles thrown out of volcanoes. Ash falls on the surrounding land to form thick sediments for soil development (*see* Fig. 9-13). Further from the volcano, ash provides a sprinkling of material that helps to rejuvenate soils.

Texture of Parent Material and Soil Properties

The textures of transported materials are related to their origin, and they may have great variability. Glacial and water-laid de-

posits range from sands to silty clays. Loessial deposits are high in silt, and many soils developed in loess have silt-loam A horizons. It is logical that the texture of the parent material will have a direct affect on the texture of the horizons of immature or young soils. The presence of resistant minerals in parent materials may have an influence on mature or old soils. Three additional ways in which the texture of the parent material influences soil development will be discussed in the following paragraphs. These are organic matter content, soil permeability (or the downward movement of water), and solum thickness.

Soils developed from fine-textured materials usually have a higher organic matter content than those formed from coarser-textured materials. The finer texture may enhance plant growth by providing a greater water and nutrient supply. This results in a greater annual addition of organic matter to the soil. Fine-textured soils also tend to be less well aerated and have slightly lower average tempera-

Figure 9-13
Diamond Head—the famous landmark near Honolulu—is an old volcanic cone composed mainly of ash. Note stratification, and that the right side is highest because of prevailing wind from the left during the ash fall.

tures. This has the effect of retarding the rate of organic matter decomposition and, thereby, aiding its accumulation. In addition, certain organic compounds may combine with clay to render soil organic-matter resistant to decomposition, as discussed in Chapter 6.

The permeability will determine, to a certain extent, the quantity of precipitation that will run off, and that which will infiltrate into the soil. In humid regions, the development of acidity can readily occur in soils developing in calcareous materials, if they are permeable. The more water that moves through soil, the more rapidly acidity develops, weathering proceeds, and colloidal materials are translocated. Certain soils of the coastal plains of the southeastern United States have developed in clay-textured marine sediments that are many thousands of years old. Where these parent materials are impermeable to water, some soils are still alkaline, even though the average annual precipitation exceeds 100 centimeters.

If the parent material is very coarsely textured or gravelly, little surface is exposed to weathering and little water is retained for weathering and plant growth. In this case, a very rapid permeability is associated with slow soil development.

It has been shown that fine-textured parent materials tend to retard leaching and the translocation of colloids. This contributes to the development of soils with thin solums. On sloping lands, fine-textured soils have greater runoff and, consequently, less water available for leaching. There is also more water active in erosion, which contributes to the development of a thin solum. Soils that develop in coarser or permeable parent materials have thicker solums. The relationships between texture and solum thickness are shown in Fig. 9-14.

Stratified Parent Materials

In the central United States, there are large areas where a thin layer of loess was deposited over glacial till. Soils in such areas frequently have A and B horizons

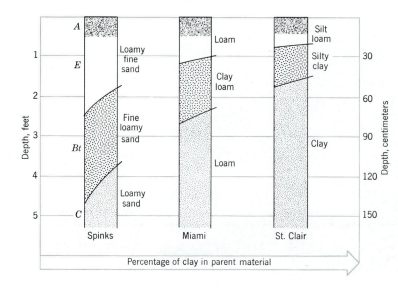

Figure 9-14
Relationship between texture of the parent material and thickness and texture of the horizons of three forest soils of the north-central United States (Alfisols).

that have developed in loess. In cases where the loess is very thin, only the A horizon may have developed in loess, whereas the B and C horizons have developed from glacial till. The horizons formed in loess, compared to those developed in glacial till, frequently have a more rapid permeability. This makes the thickness of the loess an important factor in the design of terraces and the use of lister furrows for water-erosion control.

Where streams dissect an area underlain by strata of varying composition, various parent materials will be exposed along a traverse running from the base to the top of a hill. This results in the development of a sequence of soils whose differences are due to differences in parent material. Such a sequence of soils is a *lithosequence* (see Fig. 9-15).

Nature and Source of Organic Soil Parent Material

In locations where considerable quantities of plant material grow and where decay is limited because of much water or low temperatures, a large accumulation of partially decayed vegetable matter gradually develops. Such deposits occur widely and are not restricted to any given climatic zone. They are found in Europe, Asia, Africa, Canada, South America, the United States, and various other places, including the tropics. However, accumulations of this nature are more common in northern latitudes; they occupy a larger percentage of the land surface in Norway, Sweden, Ireland, Scotland, northern Germany, Russia, and Holland than in countries lying farther south. In the tundra region, organic deposits occur frequently and extensively.

Even when a rank growth of vegetation occurs each year and remains on the soil, peat does not accumulate unless decay processes are very slow. The most common factor that limits decay is an excess of water. Accordingly, in moderate to warm climates, peat accumulates in shallow lakes and swamps. The topography, resulting from glaciation in Minnesota, Wisconsin, and Michigan, and to a lesser extent in Maine, New York, and New Jer-

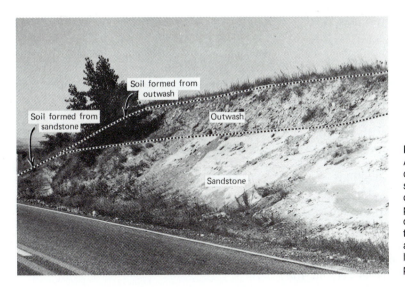

Figure 9-15
A roadcut showing outwash overlying sandstone. Strata of different materials can produce a lithosequence of soils on a hillside where the materials are exposed, as in the situation on the left side of the photograph.

sey, has given rise to innumerable small lakes and swamps. Thus, conditions have been suitable for peat accumulation. Along the Atlantic and Gulf coasts, many swamps have developed because of the slight elevation above sea level. Peat has accumulated in many of these areas. These organic deposits are the parent material for Histosols.

Soil Development in Relation to Topography

Topography modifies soil-profile development in three ways: (1) by influencing the quantity of precipitation absorbed and retained in soil, thus affecting moisture relations, (2) by influencing the rate of removal of soil by erosion, and (3) by directing movement of materials in suspension or solution from one area to another. Since moisture is essential for the action of the chemical and biological processes of weathering and effectively acts in conjunction with some of the physical forces, it is evident that a modification of moisture relationships within a soil will mate-

rially influence profile development. In a humid region, it is noteworthy that, in similar parent material of intermediate or fine texture, the soil of steep ridges or hills differs from the soil either on gentle slopes or on a level to undulating topography. In arid climates, these soil differences associated with differences in slope are much less pronounced because of the absence of water tables near the surface in more level areas.

Effect of Slope on Soil Genesis

As the steepness of a slope increases, there is greater runoff and erosion, soil creep, less water infiltration and less water available for chemical and biological activity. The net effect is a retardation in soil genesis. Generally, an increase in percent slopes is associated with a reduction in:

1. Leaching.
2. Organic matter content.
3. Clay translocation.
4. Mineral weathering.

5. Horizon differentiation.

6. Solum thickness.

Erosion on steep slopes commonly results in thin soils, especially where soils are underlain by bedrock. Mature soils are commonly found on moderately active, eroding slopes because of a quasi-equilibrium between erosion and conversion of parent material into soil. As erosion reduces the thickness of the A horizon, the upper part of the B horizon becomes incorporated into the lower part of the A horizon. The upper part of the C horizon becomes incorporated into the lower part of the B horizon. Under these conditions, the soil seems to remain the same over a long period of time as erosion and leaching evolve the landscape.

Increasing the slope length allows water, which ran off the upper part of the slope, to infiltrate in the lower part of the slope and to deposit eroded material carried in suspension. An increasing slope length on the prairies of western Iowa is associated with an increased soil thickness, depth of calcium carbonate leaching, and organic matter content, as shown in Fig. 9-16. The soils on an eastern-facing 24 percent slope, compared to a 14-percent western-facing slope, had greater organic matter content and leaching, because the eastern slope was a more humid environment because it was protected from the hot prevailing westerly winds. The aspect and nature of slope differences can create different microclimates and differences in soils in a local landscape.

Effect of Drainage on Soil Genesis

Drainage is a measure of the tendency for water to leave the soil. Well-drained soils occur on slopes where soils and parent

Figure 9-16
Percent of nitrogen (organic matter content) in an 8-60-centimeter-layer of soil, and depth of leaching in virgin soils of a prairie in western Iowa. (Adapted from Aandahl, 1948.)

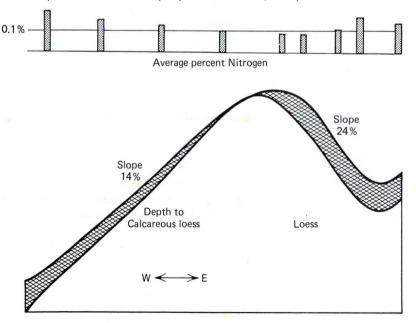

materials are permeable and the water table is far below the soil surface. Poorly drained soils tend to occur in depressional topograhy, where water accumulates and the impermeability of soil horizons or ground-water tables limit the downward movement of water. Poor drainage limits the vertical movement of water, the translocation of colloids, and the removal of soluble products. Organic matter accumulation is favored, and soils may be reduced (gray subsoils).

As a result of differences in topography or drainage or both, soil profiles developed in similar parent material of the same age and within a single zonal region vary appreciably. A group of soils developed under such conditions, and showing such variations in profile characteristics, is designated a catena or *toposequence* of soils. Locally, topography and parent material are the most frequent causes of soil differences.

Topography, Parent Material, and Time Interactions

About 500,000 years ago, the Kansan glacier ice melted and left a relatively flat surface that was underlain by till and other glacial materials in the north-central United States. During a long period of weathering, a well-developed soil with high clay content developed on the surfaces of the till, as shown in Fig. 9-17a. Loess was then deposited, and the soil developed from till was buried, becoming a paleosol (*see* Fig. 19-17b). Geologic erosion then dissected the till plain and removed the loess on the sloping areas, which reexposed the paleosol. Erosion and dissection continued until unweathered Kansan till was exposed on the lower slopes (*see* Fig. 9-17c). Today, the result is a soil of moderate age has formed on the loess of broad, nearly level upland areas. Midway

(a) Soil Developed from Kansan Glacial Till

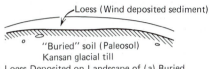

(b) Loess Deposited on Landscape of (a) Buried Soil Developed from Kansan Glacial Till

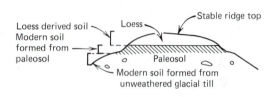

(c) Modern Soil Landscape

Figure 9-17
Steps in the formation of a soil landscape in southern Iowa. (Adapted from Oschwald, et al., 1965.)

down the slopes, the modern soil has formed from the old clayey paleosol; on the lower slopes, the modern soil has formed from unweathered Kansan till. In stream valleys, soils are very young and have developed from recent alluvium and colluvium. Within the landscape, there are four different parent materials and time factors and about six major kinds of slope conditions, as well as microclimate differences. In such a landscape in southern Iowa, these factors result in about 11 different kinds of soils, as shown in Fig. 9-18.

The alluvial soils of the Nile Valley have a depth function for nitrogen and organic matter similar to that of grassland soils on eroding slopes in California (or wherever grassland soils develop on uplands). About 1 millimeter of alluvium was deposited by the Nile River each year, producing 1 meter of material with 1,000

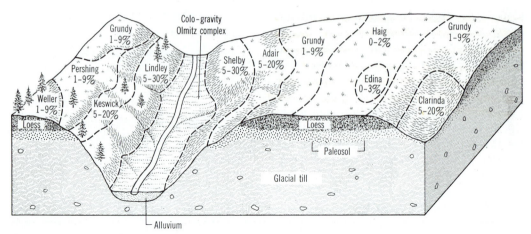

Figure 9-18
Relationship of slope, vegetation, and parent material to soils of the Adair-Grundy-Haig soil association in south-central Iowa. (From Oschwald et al., 1965.)

years. In alluvial soil, the upper material was deposited last, resulting in an increase in the age of the material with an increasing soil depth. On slopes in California, the oldest and most weathered soil is near the soil surface. Even though these two soils (Nile Valley and California soils) have similar distributions of nitrogen and organic matter, they have developed differently. The surface layers of alluvial soil contain more organic matter than the older and deeper layers, because organic matter decomposition continues after burial at a rate that is more than the addition of organic matter by the growth of primary producers. In time, the lower soil layers become depleted of most of their organic matter.

Human Beings as a Factor of Soil Formation

People's use of land for agriculture, forestry, grazing, and urbanizaton has produced extensive changes in soils. Many of these changes, including erosion, earth moving, drainage, salinity, depletion and addition of organic matter and nutrients, compaction, and flooding, have been discussed. Some activities can be judged harmful and others beneficial (*see* Fig. 9-19).

Urban soils frequently are highly modified by human beings; homeowners sometimes face huge problems in growing lawns, shrubs, and gardens. The incorporation of lime-containing mortar into soils, from the laying of bricks and other masonary activities, leaves many soils with a highly alkaline reaction that causes certain nutrient deficiencies. Sometimes, the soil is little more than an accumulation of trash. Mapping soils or making a soil survey of the area provides information useful in planning sites for gardens, buildings, and the like. Soil scientists have mapped the soil in Washington, D.C. As a result of this soil survey, it was discovered that the cherry trees lining the tidal basin around the Jefferson Memorial are difficult to maintain, because the water table is too high (poor

Figure 9-19
Earth-moving activities by humans to construct paddies and terraced areas have greatly affected the soils in many landscapes.

aeration). Many cherry trees have died and have been replaced. A soil survey before the trees were planted would have identified the problem and allowed a necessary modification of the soil for good cherry growth.

The Soil Survey

At the turn of the century, there was an increasing awareness of the bonds between land and society. In an attempt to find the underlying causes of some agricultural problems, and in an effort to build a solid foundation for future research, the United States Department of Agriculture, in cooperation with the various state experiment stations, began at that time a systematic investigation of our soil resources. This investigation assumed the form of a national inventory and survey.

Soil surveying is the process of studying and mapping the earth's surface in terms of units called *soil types*. A soil survey report consists of two parts: (1) the soil map, which is accompanied by (2) a descripiton of the area shown on the map.

Soil Mapping

The actual process of mapping or surveying consists of walking over the land at regular intervals and taking notes of soil differences and all related surface features, such as slope gradients, evidence of erosion, land use, vegetative cover, and cultural features. Boundaries are drawn directly on aerial photographs representing, in most places, changes from one soil type to another (*see* Fig. 9-20).

Types of Soil Surveys

Productive agricultural lands are inventoried in detail. A detailed map is one in which the scale is commonly 1:20,000 (5 centimeter per kilometer or 3.16 inch per mile). In addition to soil type boundaries, these maps show the location of gullies, railroads, houses, roads, streams, and

Figure 9-20
A soil mapper identifies the soil by inspecting all the soil horizons. Lines are then drawn around areas of similar soil (polypedons) on an aerial photograph to produce a soil map. (Photo courtesy of USDA.)

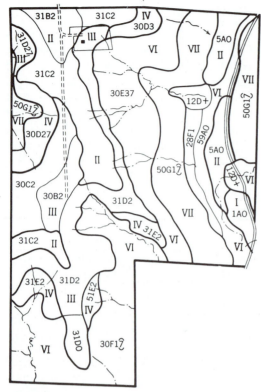

Figure 9-21
Land-capability map of a 200-acre dairy farm in Wisconsin. The Roman numerals designate the land-capability classes. The other numbers and letters indicate the land characteristics. For example, in 30E37, the 30 designates the type of soil, the E indicates the steepness of slope, and the 37 stipulates that over 75 percent of the topsoil has been lost and that there are occasional gullies. For an explanation of land-capability class, *see* Table 9-2. (Courtesy of Soil Conservation Service.)

Table 9-2
Land Capabilty Classes

Land Suited to Cultivation and Other Uses

Class I.   Soils have few limitations that restrict their uses.

Class II.   Soils have some limitations that reduce the choice of plants or require moderate conservation practices.

Class III.   Soils have severe limitations that reduce the choice of plants, require special conservation practices, or both.

Class IV.   Soils have very severe limitations that restrict the choice of plants, require very careful management, or both.

Land Limited in Use—Generally Not Suited for Cultivation

Class V.   Soils have little or no erosion hazard, but have other limitations impractical to remove that limit their use largely to pasture, range, woodland, or wildlife food and cover.

Class VI.   Soils have severe limitations that make them generally unsuited to cultivation and limit their use largely to pasture or range, woodland, or wildlife food and cover.

Class VII.   Soils have very severe limitations that make them unsuited to cultivation and that restrict their use largely to grazing, woodland, or wildlife.

Class VIII.   Soils and landforms have limitations that preclude their use for commerical plant production and restrict their use to recreation, wildlife, or water supply, or to esthetic purposes.

other details needed in planning a soil-conserving practical farm or ranch management plan. This kind of work is expensive. Large expanses of land are too poor, however, to justify this cost. These areas are mapped on scales commonly 1:200,000 or more. The resulting study, called a reconnaissance survey, shows areas and regions that are dominated by soil associations in contrast to the detailed

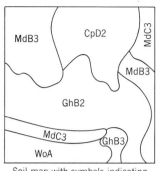

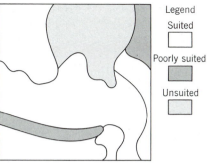

Legend

Suited

Poorly suited

Unsuited

Figure 9-22
Soil survey maps form the basis for the construction of many kinds of maps to serve a wide variety of needs; in this case for the wildlife management of a 40-acre tract. (Based on information from Allan et al., 1963.)

Soil map with symbols indicating soil type, slope, and erosion

Map constructed from the soil showing suitability of land for grain and seed crops for wildlife management

map where soil types are separated. Soil association maps are useful in studying extensive land use problems. They possess the desirable characteristics of being made quickly at a comparatively low cost.

Purpose and Value of Soil Maps

Soil survey maps contain many types of information; however, perhaps that of greatest value is the soil type (or series and phase), slope, and degree of erosion that is recorded for each area delineated on the map. These maps form the basis to develop maps for a wide variety of uses. The areas can be grouped in land capability classes. Definitions of the land capability classes are given in Table 9-2. A land capability map can be constructed, such as the one shown in Fig. 9-21. The Soil Conservation Service personnel use these maps to develop conservation plans for farmers. The use of a soil survey map in wildlife management work is shown in Fig. 9-22.

Life insurance companies, banks, and other money-lending agencies use soil surveys in determining security for loans. Real estate companies and individuals interested in buying or selling land make extensive use of soil surveys. Taxes are based on soil types in some states. They are also used by highway and drainage engineers, by various kinds of manufacturers in selecting suitable locations for factories, and by merchandising and advertising companies in selecting areas for intensive campaigns. County agricultural agents and other extension workers find soil surveys helpful in their work. And, finally, the farmers themselves are making increasing use of soil survey maps and reports in planning their management programs, and in interpreting modern agricultural research in terms of their own farm conditions.

Bibliography

Aandahl, A. R., "The Characterization of Slope Positions and Their Influence on The Total Nitrogen Content of A Few Virgin Soils of Western Iowa," *Soil Sci. Soc. Am. Proc., 13*:449–454, 1948.

Allan, P. F., L. E. Garland, and R. F. Dugan, "Rating Northeastern Soils for Their Suitability for Wildlife Habitat," *Trans. 28th Nor. Am. Wildlife and Nat. Res. Con.,* 1963.

Bidwell, O.W., and F. D. Hole, "Man as a Factor of Soil Formation," *Soil Science, 99*:65–72, 1965.

Bilzi, A. F., and E. J. Ciolkosz, "Time as a Factor in the Genesis of Four Soils Developed in Recent Alluvium in Pennsylvania," *Soil Sci. Soc. Am. Jour., 41:*122–127, 1977.

Boul, S. W., F. D. Hole, and R. J. McCracken, *Soil Genesis and Classification,* 2nd Ed., Iowa State Press, Ames, 1980.

Daniels, R. B., E. E. Gamble, and J. G. Cady, "The Relation Between Geomorphology and Soil Morphology and Genesis," *Adv. Agron. 23:*51–88, 1971.

Foth, H. D., "Properties of the Galva and Moody Series of Northwestern Iowa," *Soil Sci. Soc. Am. Proc., 18:*206–211, 1954.

Goddard, T. M., E. C. A. Runge, and B. W. Ray, "The Relationships Between Rainfall Frequency and Amount to the Formation and Profile Distribution of Clay Particles," *Soil Sci. Soc. Am. Proc., 37:*299–304, 1973.

Jenny, H., "A Study on the Influence of Climate Upon the Nitrogen and Organic Matter Content of the Soil," *Missouri Agr. Exp. Sta. Res. Bull., 152,* 1930.

Jenny, H., "Causes of the High Nitrogen and Organic Matter Contents of Certain Tropical Forest Soils," *Soil Sci., 69:*63–39, 1950.

Jenny, H., *The Soil Resource,* Springer-Verlag, New York, 1980.

Jenny, H., and C. D. Leonard, "Functional Relationships Between Soil Properties and Rainfall," *Soil Sci., 38:*363–381, 1934.

Mohr, E. C. J., and F. H. van Baren, *Tropical Soils,* Interscience, New York, 1954.

Muckenhirn, R. J., E. P. Whiteside, E. H. Templin, R. F. Chandler, and L. T. Alexander, "Soil Classification and the Genetic Factors of Soil Formation," *Soil Sci., 67:*93–105, 1949.

Oschwald, W. R., et al., *Principal Soils of Iowa,* Ames, Iowa. Special Report No. 42, Department of Agronomy, Iowa State University, 1965.

Parsons, R. B., and R. C. Herriman, "Geomorphoric Surfaces and Soil Development in the Upper Rogue Valley, Oregon," *Soil Sci. Soc. Am. Jour., 40:*933–938, 1976.

Ruhe, R. V., "Geomorphoric Surfaces and the Nature of Soils," *Soil Sci., 82:*441–455, 1956.

Ruhe, R. V., R. B. Daniels, and J. G. Cady, "Landscape Evolution and Soil Formation in Southwestern Iowa," *USDA Tech. Bull. 1349,* Washington, D. C., 1967.

Sanchez, P. A., M. P. Gichuru, and L. B. Katz, "Organic Matter in Major Soils of the Tropical and Temperate Regions," *12th Int. Congress Soil Sci., Symposia Papers Vol. 1,* New Delhi, 1982, pp. 101–114.

Simonson, R. W., "What Soils Are," in *Soil,* USDA Yearbook of Agr., Washington, D.C., 1957, pp. 17–31.

Simonson, R. W., "Outline of a Generalized Theory of Soil Genesis," *Soil Sci. Soc. Am. Proc., 23:*161–164, 1959.

Simonson, R. W., F. F. Riecken, and G. D. Smith, *Understanding Iowa Soils,* Brown, Dubuque, Iowa, 1952.

Smith, G. D., W. H. Allaway, and F. F. Riecken, "Prairie Soils of the Upper Mississippi Valley," *Advances in Agron. 2:*157–205, 1950.

Smith, Horace, "Soil Survey of District of Columbia," *Soil Con. Service, USDA,* Washington, D.C., 1976.

Smith, R. M., G. Samuels, and C. F. Cernuda, "Organic Matter and Nitrogen Buildups in Some Puerto Rican Soil Profiles," *Soil Sci., 72:*409–427, 1951.

Soil Survey Staff, *Soil Survey Manual,* USDA Handbook 18, Washington, D.C., 1951. (Also supplement May 1962.)

Soil Survey Staff, *Soil Taxonomy,* USDA Handbook 436, USDA, Washington, D.C., 1975.

Zack, A., "Soil Surveying the Nation's Capital," *Soil Cons., 41:*12–15, 1975.

CHAPTER 10

SOIL CLASSIFICATION AND GEOGRAPHY

Classification schemes of natural objects seek to organize knowledge, so that the properties and relationships of these objects may be most easily remembered and understood for some specific purpose. The ultimate purpose of soil classification is the maximum satisfaction of human wants that depend on soil use. This requires grouping soils with similar properties, so that lands can be efficiently managed for crop production. Furthermore, soils that are suitable or unsuitable for pipelines, roads, recreation, forestry, agriculture, wildlife, building sites, and so forth can be identified. This chapter is designed as an introduction to the classification, nature, distribution, and use of major soils throughout the world.

Soil Classification

A genetic classification was suggested about 1880 by the Russian scientist Dokuchaev; it has been further developed by European and American researchers. This system is based on the theory that each soil has a definite morphology that is related to a particular combination of soil-forming factors. This system reached its maximum development in 1949 and was in primary use (especially in the United

States) until 1960. In 1960, the United States Department of Agriculture published *Soil Classification, A Comprehensive System*. This classification system places a major emphasis on soil morphology and gives less emphasis to genesis or soil-forming factors, as compared to previous systems. *Soil Classification, A Comprehensive System* has been improved since 1960 and was republished as *Soil Taxonomy* in

Table 10-1
Derivation and Major Features of Diagnostic Horizons

Horizon	Derivation	Major Features
Surface diagnostic horizons-epipedons		
Mollic	L. *mollis*, soft	Thick, dark-colored, high base saturation (>50%), and strong structure so that the soil is not massive or hard when dry.
Umbric	L. *umbra*, shade	Same as mollic but highly H saturated (<50%) and may be hard or massive when dry.
Ochric	Gr. *ochros*, pale	Thin, light-colored, and low in organic matter.
Histic	Gr. *histos*, tissue	Very high organic matter content and saturated with water at some time during the year unless artificially drained.
Anthropic	Gr. *anthropos*, man	Molliclike horizon that has a very high phosphate content, 250 ppm or more P_2O_5, resulting from long time cultivation and fertilization.
Plaggen	Ger. *plaggen*, sod	Very thick, over 250 ppm P_2O_5, produced by long continued manuring.
Subsurface diagnostic horizons*		
Argillic	L. *argilla*, white clay	Illuvial horizon of silicate clay accumulation.
Natric	*Natrium*, sodium	Illuvial horizon of silicate clay accumulation, over 15% exchangeable sodium and columnar or prismatic structure.
Spodic	Gr. *spodos*, wood ash	Illuvial accumulation of free iron and aluminum oxides and organic matter.
Oxic	L. *oxide*, oxide	Altered subsurface horizon consisting of a mixture of hydrated oxides of iron, aluminum and 1:1 clays.
Cambic	L. *cambiare*, to change	An altered horizon due to movement of soil particles by frost, roots, and animals to such an extent to destroy original rock structure or aggregation into peds or both.
Agric	L. *ager*, field	An illuvial horizon of clay and organic matter accumulation just under the plow layer due to long-continued cultivation.

*Also includes other horizons as duripan, fragipan, albic, and so forth, described in *Soil Taxonomy*, 1975.

1975. *Taxonomy* is the classification of objects according to their natural relationships.

Diagnostic Horizons

Diagnostic horizons were developed to be used in defining most of the orders. Two kinds of diagnostic horizons, surface, and subsurface, are recognized. The surface diagnostic horizons are called *epipedons* (Greek *epi*, over; and *pedon*, soil). A brief description of the diagnostic horizons is given in Table 10-1.

Soil Orders of Soil Taxonomy

The order is the highest category and there are 10 orders, each ending in -sol (L. *solum* meaning soil). The orders, along with their derivation and meaning, are given in Table 10-2. Entisols are very recent soils (*see* Table 10-2). Vertisols are soils high in clay that become inverted because of alternate swelling and shrinking. Inceptisols are young soils with just the beginning of genetic horizon development. Aridisols are soils of arid regions. Mollisols are grassland soils with thick, soft, dark-colored surface horizons (mollic epipedons). Spodosols have spodic horizons (and are comparable to Podzols). Alfisol is derived from *pedalfer*—a word first used by Marbut to refer to humid region soils leached of lime, with a tendency for aluminum (Al) and iron (Fe) to accumulate in the subsoil. Ultisols are extremely leached soils, very low in bases. Oxisols are red tropical soils rich in oxides of iron and aluminum and also in 1:1 clays; that is, they have oxic horizons. Histosols are bog soils composed mainly of plant tissue. Some general relationships between the orders is given in Fig. 1-6. The world distribution of orders is given in Fig. 10-1.

Table 10-2
Formative Syllables, Derivations, and Meanings of Soil Orders

Order	Formative Syllable	Derivation	Meaning
1. Entisol	ent	Coined syllable	Recent soil
2. Vertisol	ert	L. *verto,* turn	Inverted soil
3. Inceptisol	ept	L. *inceptum,* beginning	Inception, or young soil
4. Aridisol	id	L. *aridus,* dry	Arid soil
5. Mollisol	oll	L. *mollis,* soft	Soft soil
6. Spodosol	od	Gr. *spodos,* wood ash	Ashy (Podzol) soil
7. Alfisol	alf	Coined syllable	Pedalfer (Al-Fe) soil
8. Ultisol	ult	L. *ultimus,* last	Ultimate (of leaching)
9. Oxisol	ox	F. *oxide,* oxide	Oxide soils
10. Histosol	ist	Gr. *histos,* tissue	Tissue (organic) soils

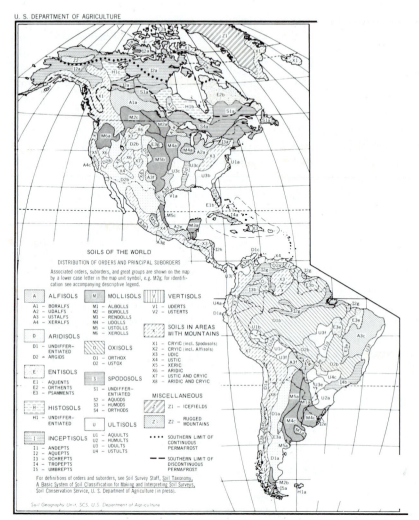

Figure 10-1
Broad schematic map of the soil orders and suborders of the world.

SOILS OF THE WORLD

Distribution of Orders and Principal Suborders and Great Groups

1:50,000,000

A **ALFISOLS**—Soils with subsurface horizons of clay accumulation and medium to high base supply; either usually moist or moist for 90

consecutive days during a period when temperature is suitable for plant growth.

A1 **Boralfs**—cold.
A1a with Histosols, cryic temperature regimes common
A1b with Spodosols, cryic temperature regimes

A2 **Udalfs**—temperate to hot, usually moist.
A2a with Aqualfs

SOIL CONSERVATION SERVICE

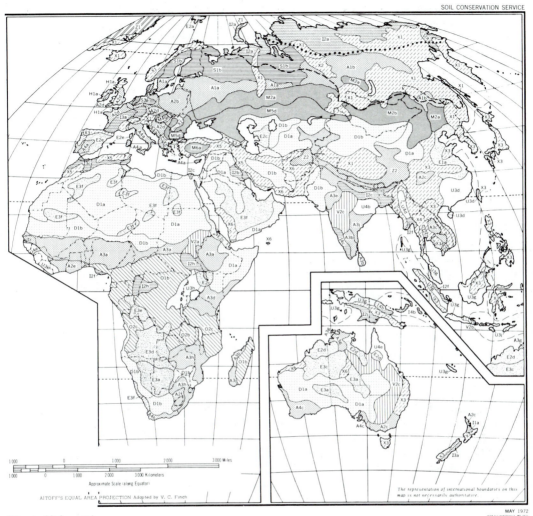

Approximate Scale (along Equator)

AITOFF'S EQUAL AREA PROJECTION Adapted by V. C. Finch

The representation of international boundaries on this map is not necessarily authoritative.

MAY 1972

Figure 10-1—cont.

A2b with Aquolls
A2c with Hapludults
A2d with Ochrepts
A2e with Troporthents
A2f with Udorthents

A3 **Ustalfs**—temperate to hot, dry more than 90 cumulative days during periods when temperature is suitable for plant growth.
A3a with Tropepts
A3b with Troporthents

A3c with Tropustults
A3d with Usterts
A3e with Ustochrepts
A3f with Ustolls
A3g with Ustorthents
A3h with Ustox
A3j Plinthustalfs with Ustorthents

A4 **Xeralfs**—temperate or warm, moist in winter and dry more than 45 consecutive days in summer.   [continued overleaf

A4a with Xerochrepts
A4b with Xerorthents
A4c with Xerults

D **ARIDISOLS**—Soils with pedogenic horizons, usually dry in all horizons and never moist as long as 90 consecutive days during a period when temperature is suitable for plant growth.

D1 **Aridisols**—undifferentiated.
D1a with Orthents
D1b with Psamments
D1c with Ustalfs

D2 **Argids**—with horizons of clay accumulation.
D2a with Fluvents
D2b with Torriorthents

E **ENTISOLS**—Soils without pedogenic horizons; either usually wet, usually moist, or usually dry.

E1 **Aquents**—seasonally or perennially wet.
E1a Haplaquents with Udifluvents
E1b Psammaquents with Haplaquents
E1c Tropaquents with Hydraquents

E2 **Orthents**—loamy or clayey textures, many shallow to rock.
E2a Cryorthents
E2b Cryorthents with Orthods
E2c Torriorthents with Aridisols
E2d Torriorthents with Ustalfs
E2e Xerorthents with Xeralfs

E3 **Psamments**—sand or loamy sand textures.
E3a with Aridisols
E3b with Orthox
E3c with Torriorthents
E3d with Ustalfs
E3e with Ustox
E3f shifting sands
E3g Ustipsamments with Ustolls

H **HISTOSOLS**—Organic soils.

H1 **Histosols**—undifferentiated.
H1a with Aquods
H1b with Boralfs
H1c with Cryaquepts

I **INCEPTISOLS**—Soils with pedogenic horizons of alteration or concentration but without accumulations of translocated materials other than carbonates or silica; usually moist or moist for 90 consecutive days during a period when temperature is suitable for plant growth.

I1 **Andepts**—amorphous clay or vitric volcanic ash or pumice.
I1a Dystrandepts with Ochrepts

I2 **Aquepts**—seasonally wet.
I2a Cryaquepts with Orthents
I2b Haplaquepts with Salorthids
I2c Haplaquepts with Humaquepts
I2d Haplaquepts with Ochraqualfs
I2e Humaquepts with Psamments
I2f Tropaquepts with Hydraquents
I2g Tropaquepts with Plinthaquults
I2h Tropaquepts with Tropaquents
I2j Tropaquepts with Tropudults

I3 **Ochrepts**—thin, light-colored surface horizons and little organic matter.
I3a Dystrochrepts with Fragiochrepts
I3b Dystrochrepts with Orthox
I3c Xerochrepts with Xerolls

I4 **Tropepts**—continuously warm or hot.
I4a with Ustalfs
I4b with Tropudults
I4c with Ustox

I5 **Umbrepts**—dark-colored surface horizons with medium to low base supply.
I5a with Aqualfs

M **MOLLISOLS**—Soils with nearly black, organic-rich surface horizons and high base supply; either usually moist or usually dry.

M1 **Albolls**—light gray subsurface horizon over slowly permeable horizon; seasonally wet.
M1a with Aquepts

M2 **Borolls**—cold.
M2a with Aquolls
M3b with Orthids
M2c with Torriorthents

M3 **Rendolls**—subsurface horizons have much calcium carbonate but no accumulation of clay.
M3a with Usterts

M4 **Udolls**—temperate or warm, usually moist.
M4a with Aquolls
M4b with Eutrochrepts
M4c with Humaquepts

M5 **Ustolls**—temperate to hot, dry more than 90 cumulative days in year.
M5a with Argialbolls
M5b with Ustalfs
M5c with Usterts
M5d with Ustochrepts

M6 **Xerolls**—cool to warm, moist in winter and dry more than 45 consecutive days in summer.
M6a with Xerorthents

O **OXISOLS**—Soils with pedogenic horizons that are mixtures principally of kaolin, hydrated oxides, and quartz, and are low in weatherable minerals.

- O1 **Orthox**—hot, nearly always moist.
  - O1a with Plinthaquults
  - O1b with Tropudults

- O2 **Ustox**—warm or hot, dry for long periods but moist more than 90 consecutive days in the year.
  - O2a with Plinthaquults
  - O2b with Tropustults
  - O2c with Ustalfs

S **SPODOSOLS**—Soils with accumulation of amorphous materials in subsurface horizons; usually moist or wet.

- S1 **Spodosols**—undifferentiated.
  - S1a cryic temperature regimes; with Boralfs
  - S1b cryic temperature regimes; with Histosols

- S2 **Aquods**—seasonally wet.
  - S2a Haplaquods with Quartzipsamments

- S3 **Humods**—with accumulations of organic matter in subsurface horizons.
  - S3a with Hapludalfs

- S4 **Orthods**—with accumulations of organic matter, iron, and aluminum in subsurface horizons.
  - S4a Haplorthods with Boralfs

U **ULTISOLS**—Soils with subsurface horizons of clay accumulation and low base supply; usually moist or moist for 90 consecutive days during a period when temperature is suitable for plant growth.

- U1 **Aquults**—seasonally wet.
  - U1a Ochraquults with Udults
  - U1b Plinthaquults with Orthox
  - U1c Plinthaquults with Plinthaquox
  - U1d Plinthaquults with Tropaquepts

- U2 **Humults**—temperate or warm and moist all of year; high content of organic matter.
  - U2a with Umbrepts

- U3 **Udults**—temperate to hot; never dry more than 90 cumulative days in the year.
  - U3a with Andepts
  - U3b with Dystrochrepts
  - U3c with Udalfs
  - U3d Hapludults with Dystrochrepts
  - U3e Rhodudults with Udalfs
  - U3f Tropudults with Aquults
  - U3g Tropudults with Hydraquents
  - U3h Tropudults with Orthox
  - U3j Tropudults with Tropepts
  - U3k Tropudults with Tropudalfs

- U4 **Ustults**—warm or hot; dry more than 90 cumulative days in the year.
  - U4a with Ustochrepts
  - U4b Plinthustults with Ustorthents
  - U4c Rhodustults with Ustalfs
  - U4d Tropustults with Tropaquepts
  - U4e Tropustults with Ustalfs

V **VERTISOLS**—Soils with high content of swelling clays; deep, wide cracks develop during dry periods.

- V1 **Uderts**—usually moist in some part in most years; cracks open less then 90 cumulative days in the year.
  - V1a with Usterts

- V2 **Usterts**—cracks open more than 90 cumulative days in the year.
  - V2a with Tropaquepts
  - V2b with Tropofluvents
  - V2c with Ustalfs

X **Soils in areas with mountains**—Soils with various moisture and temperature regimes; many steep slopes; relief and total elevation vary greatly from place to place. Soils vary greatly within short distances and with changes in altitude; vertical zonation common.

- X1 Cryic great groups of Entisols, Inceptisols, and Spodosols.

- X2 Borolfs and cryic great groups of Entisols and Inceptisols.

- X3 Udic great groups of Alfisols, Entisols, and Ultisols; Inceptisols.

- X4 Ustic great groups of Alfisols, Inceptisols, Mollisols, and Ultisols.

- X5 Xeric great groups of Alfisols, Entisols, Inceptisols, Mollisols, and Ultisols.

- X6 Torric great groups of Entisols; Aridisols.

- X7 Ustic and cryic great groups of Alfisols, Entisols, Inceptisols, and Mollisols; ustic great groups of Ultisols; cryic great groups of Spodosols.

- X8 Aridisols, torric and cryic great groups of Entisols, and cryic great groups of Spodosols and Inceptisols.

Z **MISCELLANEOUS**

- Z1 Icefields.

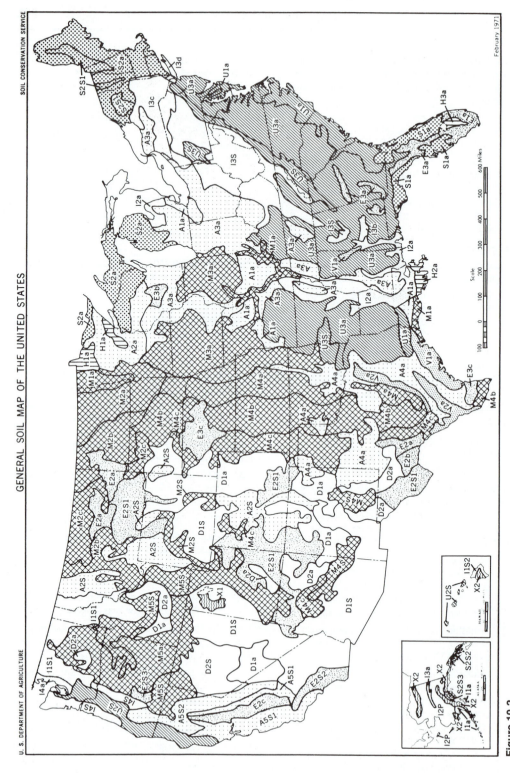

Scale

100   0   100   200   300   400   500   600 Miles

Figure 10-2
General soil map of the United States.

256

Z2 Rugged mountains—mostly devoid of soil (includes glaciers, permanent snow fields, and, in some places, areas of soil).

. . . Southern limit of continuous permafrost.

- - Southern limit of discontinuous permafrost.

Legend for general soil map of the United States.

ALFISOLS

Aqualfs

A1a–Aqualfs with Udalfs, Haplaquepts, Udolls; gently sloping.

Boralfs

A2a–Boralfs with Udipsamments and Histosols; gently and moderately sloping.

A2S–Cryoboralfs with Borolls, Cryochrepts, Cryorthods, and rock outcrops; steep.

Udalfs

A3a–Udalfs with Aqualfs, Aquolls, Rendolls, Udolls, and Udults; gently or moderately sloping.

Ustalfs

A4a–Ustalfs with Ustochrepts, Ustolls, Usterts, Ustipsamments, and Ustorthents; gently or moderately sloping.

Xeralfs

A5S1–Xeralfs with Xerolls, Xerorthents, and Xererts; moderately sloping to steep.

A5S2–Ultic and lithic subgroups of Haploxeralfs with Andepts, Xerults, Xerolls, and Xerochrepts; steep.

ARIDISOLS

Argids

D1a–Argids with Orthids, Orthents, Psamments, and Ustolls; gently and moderately sloping.

D1S–Argids with Orthids, gently sloping; and Torriorthents, gently sloping to steep.

Orthids

D2a–Orthids with Argids, Orthents, and Xerolls; gently or moderately sloping.

D2S–Orthids, gently sloping to steep, with Argids, gently sloping; lithic subgroups of Torriorthents and Xerorthents, both steep.

ENTISOLS

Aquents

E1a–Aquents with Quartzipsamments, Aquepts, Aquolls, and Aquods; gently sloping.

Orthents

E2a–Torriorthents, steep, with borollic subgroups of Aridisols; Usterts and aridic and vertic subgroups of Borolls; gently or moderately sloping.

E2b–Torriorthents with Torrerts; gently or moderately sloping.

E2c–Xerorthents with Xeralfs, Orthids, and Argids; gently sloping.

E2S1–Torriorthents; steep, and Argids, Torrifluvents, Ustolls, and Borolls; gently sloping.

E2S2–Xerorthents with Xeralfs and Xerolls; steep.

E2S3–Cryorthents with Cryopsamments and Cryandepts; gently sloping to steep.

Psamments

E3a–Quartzipsamments with Aquults and Udults; gently or moderately sloping.

E3b–Udipsamments with Aquolls and Udalfs; gently or moderately sloping.

E3c–Ustipsamments with Ustalfs and Aquolls; gently or moderately sloping.

HISTOSOLS

Histosols

H1a–Hemists with Psammaquents and Udipsamments; gently sloping.

H2a–Hemists and Saprists with Fluvaquents and Haplaquepts; gently sloping.

H3a–Fibrists, Hemists, and Saprists with Psammaquents; gently sloping.

INCEPTISOLS

Andepts

I1a–Cryandepts with Cryaquepts, Histosols, and rock land; gently or moderately sloping.

I1S1–Cryandepts with Cryochrepts, Cryumbrepts, and Cryorthods; steep.

I1S2–Andepts with Tropepts, Ustolls, and Tropofolists; moderately sloping to steep.

Aquepts

I2a–Haplaquepts with Aqualfs, Aquolls, Udalfs, and Fluvaquents; gently sloping.

I2P–Cryaquepts with cryic great groups of Orthents, Histosols, and Ochrepts; gently sloping to steep.

Ochrepts

I3a–Cryochrepts with cryic great groups of Aquepts, Histosols, and Orthods; gently or moderately sloping.

I3b–Eutrochrepts with Uderts; gently sloping.

I3c–Fragiochrepts with Fragiaquepts, gently or moderately sloping; and Dystrochrepts, steep.

I3d–Dystrochrepts with Udipsamments and Haplorthods; gently sloping.

I3S–Dystrochrepts, steep, with Udalfs and Udults; gently or moderately sloping. *[continued overleaf*

Umbrepts

14a—Haplumbrepts with Aquepts and Orthods; gently
or moderately sloping.

I4S—Haplumbrepts and Orthods; steep, with Xerolls
and Andepts; gently sloping.

MOLLISOLS

Aquolls

M1a—Aquolls with Udalfs, Fluvents, Udipsamments,
Ustipsamments, Aquepts, Eutrochrepts, and
Borolls; gently sloping.

Borolls

M2a—Udic subgroups of Borolls with Aquolls and
Ustorthents; gently sloping.

M2b—Typic subgroups of Borolls with
Ustipsamments, Ustorthents, and Boralfs; gently
sloping.

M2c—Aridic subgroups of Borolls with Borollic
subgroups of Argids and Orthids, and
Torriorthents; gently sloping.

M2S—Borolls with Boralfs, Argids, Torriorthents, and
Ustolls; moderately sloping or steep.

Udolls

M3a—Udolls, with Aquolls, Udalfs, Aqualfs, Fluvents,
Psamments, Ustorthents, Aquepts, and Albolls;
gently or moderately sloping.

Ustolls

M4a—Udic subgroups of Ustolls with Orthents,
Ustochrepts, Usterts, Aquents, Fluvents, and
Udolls; gently or moderately sloping.

M4b—Typic subgroups of Ustolls with Ustalfs,
Ustipsamments, Ustorthents, Ustochrepts, Aquolls,
and Usterts; gently or moderately sloping.

M4c—Aridic subgroups of Ustolls with Ustalfs,
Orthids, Ustipsamments, Ustorthents, Ustochrepts,
Torriorthents, Borolls, Ustolls, and Usterts, gently
or moderately sloping.

M4S—Ustolls with Argids and Torriorthents;
moderately sloping or steep.

Xerolls

M5a—Xerolls with Argids, Orthids, Fluvents,
Cryoboralfs, Cryoborolls, and Xerorthents; gently
or moderately sloping.

M5S—Xerolls with Cryoboralfs, Xeralfs, Xerorthents,
and Xererts; moderately sloping or steep.

SPODOSOLS

Aquods

S1a—Aquods with Psammaquents, Aquolls, Humods,
and Aquults; gently sloping.

Orthods

S2a—Orthods with Boralfs, Aquents, Orthents,
Psamments, Histosols, Aquepts, Fragiochrepts,
and Dystrochrepts; gently or moderately sloping.

S2S1—Orthods with Histosols, Aquents, and
Aquepts; moderately sloping or steep.

S2S2—Cryorthods with Histosols; moderately sloping
or steep.

S2S3—Cryorthods with Histosols, Andepts and
Aquepts; gently sloping to steep.

ULTISOLS

Aquults

U1a—Aquults with Aquents, Histosols,
Quartzipsamments, and Udults; gently sloping.

Humults

U2S—Humults with Andepts, Tropepts, Xerolls,
Ustolls, Orthox, Torrox, and rock land; gently
sloping to steep.

Udults

U3a—Udults with Udalfs, Fluvents, Aquents,
Quartzipsamments, Aquepts, Dystrochrepts, and
Aquults; gently or moderately sloping.

U3S—Udults with Dystrochrepts; moderately sloping
or steep.

VERTISOLS

Uderts

V1a—Uderts with Aqualfs, Eutrochrepts, Aquolls, and
Ustolls; gently sloping.

Usterts

V2a—Usterts with Aqualfs, Orthids, Udifluvents,
Aquolls, Ustolls, and Torrerts; gently sloping.

Areas With Little Soil

X1—Salt flats.

X2—Rock land (plus permanent snow fields and
glaciers).

Slope Classes

Gently sloping—Slopes mainly less than 10 percent,
including nearly level.

Moderately sloping—Slopes mainly between 10 and
25 percent.

Steep—Slopes mainly steeper than 25 percent.

Suborders

The orders are divided into suborders, primarily on the basis of chemical and physical properties that reflect (1) the presence or absence of waterlogging or (2) the genetic differences caused by climate and its partially associated variable, vegetation. The distribution of suborders in the United States is given in Fig. 10-2. For example, the Aqualfs are "wet" (aqu- for aqua) Alfisols saturated with water sometime during the year. Borolls are Mollisols of the cool regions. The suborder names all have two syllables, with the last syllable indicating the order, such as *alf* for Alfisol and *oll* for Mollisol. Formative elements in names of suborders are given in Table 10-3.

Soil Orders—Properties, Distribution, and Use

Most of the world's population make their living by tilling the soil. The soil directly influences their lives everyday in that it determines how they build their houses and roads, and how they grow their crops. By affecting the amount and kinds of food they eat, soils affect their health. The order category is sufficiently general and, yet, sufficiently well defined to make a discussion of these and other aspects of the world's soils possible in one chapter. A broad schematic map of the soil orders of the world is presented in Fig. 10-1, and suborders for the United States appear in Fig. 10-2.

Entisols

Entisols are soils that tend to be of recent origin. They are characterized by youthfulness and are without natural genetic horizons, or else they have only the beginnings of horizons. The central concept of Entisols is soils in deep regolith or earth with no horizons, except perhaps a plow layer. Some Entisols, however, have plaggen, agric, or E (albic) horizons; some have hard rock close to the surface.

Soils that develop on alluvium of recent origin and have very weakly developed profiles are commonly *Fluvents*. In many of them, the color change from the A to C horizon is hard to see or is nonexistent. They are, in large part, soils in which most of the properties have been inherited. They are usually characterized by stratification. The texture is related to the rate at which water deposited the alluvium. For this reason, they tend to be coarse textured near the stream and finer textured near the outer edges of floodplains. Mineralogically, they are related to soils that served as a source of alluvium.

Periodic flooding brings fresh minerals to the soils, and they tend to remain fertile (*see* Fig. 10-3). The soil remains youthful, because it is buried before maturity is reached.

Most soils throughout the world that developed from unconsolidated sediments were Entisols when they were young. Steep slopes (where erosion occurs rapidly), an insufficient length of time, or movement of the material (as in the case of sand dunes), are the major causes for their existence. Unstabilized sand dunes develop into Entisols after vegetation has become established. Entisols *(Psamments)* are the dominant soils on stabilized sand dunes in the Nebraska Sand Hills (*see* Fig. 10-4).

Entisols that developed from sand dunes have a limited agricultural-cropping value. The moderating influence of lakes on climate has made it profitable to

Table 10-3
Formative Elements in the Names of Suborders

Formative Elements	Derivation of Formative Element	Connotation of Formative Element
alb	L. *albus,* white	Presence of albic horizon (a bleached eluvial horizon)
and	Modified from Ando	Ando-like (containing volcanic ash)
aqu	L. *aqua,* water	Characteristics associated with wetness
ar	L. *arare,* to plow	Mixed horizons
arg	Modified from argillic horizon; L. *argilla,* white clay	Presence of argillic horizon (a horizon with illuvial clay)
bor	Gr. *boreas,* northern	Cool
ferr	L. *ferrum,* iron	Presence of iron
fibr	L. *fibra,* fiber	Least decomposed stage
fluv	L. *fluvius,* river	Floodplains
hem	Gr. *hemi,* half	Intermediate stage of decomposition
hum	L. *humus,* earth	Presence of organic matter
lept	Gr. *leptos,* thin	Thin horizon
ochr	Gr. base of *ochros,* pale	Presence of ochric epipedon (a light colored surface)
orth	Gr. *orthos,* true	The common ones
plag	Modified from Ger. *plaggen,* sod	Presence of plaggen epipedon
psamm	Gr. *psammos,* sand	Sand textures
rend	Modified from Rendzina	Rendzina-like (thin soil over limestone)
sapr	Gr. *sapros,* rotten	Most decomposed stage
torr	L. *torridus,* hot and dry	Usually dry
trop	Modified from Gr. *tropikos,* of the solstice	Continually warm
ud	L. *udus,* humid	Of humid climates
umbr	L. *umbra,* shade	Presence of umbric epipedon (a dark colored surface)
ust	L. *ustus,* burnt	Of dry climates, usually hot in summer
xer	Gr. *xeros,* dry	Annual dry season

raise fruit on some Entisols in Michigan. Small areas of Entisols frequently exist on the steepest parts of cultivated fields, and they are effectively used with the surrounding soils that may be older and comprise the major portions of the fields. They are low in organic matter content and are generally responsive to nitrogen fertilization. Many of them are neutral in reaction or calcareous at the surface.

Some Entisols have an A horizon resting directly on hard rock. An AR type of Entisol is shown in Fig. 1-3. Two very important factors that contribute to their development are the hardness of the rock and the steepness of the slope. The rate

Figure 10-3
Lettuce being grown on highly productive Entisols (alluvial soils) with irrigation. Water from the river in the background (located where the line of trees are growing) is distributed over the field by gravity flow in the furrows between the rows. (Photo courtesy of USDA.)

of rock disintegration does little more than keep pace with the removal of material via erosion. Cracks in the underlying rock may enable roots to penetrate much deeper than the A horizon. Where the A horizon is 20 to 50 centimeters thick, the land is profitably used for pasturing.

AR Entisols *(Orthents)* are common in mountainous areas and give evidence to the fact that deep soils did not cover the land everywhere before agriculture be-

Figure 10-4
Cattle grazing is the dominant land use of Entisols (Psamments) of the Nebraska Sandhills. A grass cover is needed to prevent wind erosion.

gan. Many deep, productive soils were once Entisols and, in a sense, these soils may be transitory in the development of well-differentiated profiles.

Inceptisols

Inceptisol is derived from the Latin *inceptum*, meaning beginning. The development of genetic horizons is just beginning in Inceptisols, but they are considered older than Entisols. Typically, Inceptisols have ochric epipedons and cambic subsurface horizons. They may have other diagnostic horizons, but show little evidence of eluviation or illuviation. Evidence of extreme weathering is generally lacking. They lack sufficient diagnostic features to be placed in any of the remaining eight soil orders.

Inceptisols occur in all climatic zones where there is some leaching in most years. On the soil order map in Fig. 10-1, two large areas are shown that include the tundras of North America and Europe-Asia. Soils of the tundra are characterized by a high organic matter content. The vegetation consists mostly of low-growing mosses, lichens, and sedges (*see* Fig. 10-5). These plants grow slowly, but the low soil temperature inhibits organic matter decomposition, resulting in soils with a high content of organic matter. They usually have permafrost, are slightly to strongly acidic and have a surface microrelief caused by freezing and thawing. Most soils on the tundra show evidence of wetness or poor drainage. They are *Aquepts.*

Inceptisols of the tundra support a sparse population of nomadic hunters that live almost entirely on the products of caribou. In recent years, some of these Eskimos have been in the news for having high radioactivity. Testing of nuclear bombs in the Arctic produced radioactive fallout that was absorbed by lichens. Caribou ate the lichens and the radioactivity was transferred to humans via the meat. Eskimos of the Brooks Range in Alaska were found to have about 100 times more

radioactivity than persons in the "lower states."

Many Inceptisols are volcanic ash soils or *Andepts.* They represent a stage in the ultimate development of Ultisols and Oxisols in humid tropics. They have amorphous clays and are usually very acidic. Many are used intensively for production of sugar cane, coffee, and other crops.

Inceptisols are widely distributed throughout the world and some make good agricultural and grazing lands. The data in Table 10-4 show that Inceptisols occupy 15.8 percent of the land surface of the world and rank second in order of abundance. The wet Inceptisols (Aquepts) located in large river valleys in Asia are the most extensive soil used for paddy (rice) production.

Aridisols

Aridisols have an aridic soil moisture regime and are the dominant soils of desert regions. They are the most abundant soil order, making up nearly 20 percent of the world's soils (*see* Table 10-4). Desert shrubs dominate the most arid regions, with shrubs giving way to bunch grasses with increasing moisture. Plants are widely spaced and use whatever soil moisture there is quite effectively. Perhaps one of the surprising things for persons who have lived all their lives in humid regions and then taken a trip to the desert is the great diversity of plants and the considerable amount of vegetation. Many desert plants grow and function during the wetter seasons of the year and go dormant during the driest seasons. On the more humid or eastern edge of the arid region of the western United States, the desert bunch grasses give way to taller and more

Figure 10-5
Dominant soils on the tundra above timberline in mountains are Inceptisols (Aquepts). Vegetation is sparse and the environment is very fragile.

Table 10-4

Area of Soils of the World by Soil Order

Soil Order	Area in Thousands of Square Miles	World Total, %	Rank
Alfisols	7,600	14.7	3
Aridisols	9,900	19.2	1
Entisols	6,500	12.5	4
Histosols	400	0.8	10
Inceptisols	8,100	15.8	2
Mollisols	4,600	9.0	6
Oxisols	4,800	9.2	5
Spodosols	2,800	5.4	8
Ultisols	4,400	8.5	7
Vertisols	1,100	2.1	9
Ice fields and rugged mountains	1,200	2.4	—
Islands, unclassified	200	0.4	—
Grand total	51,600	100.0	—

vigorous grasses; Aridisols merge with Mollisols.

Development and Properties of Aridisols. In arid regions, the soil-forming processes are similar to those of humid regions; however, the rate of soil development is much slower in arid regions. The lesser amount of plant growth and the potential for organic matter decomposition produce soils with low organic matter contents. Winds play a major role in the development of Aridisols. Winds move dust about; occasional rains wash soluble nutrients from the dust on its transient journey across the desert. A more obvious role of wind is the blowing away of fine soil particles, resulting in the formation of a concentration of gravel or the formation of desert pavement.

Water is less effective in leaching soluble salts and translocating colloidal material in arid regions because of the low amount of precipitation. Another factor is the torrential nature of much of the rainfall, which results in considerable runoff. A striking feature of most Aridisols is a zone at varying distances below the surface, where calcium carbonate has been deposited by percolating water (calcic horizon). Many Aridisols have well-developed argillic (Bt) horizons, which is evidence of considerable clay movement. The widespread occurrence of argillic horizons in many Aridisols suggests that many years ago a more humid climate existed than exists today. Mohave is a common Aridisol from the southwestern United States; some data of a Mohave is given in Table 10-5 to illustrate features or properties commonly found in Aridisols. Note the presence of an argillic horizon (Bt), a low content of organic matter and low carbon-nitrogen ratio of the or-

Table 10-5

Some Properties of Mohave Sandy Clay Loam—an Aridisol

Horizon	Depth, cm	Clay, %	Organic Matter, %	C/N	CEC mEq/ 100 grams	Exch. Na, %	pH	CaCO$_3$, %
A	0–10	11	0.25	6	8	1.2	7.8	—
Bt1	10–25	14	0.19	6	15	2.0	7.4	—
Bt2	25–69	25	0.24	7	22	2.5	8.5	—
Bk	69–94	21	0.25	8	17	4.1	8.9	10
Ck	94–137	17	0.08	—	6	12.7	9.2	22

Adapted from profile 62 of *Soil Classification, a Comprehensive System*, USDA, 1960.

ganic matter, the presence of significant exchangeable sodium, high pH values, and the accumulation of calcium carbonate (k) in the lower part of the profile (*see* Color Plate 2 in Chapter 1).

Aridisols are placed in suborders on the basis of the presence or absence of argillic horizons. The suborder *Orthids* includes Aridisols without argillic horizons; by contrast, the suborder *Argids* includes Aridisols with argillic horizons. As we have already noted, Mohave has an argillic horizon; therefore, Mohave is an Argid. The general distribution of Orthids and Argids in the United States is shown in Fig. 10-2.

Relationship of Land Surfaces to Age of Aridisols.

It is believed that *Orthids* are the "younger" Aridisols and that the *Argids* are the "older" Aridisols. There is evidence that Orthids, in the United States, have developed largely within the past 25,000 years in an arid climate. Orthids are mostly located where recent alluvium has been deposited. Argids are common on older land surfaces in any landscape where there has been more time for the development of argillic horizons and a greater likelihood that the soil has been influenced by a more humid climate over 25,000 years ago. On the most recent sediments or land surfaces, Entisols are abundant. Fig. 10-6 shows a desert landscape with land surfaces of greatly different ages.

Land Use on Aridisols.

Aridisols of the western United States occur almost entirely within a region called the "western range and irrigated region." As the name implies, grazing of sheep and cattle and the production of crops by irrigation are the two major uses of land. The use of land for grazing is closely related to precipitation, which largely determines the amount of forage produced. Some areas are too dry for grazing, while other areas are more favorable and may also be able to take advantage of summer grazing on mountain meadows. As much as 35 hectares or 75 acres or more in the drier areas is required per head of cattle, thus making large farms or ranches a necessity. Most ranchers supplement range forage by producing some crops on a small acreage favorably located for irrigation (*see* Fig. 10-6). The major hazard in the use of grazing lands is overgrazing, which results in the invasion of less desirable

Figure 10-6
Grazing lands dominated by Aridisols. Soils in the area are closely related to the age of land surface with Aridisols on the older surfaces and Entisols on the youngest surfaces. Note the small irrigation reservoir and irrigated cropland near ranch headquarters. (USDA photo by B. W. Muir.)

plant species and an increased soil erosion.

Only about 1 or 2 percent of the land is irrigated, because the production of crops by irrigation depends on a water supply. Most of the irrigated land is located on alluvial soils, or Entisols, along streams and rivers where the land is nearly level and the irrigation water can be distributed over the field by gravity (*see* Fig. 10-7). In addition, nearby rivers serve as a source of water from natural river flow and carry irrigation water released from water storage reservoirs. The alkaline nature of Aridisols may cause deficiencies of various micronutrients on certain crops. Major crops include alfalfa, cotton, citrus fruits, vegetables, and grain crops. In Arizona, crop production on only 2 percent of the land that is irrigated accounts for 60 percent of the total farm income. Grazing, by contrast, uses 80 percent of the land and accounts for only 40 percent of the total farm income.

Mollisols

Bordering many desert regions are areas of higher rainfall that support grasses that tend to cover the ground completely and produce an abundance of organic matter, which decomposes within the soil. The rainfall, however, is sufficiently limited to prevent excessive leaching, and base saturation remains high. Decomposition of abundant organic matter within soil in the presence of calcium leads to the formation of mollic epipedons. The well-aggregated soil structure gives rise to the softness of the soil, which is neither massive nor very hard when dry. All Mollisols have mollic epipedons. Features of mollic horizons include: (1) softness when dry, (2) dark color and at least 1 percent organic matter (unless very high in lime), and (3) base saturation 50 percent or more (measured at pH 7). Most mollic horizons are 18 or more centimeters thick. A soil with a mollic epipedon is shown in Fig. 10-8. In many Mollisols, there has

Figure 10-7
Aridisol landscapes on the Sonoran Desert of southern Arizona. *(Left)* Natural vegetation and the presence of desert pavement in the foreground. *(Right)* Similar land used for irrigated agriculture.

been sufficient clay migration to form a Bt or argillic horizon. As a group, Mollisols combine high soil fertility and fair to adequate rainfall, so that they comprise perhaps the world's most productive agricultural soils.

Geographic Relationships of Mollisols.   Geographically, large areas of Mollisols and Aridisols share a common boundary. Illustrations shown in Fig. 10-1 include the Great Plains of North America, and the plains of southern South America (Argentina), the southern Soviet Union, and northeastern China. The drier Mollisols of the Great Plains, near the Aridisol-Mollisol border, have ustic moisture regimes and are called *Ustolls*. Ustolls typically have lime accumulation (k) layers. In eastern Nebraska, Ustolls border with *Udolls*-Mollisols that have udic moisture regimes. Humid Mollisols or Udolls lack lime accumulation layers. In

these areas, increasing precipitation from Aridisols to the Mollisols results in gradual changes in soil properties, as illustrated in Fig. 10-9. With increasing precipitation, there is a gradual increase in solum thickness, organic matter content, development of Bt horizon, and depth to the k layer. A soil profile near the Mollisol-Aridisol border in Colorado is shown in Fig. 10-10 and illustrates many of the dominant features of Mollisols (*see also* Color Plate 2).

Mollisols in the Palouse country of Washington, Idaho, and Oregon developed in loess (plus some volcanic ash) and are similar to Udolls in Iowa and Illinois. These western soils have a Mediterranean climate, so that they are wet in the winter and dry in the summer. The soils have a xeric soil moisture regime and are *Xerolls*. The cool Mollisols of the northern Great Plains are *Borolls*. Wet Mollisolls are *Aquolls* and are very important on the re-

Mollic epipedon

Figure 10-8
A Mollisol with a mollic epipedon. The soil is the Aguilita clay loam from southwestern Puerto Rico and has developed from soft limestone. It is a *Rendoll.*

cently glaciated plains of the Midwest. Fig. 10-2 shows the distribution of the major areas of Mollisol suborders.

Land Use on Mollisols. The world's major grasslands lacked trees for lumber, readily available water supplies, and nat-

ural sites for protection against invaders. As a result, the areas were inhabited mainly by nomadic people until about 150 years ago. In fact, the virgin lands of the Soviet Union were opened to settled agriculture in 1954. Today, these areas are characterized by having excellent soil for

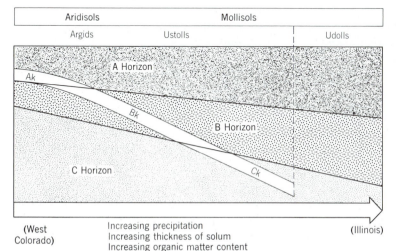

Aridisols	Mollisols	
Argids	Ustolls	Udolls

A Horizon
Ak
Bk
B Horizon
C Horizon
Ck

(West Colorado)

Increasing precipitation
Increasing thickness of solum
Increasing organic matter content
Increasing development of Bt horizon
Increasing depth to k horizon

(Illinois)

Figure 10-9
Generalized relationships between some suborders of Mollisols and Aridisols in many parts of the world, showing a gradual change in soil properties with increasing precipitation. Specifically, the diagram illustrates soil relationships on the Great Plains of the United States.

Figure 10-10
Ustoll near the Mollisol-Aridisol border in eastern Colorado. Obvious lime accumulation zone can be seen. (Caliche is a more or less cemented deposit of calcium carbonate.)

crops and a low population density. Land use is importantly related to moisture supply and temperature.

Grazing is a major land use of Ustolls (*see* Fig. 10-11). Dryland farming is practiced extensively for wheat production. Ir-

rigation is used, particularly along the river valleys; a wide variety of irrigated crops are grown. Fruits, vegetables, and sugar beets occupy smaller irrigated acreages.

As one travels east across the Great

Figure 10-11
Beef cattle grazing and wheat production are the two major land uses on Ustolls. (Photo courtesy of SCS, USDA.)

Plains of the United States, the rainfall increases; there is a decrease in grazing and an increase in wheat production. Near the Udoll border, corn becomes a major crop. Udolls combine high natural soil fertility, adequate moisture, and warm summers that are ideal for corn and soybean production. The two most famous Udoll areas are the Corn Belt of the United States and the humid pampa of Argentina.

Most Borolls of the northern Great Plains and Canada have cryic soil temperature regimes. Winters are too cold for winter wheat (and summers are too cool for corn and soybeans). Winter wheat is planted in the fall and spring wheat is planted in the spring, giving winter wheat a longer growing season. Consequently, yields are greater in the winter wheat region. Winter wheat is also the major crop of the Xerolls of the Palouse Hills (*see* Fig. 10-12).

A very significant amount of the soil in the upper Mississippi River Valley developed under the influence of poor drainage and has aquic moisture regimes. The soils are Aquolls. Mollic horizons of these soils tend to be very thick, very dark colored, and contain a high content of organic matter. Leaching has been minimal because of the presence of a high water table much of the time. They require

Figure 10-12
Landscape of Xerolls in the Palouse, where winter wheat production is the dominant land use. (Photo courtesy Dr. H. W. Smith.)

drainage for crop production and, when properly drained, become some of the most productive soils for agriculture. In Illinois, the most abundant single soil—the Drummer silty clay loam—is an Aquoll. Much of the Corn Belt's reputation of corn production has resulted from the large acreages of Mollisols developed on flat lands under poor drainage (*see* Fig. 10-13). Aquolls are also the dominant soils in the Red River Valley on the Minnesota–North Dakota border.

Wheat is the most important food crop in the world, in terms of acreage and production. Corn is perhaps third in importance after rice. Compare the maps in Fig. 10-14, which show the production of wheat and corn throughout the world. The Soviet Union is the major wheat producer and the United States is the major corn producer. This correlates with the abundance of Udoll-type Mollisols in the United States and their general absence in the Soviet Union. In the United States, corn production is centered in the Corn Belt on moist Udolls and the wheat production is concentrated on the drier Ustolls and cool Borolls. A similar comparison can be made for the pampa of Argentina. Over 50 percent of the world's production of corn and about 15 percent of the world's wheat is produced in the United States, which is a partial indication that a considerable amount of the Mollisols in the world are located in the United States. Mollisols occupy only 9 percent of the world's surface (*see* Table 10-4), but they produce a much larger percentage of the world's food.

Spodosols

All Spodosols have a spodic horizon. Spodic horizons are illuvial subsurface horizons, where amorphous materials composed of organic matter, aluminum, and iron have accumulated; they are comparable to Bhs horizons. All Spodosols form in a humid climate, mostly from sandy siliceous or quartzitic parent material. They are found from the tropics to the boreal regions, but the major areas in the world are just south of the tundras of North America and Europe-Asia. Here, glaciation left large areas of sandy parent materials, and the climate is humid. No major area of Spodosols is shown in the southern hemisphere on the map in Fig. 10-1. Trees are the common vegetation (*see* Fig. 10-15), although some of the most intensely developed Spodosols develop under heath vegetation. Ashy-gray E horizons (albic) are a major feature of most Spodosols, but not a requirement (*see* Fig. 10-16).

Properties of Spodosols. Spodosols have solums that are very acidic throughout. They have a low cation exchange capacity (except where humus has accumulated) and low base saturation percentages. The base saturation of some horizons is frequently less than 10 percent. Spodosols have a limited capacity to store water and are naturally infertile for most

Figure 10-13
Typical Aquoll-Udoll landscape in central Illinois with corn and soybeans as the major crops.

crops. Properties of a typical Spodosol are presented in Table 10-6 and Color Plate 1.

Land Use on Spodosols.

The unsuitability of Spodosols for agriculture in colonial New England was described by C. L. W. Swanson, the former chief soil scientist of Connecticut:

The virgin soil under a long-established forest is not always good. When the settlers cleared New England forests 300 years ago, the topsoil they found was only 2 to 3 inches thick. Below this was sterile subsoil and when the plow mixed the two together, the blend was low in nearly everything a good soil should have. It was not the lavish virgin soil of popular fancy. Such soil could not support extractive agriculture which takes nutrients out of the soil and does not replace them. Many New England lands that were treated in this way soon went back to forest.

Similar conditions existed in other places, such as the Lake States. Large acreages of Spodosols were settled by farmers after logging removed the timber; however, low soil fertility and droughtiness caused many farmers to abandon the land (*see* Fig. 10-17). The first bulletin published by the Michigan Agricultural Experiment Station was devoted to solving the problems of farmers on the "sand plains" of north-central Michigan. Today, the northern Spodosol regions are characterized generally by sparse farming, but in localized areas intensive production of fruits and vegetables occurs. The cool summers of the region attract many summer tourists. Many cities owe their existence to mining and lumbering.

Many Spodosols have developed from sandy parent material of marine origin along the southeastern coast of the United States and on the Florida peninsula. The use of modern technology has resulted in a successsful use of these soils for vegetable and cattle production.

Alfisols

The moister Mollisols (Udolls) occur in humid regions, where trees are the natural vegetation. Many theories have been advanced to account for the extensive grasslands that exist in Iowa and Illinois.

Table 10-6
Some Properties of the Horizons of a Spodosol (Orthod)

Depth, cm	Horizon	Cation Exchange Capacity, mEq/100 grams	Percentage Base Saturation	pH	Percentage Organic Matter	Percentage Clay
5–0	O2	—	—	3.6	45.6	—
0–10	E	7.1	10	3.8	0.8	2
10–23	Bhs	14.3	4	4.4	7.2	4
23–38	Bs	4.1	22	4.8	2.0	3
38–71	BC	6.5	9	5.2	0.8	2
71–127	C	4.6	4	5.0	0.1	7

Adapted from *Soil Survey Laboratory Memorandum* 1, 1952. Soil is profile No. 39, Worthington loam from Coos County, New Hampshire.

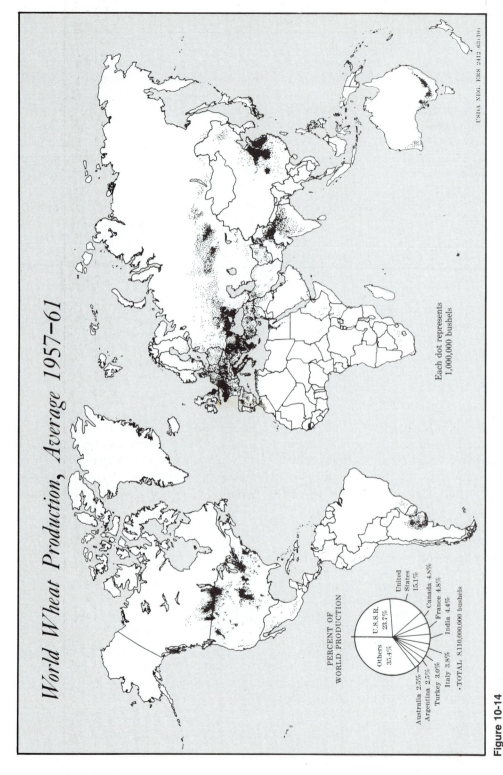

World Wheat Production, Average 1957–61

Each dot represents
1,000,000 bushels

PERCENT OF
WORLD PRODUCTION

U.S.S.R. 23.7%

United States 15.1%

Canada 4.8%

France 4.8%

India 4.4%

Italy 3.8%

Turkey 3.0%

Argentina 2.5%

Australia 2.5%

Others 35.4%

TOTAL 8,110,000,000 bushels

USDA NEG. ERS 2412-63(10)

Figure 10-14
World production of wheat and corn. The Soviet Union is the major producer of wheat and the United States is the major producer of corn. This reflects the large acreages of drier and cooler Mollisols in the Soviet Union and the large acreages of more humid and warmer Mollisols in the United States.

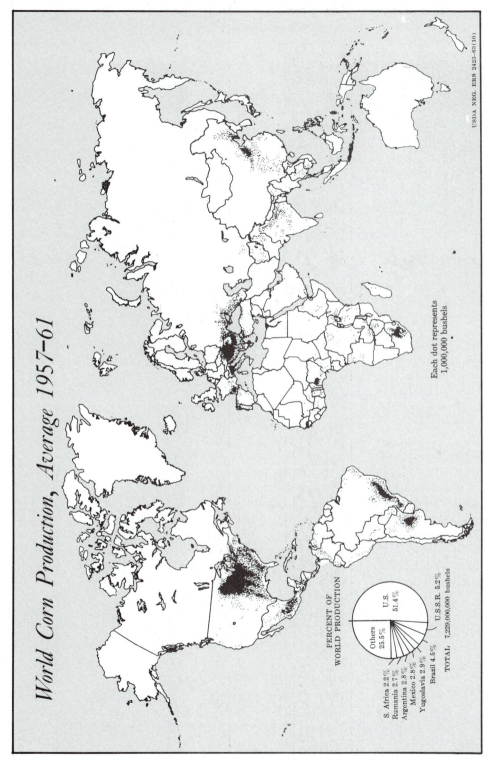

World Corn Production, Average 1957–61

Each dot represents
1,000,000 bushels

PERCENT OF
WORLD PRODUCTION

U.S.
51.4%

Others
25.5%

S. Africa 2.2%
Rumania 2.7%
Argentina 2.8%
Mexico 2.8%
Yugoslavia 2.9%
Brazil 4.5%

U.S.S.R. 5.2%

TOTAL 7,229,000,000 bushels

USDA NEG. ERS 2423–63(10)

Figure 10-14—(continued)

273

Figure 10-15
High rainfall, as found in forest regions, and quartz sand-parent material are common where soils are Spodosols. Forestry and recreation are important land uses. (Photo courtesy of W. Schmidt.)

Along this wetter Mollisol boundary are large areas of soils developed under trees with ochric epipedons, argillic subsurface horizons, and a slightly lower base saturation than nearby Mollisols. These soils are Alfisols.

Properties of Alfisols. Alfisols have argillic horizons and occur in regions where the soil is moist at least part of the year. The requirement for 35 percent or more base saturation in the lower horizons of Alfisols means that bases are being released in the soil by weathering about as fast as they are being leached out. Thus, Alfisols rank only slightly lower than Mollisols for agricultural use, and both kinds of soils are considered high base-status soils. Some physical properties of an Alfisol, *Udalf*, are shown in Figs. 2-8 and 2-11 and Color Plate 1.

Land Use of Alfisols. Alfisols rank third in abundance in the world (*see* Table 10-4). The largest Alfisol area is south of the Sahara Desert in Africa, where *Ustalfs* are common (*see* Fig. 10-1). These

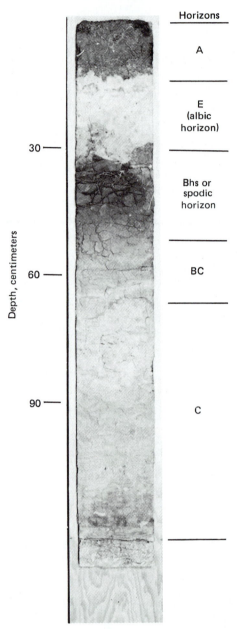

Figure 10-16
A Spodosol profile showing a nearly "white" E horizon and an ortstein—a cemented Bhs or spodic horizon. All mineral horizons have a sand texture, which is a common characteristic of Spodosols.

Figure 10-17
These disintegrating buildings tell the story of forest removal, farming, and then abandonment of land in the Lake States with Spodosols of very low, inherent fertility. It was not that the land was misused, but that high production costs have made it noncompetitive for general farming.

soils have a distinct dry season and, when cultivated, are subject to serious erosion by intense rainy season storms. The major agricultural land uses are cattle raising and mixed farming. Other large areas dominated by Ustalfs are located in east and southeastern Africa, Brazil, and India.

Udalfs have udic soil moisture regimes. They are dominant in the northeastern United States, southeastern Canada, and northwestern Europe. Favorable climate and soils, with fairly good fertility and physical properties, make Alfisols one of the most productive of the soil orders for agriculture (*see* Fig. 10-18). In all areas, farm animals have played a very important role for power, food, and the production of manure, which was carefully conserved and applied to the land to maintain soil fertility. Today, the availability of fertilizers has greatly reduced the farmers dependence on farm animals for manure. Wheat is a major crop of Alfisols in all three areas, but only in the United States is the climate warm enough for good corn production (*see* Fig. 10-14). Potatoes are a major crop in Europe; rice and wheat are important crops in Asia. Cotton is grown on Udalfs developed from loess in western Tennessee and Louisiana in the United States.

When settlers moved across the United States in the eighteenth and nineteenth centuries, the farmers who settled on the Alfisols and Mollisols of the then "Northwest" found good soil. A surplus of agricultural products quickly followed. In 1827, Timothy Flint wrote:

Everyone who was willing to work had an abundance of the articles which the soil produced, far beyond the needs of the country, and it was a prevalent complaint in the Ohio Valley that this abundance greatly exceeded the chances for a profitable sale.

The farmers became agitators for the development of waterways that would per-

Figure 10-18
A landscape in the Alfisol region of the eastern United States, where much of the land is used for general farming and livestock production.

mit shipment of excess food to markets. By the time the population of the United States reached 50,000,000 (about 1880), enough food was being shipped to Europe to feed 25,000,000 people.

A large area of Alfisols with a ustic moisture regime exists in the southwest, particularly in Texas and New Mexico. They occur in natural grasslands, but do not have a sufficient organic matter or dark color in epipedons to be Mollisols. Their use is similar to that of Ustolls in the same region. Alfisols with a cryic temperature regime are *Boralfs* and are extensive east of the Mollisols on the northern Great Plains of the United States and Canada. Spring wheat is a major crop of Boralfs.

Xeralfs are Alfisols with xeric moisture regimes. They are important in the foothills and mountain ranges surrounding the Central Valley of California. Some, such as the San Joaquin on old terraces and alluvial fans, are highly developed (*see* Fig. 2-6). Where slopes are favorable and irrigation water is available, a wide variety of crops are grown, including fruits, nuts, and cotton. A major citrus area of California is located on Xeralfs east of Fresno. A landscape with Xeralfs on the drier lower slopes and *Xerults* (Ultisols with xeric moisture regime) on the higher, more humid slopes is shown in Fig. 10-19.

Ultisols

All soil orders discussed so far have *not* shown evidence of extreme weathering or age. Recent glaciation has played an important role in the existence of vast areas of slightly, or only moderately, weathered soils in the northern part of the northern hemisphere. As one approaches the humid tropics, ancient landscapes with very long periods of weathering, coupled with

Figure 10-19
Grazing is important on Xeralfs on the lower slopes of the Sierra Nevada Mountains in California. On the cooler and more humid upper slopes are Xerults (Ultisols with a xeric moisture regime).

an abundant rainfall and high temperature, have created two unique soil orders found in humid tropics. These orders are Ultisols and Oxisols.

Properties of Utisols. The word Ultisol comes from the latin, *ultimus,* meaning last or, in the case of Ultisol, soils that are the most weathered and that show the ultimate effects of leaching. Ultisols have argillic horizons with low base saturation, it being less than 35 percent in the lower soil horizons. Large amounts of exchangeable aluminum are usually present. They occur in warmer parts of the world where the mean annual soil temperature is 47°F (8°C) or more, and they have a period each year when rainfall is considerably in excess of evapo-transpiration. Few weatherable minerals usually exist in soil to release bases, and trees play a major role in transporting nutrients from the lower part of the soil to the upper part of the solum. Agriculture can be maintained only by shifting cultivation, or by the use of fertilizers.

The gross morphology and horizon sequence is similar to Alfisols (*see* Fig. 10-20). Evidence points to the fact that Ultisols can be Alfisols before they become sufficiently weathered to become Ultisols. Some properties of Ultisols located in the southeastern United States are presented in Table 10-7. Several things can be noted from the data: (1) the clay content shows the development of an argillic horizon, (2) the organic matter content of all horizons, except the very thin A, is quite low, (3) cation exchange capacity is relatively low, expressing the low organic matter content and presence of low cation exchange capacity clays such as kaolinite, and (4) the amount of exchangeable bases and the percentage base saturation are very low, except for the very thin A horizon. The addition and decomposition of residues from vegetation play an important role in maintaining the higher base saturation in the upper part of the solum (A and E horizons). High aluminum saturation is common. Generally, Ultisols have a very low

Figure 10-20
Ultisol on coastal plain of the southeastern United States. Dark-colored (reddish) areas in lower part of profile are discontinuous plinthite. Scale in centimeters.

Table 10-7
Selected Characteristics of an Ultisol

Horizon	Depth, cm	Clay, %	Organic Matter, %	Cation Exchange Capacity	Exchangeable Bases, mEq/ 100 grams	Base Saturation, %
A	0–2	9	5.2	14	4.1	29
E	2–8	8	1.8	6	0.9	15
EB	8–19	8	1.2	4	0.4	9
Bt1	19–30	17	0.4	5	0.4	8
Bt2,3	30–81	35	0.3	10	0.8	8
2C	81–119	48	0.1	13	0.8	6

Data from profile 92 of *Soil Classification, a Comprehensive System,* USDA, 1960. Soil from Jenkins County, Georgia that developed from coastal plain alluvium, under a mixed forest of pines, oak, sweetgum, and blackgum.

fertility level for food crops, but respond well to fertilization because of their desirable physical properties (*see* Color Plate 1 in Chapter 1).

Land Use on Ultisols. Interestingly, the Ohio River is an approximate boundary between the Alfisolic and Mollisolic soils of the North and the Ultisolic soils of the South. An entirely different fortune befell the early settlers who moved south and east, where Ultisols dominated the landscape. The plight of early Virginians and one farmer in particular, Edmund Ruffin, has been aptly described as follows:

Virginians by the thousands, seeing only a dismal future at home, emigrated to newer states of the South and Middle West. The rate of population growth dropped from 38 percent in 1820 to less than 14 percent in 1830, and then to a mere two percent in 1840. As Ruffin himself later described it, "All wished to sell, none to buy." So poor and exhausted were his lands and those of his neighbors that they averaged only ten bushels of corn per acre and even the better lands a mere six bushels of wheat.

Ruffin was determined to make good on the plantation he inherited from his father in Virginia. By chance, he read Sir Humphrey Davy's book, *Elements of Agricultural Chemistry,* published in London in 1813. He was particularly intrigued by Davy's statement to the effect that a soil that contains salts of iron or any other acidic matter could be ameliorated by the application of quick lime. Ruffin experimented with marl and found it beneficial. Not only were Ultisols very low in fertility, the low pH made manure ineffective through pH influence on nutrient availability; aluminum toxicity was a problem in the argillic horizon. Ruffin's discovery was a turning point in the use of very weathered Ultisols, and it led to the move away from soil exhaustion that was followed by abandonment to permanent agriculture. With the long, year-round growing season and rainfall, this region that proved so troublesome at first is now one of the most productive agricultural and forestry regions in the United States. Ultisols of the nearly level coastal plain have an ideal topography for agricultural use. One big problem remaining to be solved is the development of low-cost methods of incorporating lime into subsoils to raise the pH and to reduce the Al toxicity.

One other major area of Ultisols is shown in Fig. 10-1 in southeastern Asia. Many smaller areas of Ultisols occur in regions where Oxisols are common. On many of these Ultisols, shifting cultivation is still practiced.

Oxisols

All soils with oxic horizons belong to the Oxisol order. Oxic horizons are subsurface horizons consisting of a mixture of hydrated oxides of iron and/or aluminum, along with variable amounts of 1:1 lattice clays. Few other minerals exist in oxic horizons, except some that are highly insoluble. More specifically, the oxic horizon has: (1) a thickness of 30 centimeters or more, (2) a cation exchange capacity of 16 milliequivalents or less for each 100 grams of clay, (3) none or only a trace of minerals that can weather to release bases, (4) little if any water-dispersible clay, and (5) diffuse boundaries with adjacent horizons. Oxisols with very diffuse horizon boundaries are shown in Fig. 8-5 and Color Plate 2 in Chapter 1.

Oxisols exist on ancient land surfaces in

humid tropics and contain no or few reserve of bases beyond those on the exchange sites. Agriculture on Oxisol soils uses shifting cultivation similar to that on Ultisol soils. Ultisols and Oxisols commonly exist in the same landscapes. They probably owe at least part of their differences to the tendency for Oxisols to develop from more basic rocks, in which the minerals are more weatherable and have less tendency for silicate clays to form. As a result, Oxisols are richer in iron and have fewer weatherable minerals still remaining in the soil. Soil aggregates are very stable and soils are very erosion resistant. Liming, fertilizers, irrigation, and other management practices have made some Oxisols some of the world's most productive soils, as illustrated in Fig. 10-21.

Two very large areas of Oxisols exist: one in South America (including the Amazon Valley) and the other in Africa (including the Congo). Other large acreages exist in eastern India, Burma, and surrounding regions. The two major suborders are *Orthox* and *Ustox*. Orthox soils tend to straddle the equator and have high rainfall every month. Ustox tend to occur further from the equator and have ustic soil moisture regime (*see* Fig. 10-1). The total acreage is comparable to that of Mollisols, but few people, by comparison, live on Oxisol lands. The great Amazon Basin, the Sahara Desert, and the tundra region all have very low population densities of less than one person per square kilometer or 2 persons per square mile.

Plinthite. A soil horizon associated with some Oxisols (also Ultisols and some Alfisols) is an iron-rich, humus-poor mixture of clay with quartz and other diluents. This material, called *plinthite*, dries irreversibly with repeated wetting and drying. The hard, dried rock-like material is hardened plinthite, and it is commonly referred to as ironstone or laterite.

Plinthite can perhaps form from the concentration of iron in *situ,* but it more likely forms via the influx of iron from some other place. Plinthite formation is characteristically associated with a fluctuating water table in areas with at least a short dry season. Reduced iron moves in the water and precipitates on encounter with a good supply of oxygen. Plinthite normally forms as a subsurface layer that is saturated at some season, but it also forms at the base of slopes where water seeps out. Plinthite is soft to the extent that it can be cut with a spade and readily mined. On exposure to the sun and repeated cycles of wetting and drying, the material hardens to a brick-like material, which is used as a building material (*see* Fig. 10-22).

There is a great diversity in plinthite, ranging from continuous thick layers to thin, discontinuous nodular forms. Small amounts of nodular plinthite or deeply buried, continuous layers below the root zone have little influence on plant growth. Where continuous plinthite is exposed by

Figure 10-21
Pineapple production on Oxisols in Hawaii.

Figure 10-22
Mining plinthite in Orissa state, India. On drying, the bricks harden and are used for construction purposes.

erosion, drying and hardening can seriously limit plant growth. As long as the layer remains moist and soft, and is below the root zone, soils can be effectively used for a wide variety of crops. In some cases, holes are dug in hardened plinthite and then filled with soil to grow high-value tree crops. It is estimated that hardened plinthite (ironstone or laterite) occurs on 2 percent of the land in tropical America, 5 percent in Brazil, 7 percent in the tropical part of the Indian sub continent, and 15 percent for sub-Saharan West Africa.

Vertisols

Vertisols are mineral soils that: (1) are over 50 centimeters thick, (2) have 30 percent or more clay in all horizons, and (3) have cracks at least 1-centimeter wide to a depth of 50 centimeters (unless irrigated) at some time in most years. Conditions that give rise to the development of Vertisols are: (1) parent materials high in, or that weather to form, large amounts of montmorillonitic (expanding) clay and (2)

a climate with a wet and dry season. The typical vegetation in natural areas is grass or herbaceous annuals, although some Vertisols support drought-tolerant woody plants (see Color Plate 1).

Properties of Vertisols. The central concept of Vertisols is one of soils that crack widely in dry seasons. After the cracks develop in the dry season, surface soil material sloughs off into cracks. The soil rewets in the wet season from water that quickly runs into the cracks, and it is held in the soil by impermeable underlying layers. Repeated drying or rewetting periods cause a "humping up" of areas between the cracks to produce a microrelief called *gilgai*. Repeated cycles of expansion and contraction cause a gradual inverting of soil; thus, they are called Vertisols. Expansion and contraction in subsoil, with wetting and drying, produces shiny ped surfaces, called *slickensides* (*see* Fig. 10-23). The expansion and contraction causes a misalignment of fence and telephone posts. Pipelines may be broken

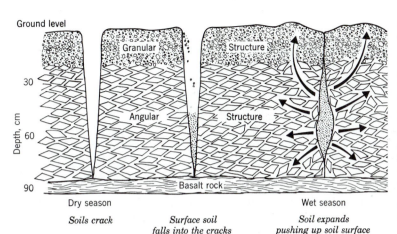

Figure 10-23
Schematic drawing showing the formation of Vertisols. From left to right: (1) Cracks develop in dry season, (2) Loose material falls into cracks, and (3) Wetting of the soil in the wet season causes expansion and movement of soil in the lower part of the soil to produce angular or wedge-shaped peds with shiny surfaces (slickensides) and a microrelief called gilgai. (Adapted from Boul, 1966.)

and road and building foundations destroyed (*see* Fig. 10-24). G. W. Olson, a soil scientist at Cornell University, found that the Mayans in Guatemala avoided Vertisol areas in the construction of temples.

Some properties of Houston clay from the Blackland prairie region of Texas are presented in Table 10-8 to illustrate some of the important properties of Vertisols relative to their nature and use. There is no evidence of clay migration, and the content of clay is very high in all horizons. All horizons have a clay texture. The high content of expanding clay makes the soil very sticky when wet and very hard when dry. Hydraulic conductivity of the soil is very low when the soil is wet. Organic matter decreases gradually with increasing soil depth. Lime was present in the parent material, and the large amount of

Figure 10-24
Landslide on Vertisols produced by wet soil. Note that pipes are on top of ground to prevent breaking caused by expansion and contraction of soil.

Table 10-8

Some Properties of Houston Clay—a Vertisol

Horizon	Depth, cm	Clay, %	Organic Matter, %	CaCO₃, %	Cation Exchange Capacity, mEq/100 grams
A1	0–46	58	4.1	17	64
A2	46–100	58	2.1	20	58
AC	100–152	58	1.0	26	53
C	152–198	59	0.4	32	47

Table is based on Kunze and Templin, 1956. Average of 5 profiles.

lime still remaining in the upper soil horizons is evidence of the closed-system nature of the soil and the limited opportunity for any soluble material to be leached out of the bottom of the profile. The high cation exchange capacity reflects the high content of smectite clay.

Land Use on Vertisols. Vertisols are widely distributed throughout the world between 45 degrees north and south latitude (*see* Fig. 10-1). The three largest areas of Vertisols in the world are in Australia (70,000,000 acres), India (60,000,000 acres), and the Sudan (40,000,000 acres). Agriculturally, the soils have a great potential where power tools, fertilizers, and irrigation are available (*see* Fig. 10-25). Vertisols are high base-status soils. The natural fertility level can be considered quite high, although the use of nitrogen and phosphorus is beneficial. Tillage of the soil is difficult with primitive tillage tools. The blacklands of Texas and Alabama are some of the best agricultural lands in the United States. Worldwide, Vertisols are used mainly for cotton, wheat, corn, sorghum, rice, sugar cane, and pasture.

Histosols

Organic soils are classified as Histosols. Most Histosols are recognized by: (a) a histic epipedon, (2) saturation with water at least 30 consecutive days a year, and (3) at least 20 percent organic matter. Alaska and Minnesota are two states in the United States that have large areas of these soils. Perhaps, the most famous area of Histosols in the United States is the Everglades of Florida. Histosols, however, are found throughout the world. Their

Figure 10-25

Cotton growing on Vertisols (Black Cotton soils) on the Deccan Plateau of India. Note the large cracks and shiny slickenside surfaces exposed in the soil pit.

Figure 10-26
Histosol or organic soil landscape. Crop on the left is grass sod to be used for landscaping. The lightness of the soil makes it ideal for sod production, but this also contributes to its susceptibility to wind erosion. Note the windbreak of trees at the far end of the field.

total extent is less than 1 percent of the land surface of the world (*see* Table 10-4).

Development of Histosols.
Histosols develop where the soil is saturated ranging from at least 1 month each year to continuous saturation. The characteristics of Histosols depend primarily on the na-

ture of the vegetation that was deposited in water and the degree of decomposition. In relatively deep water, remains of algae and other aquatic plants give rise to highly colloidal material that shrinks greatly on drying. As the lake gradually fills, rushes, wild rice, water lilies, and similar plants flourish. The partially decayed remains of these plants are less slimy and colloidal. Gradually, sedges, reeds, and eventually grasses are able to grow. Peat from such plants is much more fibrous than that produced from plants growing in deeper water. Shrubs and trees follow, in time, and produce a woody type of peat. Changes in water depth may cause a recurrence of deeper-water plants; hence, layers of a more pulpy material may occur over fibrous peat and the like. The following plant succession in the filling of a Minnesota lake has been suggested by Soper:

1. Stonewort: waterweed stage.
2. Pondweed: water lily stage.
3. Rush: wild rice stage.
4. Bog: meadow stage.

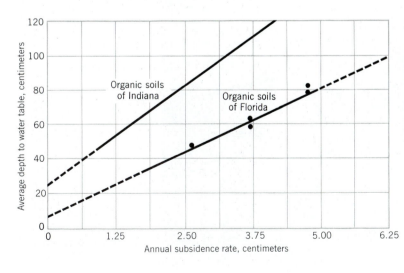

Figure 10-27
This shows that the subsidence rate for organic soils depends on the depth to the water table; the lower the water table, the greater the soil loss. Subsidence is greatest in Florida, where temperature is highest.

5. Bog: heath stage.
6. Tamarack: spruce stage.
7. Pine association.

Land Use on Histosols. When organic soils are drained in such a way as to remove excess water rapidly, yet maintaining the water level at a relatively shallow depth, they may be used for very intensive types of crop production. In the northern United States, these soils are used for the production of onions, celery, mint, potatoes, cabbage, cranberries, carrots, and other root crops. Corn is produced to some extent, and considerable areas are used as pasture. Late spring or summer and early fall frosts are the greatest hazard to crop growth in the temperate region. Other hazards include fires and wind erosion (*see* Fig. 10-26). A great variety of special crops is grown on the organic soils of the South and East. Special methods of tillage, coupled with careful

application of fertilizer, are required to bring these soils to their highest state of productivity. Drainage is required. This results in subsidence because of oxidation and dehydration of organic matter. Comparative rates of subsidence in Florida and Indiana are shown in Fig. 10-27. Subsidence not only results in an eventual loss of Histosols for crop production, but also creates engineering problems (*see* Fig. 10-28).

Soil Classification Categories Below the Suborders

The discussion of soil orders and suborders permitted an introduction to soil geography on a worldwide basis. Within each suborder, there is much diversity and a need for a more precise definition and classification to permit wise usage. These additional categories will be considered next.

Figure 10-28
Organic soil being removed and replaced with sand to create a stable roadbed.

Great Groups and Subgroups

Suborders are divided into great groups. Great group names are coined by prefixing one or more additional formative elements to the appropriate suborder name. The prefixes are used to indicate the presence or absence of certain diagnostic horizons. The formative elements, with their meanings and connotations, are shown in Table 10-9. As an example, a Fragiaqualf is a wet Alfisol with a fragipan. Subgroup names indicate to what extent the central concept of the great group is expressed. A typic Fragiaqualf is a soil that is typical for the Fragiaqualf great group.

Table 10-9
Formative Elements for Names of Great Groups

Formative Element	Connotation
acr	Extreme weathering
agr	Agric horizon
alb	Albic horizon
and	Ando-like
anthr	Anthropic epipedon
aqu	Wetness
arg	Argillic horizon
calc	Calcic horizon
camb	Cambic horizon
chrom	High chroma
cry	Cold
dur	Duripan
dystr, dys	Low base saturation
eutr, eu	High base saturation
ferr	Iron
frag	Fragipan
fragloss	See *frag* and *gloss*
gibbs	Gibbsite
gloss	Tongued
hal	Salty
hapl	Minimum horizon
hum	Humus
hydr	Water
hyp	Hypnum moss

Table 10-9 (continued)
Formative Elements for Names of Great Groups

Formative Element	Connotation
luv	Illuvial
med	Temperate climate
nadur	See *Natr* and *Dur*
natr	Natric horizon
ochr	Ochric epipedon
pale	Old development
pell	Low chroma
plac	Thin pan
plagg	Plaggen horizon
plinth	Plinthite
quartz	High quartz
rend	Rendzina-like
rhod	Dark-red colors
sal	Salic horizon
sider	Free iron oxides
sombr	A dark horizon
sphagn	Sphagnum-moss
torr	Usually dry
trop	Continually warm
ud	Humid climates
umbr	Umbric epipedon
ust	Dry climate, usually hot in summer
verm	Wormy, or mixed by animals
vitr	Glass
xer	Annual dry season

Family and Series

Families indicate features that are important to plant growth, such as texture, mineralogical composition, or temperature. Series get down to the individual soil, and the name is that of a natural feature or place near where the soil was first recognized. Familiar series names include Amarillo, Carlsbad, and Fresno; they obviously refer to soils located in Texas, New Mexico, and California, respectively.

Bibliography

Alexander, L. T., and J. G. Cady, "Genesis and Hardening of Laterite in Soils," *USDA Tech. Bull. No. 1282*, Washington, D.C., 1962.

Anonymous, "Fallout in the Food Chain," *Time*, September 13, 1963, p. 63.

Buol, S. W., "Soils of Arizona," *Ariz. Agr. Exp. Sta. Tech. Bull., 117*, 1966.

Dudal, R., and D. L. Bramao, "Dark Clay Soils of Tropical and Subtropical Regions," *FAO Agr. Dev. Paper, 83*, 1965.

Foth, H. D., and J. W. Schafer, *Soil Geography and Land Use*, Wiley, New York, 1980.

Kellogg, C. E., "Soils," *Sci. Am., 183:*30–39, July 1950.

Kunze, G. W., and E. H. Templin, "Houston Black Clay, the Type Grumusol: II. Mineralogical and Chemical Characterization," *Soil Sci. Soc. Am. Proc., 20:*91–96, 1956.

Lotspeich, F. B., and H. W. Smith, "Soils of the Palouse Loess: I. The Palouse Catena," *Soil Science, 76:*467–480, 1953.

Oakes, H., and J. Thorp, "Dark-Clay Soils of Warm Regions Variously Called Rendzina, Black Cotton Soils, Regur and Tirs," *Soil Sci. Soc. Am. Proc., 15:*347–354, 1951.

Rourke, J. D., *Soils of the World—Probable Occurrence of Orders and Suborders*, USDA, Washington, D.C., May 1968.

Ruffin, Edmund, *An Essay on Calcareous Manures*, Belknap Press, Cambridge, Mass., 1961. Original book published in 1832.

Sanchez, P. A., and S. W. Buol, "Soils of the Tropics and the World Food Crisis," *Science, 188:*598–603, 1975.

Smith, H. N., *Virgin Land*, Vintage, New York, 1957.

Soil Survey Staff, *Soil Survey Laboratory Memorandum I*, USDA, Washington, D.C., 1952.

Soil Survey Staff, *Soil Classification, A Comprehensive System*, USDA, Washington, D.C., 1960.

Soil Survey Staff, *Soil Taxonomy*, USDA Handbook 436, Washington, D.C., 1975.

Soper, E. K., "The Peat Deposits of Minnesota," *Minn. Geol. Sur. Bull., 16*, 1919.

Stephens, John C., "Drainage of Peat and Muck Lands," in *Water*, USDA Yearbook, Washington, D.C., 1955, pp. 539–557.

Swanson, C. L. W., "The Road to Fertility," *Time*, January 18, 1954.

Tuan, Yi-Fu, *China*, Aldine, Chicago, 1969.

Young, A., *Tropical Soils and Soil Survey*, Cambridge, London, 1976.

CHAPTER 11

SOILS AND MINERAL NUTRITION OF PLANTS

The growth and development of plants are determined by numerous factors of soil and climate, and by factors inherent in plants themselves. Some of these factors are under the control of human beings, but many are not. People have little control over air, light, and temperature, for example, but they can influence the supply of nutrients in soil. Anyone dealing directly with the growth of plants is particularly concerned with their nutrient requirements. The emphasis in this chapter will be on the amounts and forms of nutrients in soils and their uptake by plants.

The Essential Elements or Nutrients

If a soil is to produce crops successfully, it must have, among other things, an adequate supply of all the essential elements or nutrients. Not only must required nutrient elements be present in forms that plants can use, but there should also be a rough balance between them in accordance with the amounts needed by plants. If any of these elements is lacking or if it is present in improper proportions, normal plant growth will not occur.

Characteristics of an Essential Element

For many centuries, people knew that substances such as manure, ashes, and blood had a stimulating effect on plant growth. The effect was found to result basically from the essential elements contained in the materials. As recently as 1800, however, it was not known which elements removed from the soil were indispensable. Discovery of chemical elements and techniques for their determination were prerequisites in determining which nutrients were essential for plant growth. Two criteria commonly used in establishing the essentiality of a plant nutrient are: (1) its necessity for the plant to complete its life cycle, and (2) its direct involvement in the nutrition of the plant, apart from the possible effects in correcting some unfavorable condition in the soil or culture medium.

Concentrations of Essential Elements in Plants and Functions

The elements generally required by plants are divided into two groups, based on the *amount* that plants require. *Macronutrients* are *necessary* in relatively large amounts, usually over 500 parts per million in a plant. *Micronutrients* are necessary only in extremely small amounts, usually less than 50 parts per million in a plant. A list of the macro- and micronutrients and their major roles in plant growth are given in Table 11-1.

Quantities of Nutrients in Crops

The nutrient content of several crops is given in Table 11-2. The quantities of elements do not represent the total quantities that crops require during growth; they represent, instead, the quantities contained in harvested material. Roots and other portions of a plant that may not be harvested require considerable quantities of nutrients.

Nutrient Deficiency Symptoms

When plants are starving for any particular nutrient, characteristic symptoms usually appear on these plants. If crops are not vigorous and healthy, it is important to know and understand the cause. If the unhealthy appearance is due to disease, it may be possible to save the crop by spraying. If it is a nutrient deficiency, fertilizers may be applied as a top-dressing on soil or on foliage, as a spray, in time to save the crop. These deficiency symptoms appear only when the supply of a particular element is so low that the plant can no longer function normally.

Nutrient Mobility in Plants and Deficiency Symptoms

Translocation of nutrients within a plant is an ever-continuing process. In this regard, there is a considerable difference in the mobility of various nutrients. When a shortage of a mobile nutrient occurs, it is removed from the older, first-formed tissues and translocated to the growing points. This causes symptoms to appear on the lower leaves. Nitrogen is very mobile in plants, and deficient plants have yellow-colored lower leaves and green upper leaves. Other nutrients that are mobile in a plant include phosphorus, potassium, and magnesium. Those nutrients with a limited mobility that produce symptoms on new leaves or growing points include calcium, boron, iron, copper, and manganese (*see* Fig. 11-1).

It should be pointed out that nutrient deficiency symptoms are not always easily diagnosed. Some of them might be mistaken for discoloration or abnormal char-

Table 11-1

Essential Mineral Elements and Role in Plants

Element	Role in Plants
Macronutrients	
Nitrogen (N)	Constituent of all proteins, chlorophyll, and in coenzymes, and nucleic acids.
Phosphorus (P)	Important in energy transfer as part of adenosine triphosphate. Constituent of many proteins, coenzymes, nucleic acids, and metabolic substrates.
Potassium (K)	Little if any role as constituent of plant compounds. Functions in regulatory mechanisms as photosynthesis, carbohydrate translocation, protein synthesis, etc.
Calcium (Ca)	Cell wall component. Plays role in the structure and permeability of membranes.
Magnesium (Mg)	Constituent of chlorophyll and enzyme activator.
Sulfur (S)	Important constituent of plant proteins.
Micronutrients	
Boron (B)	Somewhat uncertain, but believed important in sugar translocation and carbohydrate metabolism.
Iron (Fe)	Chlorophyll synthesis and in enzymes for electron transfer.
Manganese (Mn)	Controls several oxidation-reduction systems, formation of O_2 in photosynthesis.
Copper (Cu)	Catalyst for respiration, enzyme constituent.
Zinc (Zn)	In enzyme systems that regulate various metabolic activities.
Molybdenum (Mo)	In nitrogenase needed for nitrogen fixation.
Cobalt[a] (Co)	Essential for symbiotic nitrogen fixation by *Rhizobium.*
Chlorine (Cl)	Activates system for production of O_2 in photosynthesis.

*Compiled from many sources.

[a]Not essential for all vascular plants according to the definition of an essential element by Arnon.

acteristics produced by diseases; they may be due to a deficiency of some other element or factor of plant growth. These symptoms have become a valuable aid in determining the need for certain nutrients, especially when used in conjunction with tissue and soil tests.

Nutrient Uptake from Soils

During seed germination, nutrients are supplied from the supply contained in the plant seed. Nutrients in the seed are eventually depleted, and the plant becomes dependent on the soil for nutrients. As roots

Table 11-2
Approximate Pounds per Acre of Nutrients Contained in Crops

Crop	Acre Yield	Nitrogen	Phosphorus Potassium as P	as K	Calcium	Magnesium	Sulfur	Copper	Manganese	Zinc
Grains										
Barley (grain)	40 bu.	35	7	8	1	2	3	0.03	0.03	0.06
Barley (straw)	1 ton	15	3	25	8	2	4	0.01	0.32	0.05
Corn (grain)	150 bu.	135	23	33	16	20	14	0.06	0.09	0.15
Corn (stover)	4.5 tons	100	16	120	28	17	10	0.05	1.50	0.30
Oats (grain)	80 bu.	50	9	13	2	3	5	0.03	0.12	0.05
Oats (straw)	2 tons	25	7	66	8	8	9	0.03	—	0.29
Rice (rough)	80 bu.	50	9	8	3	4	3	0.01	0.08	0.07
Rice (straw)	2.5 tons	30	5	58	9	5	—	—	1.58	—
Rye (grain)	30 bu.	35	5	8	2	3	7	0.02	0.22	0.03
Rye (straw)	1.5 tons	15	4	21	8	2	3	0.01	0.14	0.07
Sorghum (grain)	60 bu.	50	11	13	4	5	5	0.01	0.04	0.04
Sorghum (stover)	3 tons	65	9	79	29	18	—	—	—	—
Wheat (grain)	40 bu.	50	11	13	1	6	3	0.03	0.09	0.14
Wheat (straw)	1.5 tons	20	3	29	6	3	5	0.01	0.16	0.05
Hay										
Alfalfa	4 tons	180	18	150	112	21	19	0.06	0.44	0.42
Bluegrass	2 tons	60	9	50	16	7	5	0.02	0.30	0.08
Coastal Bermuda	8 tons	185	31	224	59	24	—	0.21	—	—
Cowpea	2 tons	120	11	66	55	15	13	—	0.65	—
Peanut	2.25 tons	105	11	79	45	17	16	—	0.23	—
Red clover	2.5 tons	100	11	83	69	17	7	0.04	0.54	0.36
Soybean	2 tons	90	9	42	40	18	10	0.04	0.46	0.15
Timothy	2.5 tons	60	11	79	18	6	5	0.03	0.31	0.20

Fruits and vegetables										
Apples	500 bu.	30	5	37	8	5	10	0.03	0.03	0.03
Beans, dry	30 bu.	75	11	21	2	2	5	0.02	0.03	0.06
Cabbage	20 tons	130	16	108	20	8	44	0.04	0.10	0.08
Onions	7.5 tons	45	9	33	11	2	18	0.03	0.08	0.31
Oranges (70 pound boxes)	800 boxes	85	13	116	33	12	9	0.20	0.06	0.24
Peaches	600 bu.	35	9	54	4	8	2	—	—	0.01
Potatoes (tubers)	400 bu.	80	13	125	3	6	6	0.04	0.09	0.05
Spinach	5 tons	50	7	25	12	5	4	0.02	0.10	0.10
Sweet potatoes (roots)	300 bu.	45	7	62	4	9	6	0.03	0.06	0.03
Tomatoes (fruit)	20 tons	120	18	133	7	11	14	0.07	0.13	0.16
Turnips (roots)	10 tons	45	9	75	12	6	—	—	—	—
Other crops										
Cotton (seed and lint)	1,500 lbs.	40	9	13	2	4	2	0.06	0.11	0.32
Cotton (stalks, leaves, and burs)	2,000 lbs.	35	5	29	28	8	—	—	—	—
Peanuts (nuts)	1.25 tons	90	5	13	1	3	6	0.02	0.01	—
Soybeans (grain)	40 bu.	150	16	46	7	7	4	0.04	0.05	0.04
Sugar beets (roots)	15 tons	60	9	42	33	24	10	0.03	0.75	—
Sugarcane	30 tons	96	24	224	28	24	24	—	—	—
Tobacco (leaves)	2,000 lbs.	75	7	100	75	18	14	0.03	0.55	0.07
Tobacco (stalks)	—	35	7	42	—	—	—	—	—	—

[a]These values are about the same as kilograms per hectare (multiply pounds per acre by 1.121).
Reprinted from *Plant Food Review*, 1962, pp. 22–25, publication of The Fertilizer Institute.

Figure 11-1
Manganese-deficient bean plants. Note the healthy lower leaves and the light-colored intervein areas of the upper leaves. A progression of more severe symptoms occurs from the bottom to the top of the plant, which is related to the immobility of manganese in plants.

elongate through soil, an increasing amount of nutrients becomes *positionally available* to the plant. As the root system enlarges and ramifies a greater soil volume, there is an increase in the ability of a plant to absorb nutrients from soil. The extent and distribution of roots in soils are discussed in Chapter 1. This discussion will emphasize the processes that are important in the movement of nutrients to root surfaces, the nutrient absorption process, and the factors that affect nutrient absorption from soils.

Role of Mass Flow and Diffusion

Two important points must be kept in mind to understand how plants use nutrients in soils so effectively. As we have noted, root extension through soil continuously exposes roots to new supplies of nutrients (and water). Second, after roots have invaded a soil region, *mass flow* and

diffusion play an important role in moving nutrients over short distances to root surfaces. As water is absorbed, a water potential gradient is established, and water slowly moves to root surfaces. Nutrients dissolved in water are carried along, by mass flow. The amount of nutrients moved to roots via mass flow depends on the amount of water moved to the root and the concentration of nutrients in the water.

The range of concentration for some nutrients in soil water is given in Table 11-3. Calcium concentration ranges from 8 to 450 parts per million. For a concentration of calcium of 8 parts per million in soil water and 2,200 parts per million of calcium in a plant, it would require the plant to absorb 275 times more water than the plant's weight to move the amount of calcium needed to the roots via mass flow. Stated in another way, if the transpiration ratio is 275, and the concentration of calcium in soil water is 8 parts per million, enough calcium will be moved to root surfaces to supply plant needs. Transpiration ratios are more commonly about 500, resulting in the expectation that more calcium is moved to root surfaces by mass flow than plants need.

The situation for phosphorus is very different. Phosphorus concentration in soil solution is usually low, and a transpiration ratio of over 60,000 is sometimes needed to move enough phosphorus to the roots, according to the data in Table 11-3. From this illustration and others that could be drawn from the data, it is evident that some situations exist where a mechanism other than mass flow is needed to account for the movement of nutrients to the roots. This mechanism or process is *diffusion*.

Diffusion includes the movement of nutrient ions through soil water. When there

Table 11-3

Relation Between Concentration of Ions in the Expressed
Soil Solution and Concentration within the Corn Plant

	Concentration, parts per million			Ratio of Corn Plant Content to Lowest and Highest Soil Solution Contents	
	Soil Solution		Corn Plant[a]		
	Low	High	Average	Low	High
Calcium	8	450	2,200	275	4.9
Potassium	3	156	20,000	6,666	128.0
Magnesium	3	204	1,800	600	8.8
Nitrogen	6	1,700	15,000	2,500	8.8
Phosphorus	0.03	7.2	2,000	66,666	278.0
Sulfur	118	655	1,700	155	2.6

Adapted from S.A. Barber, "A Diffusion and Mass Flow Concept of Soil
Nutrient Availability," *Soil Sci.*, 93:39–49, 1962.
Used by permission of the author and The Williams and Wilkins Co.,
Baltimore.
[a]Dry weight basis.

is an insufficient movement of nutrients to the root surface via mass flow, diffusion plays an important role. In these cases, the uptake of the ion reduces the concentration of that ion at the root surfaces, establishing a diffusion gradient outward from the root surface and causing the diffusion of ions toward the root. From the data in Table 11-3, it can be concluded that (1) mass flow usually plays the dominant role in the movement of calcium and sulfur to root surfaces and (2) that diffusion plays the dominant role in the movement of phosphorus to root surfaces. For potassium, magnesium, and nitrogen, it appears that both mass flow and diffusion are important; this depends, of course, on the particular concentration in the soil solution and the transpiration ratio. At certain points along root surfaces, there is an intimate contact with soil surfaces so that ions are exchanged directly from the soil particle to the root surface by a process called *contact exchange*.

The Process of Nutrient Uptake

Two well-established phenomena are known about nutrient absorption by plants. First, metabolic energy is required. If root respiration is curtailed, the net uptake of nutrients is minor, even from a concentrated solution. Second, the process is selective. Plants have a capacity for selectively absorbing certain ions over a wide range of conditions, while effectively excluding others. Any theory of ion absorption must take into account these two phenomena.

Plant roots are more or less surrounded by the soil solution and are in intimate contact with soil particles at many points.

Root cells have an "outer" space into which ions from both the soil solution and from the exchange sites of the soil can diffuse. Diffusion of ions into these spaces is reversible. It occurs without regard for a plant's metabolic activities. It is a passive activity so far as a plant is concerned. This is a prelude to the irreversible transport of ions across a seemingly impermeable membrane that requires an expenditure of energy.

The interior surface of the outer space has binding sites, where carrier molecules are believed to be located. Carriers combine with ions; together, they migrate across a membrane that is impermeable to the ion alone. Once across the membrane, the carrier molecule and ion separate as the ion is deposited in the "inner" space of the cell, which is commonly called the vacuole. The carrier transport requires energy that is derived from respiration, and the process enables the cell to achieve an ionic concentration (in the cell) that may be many times that of the external soil solution (*see* Fig. 11-2).

This theory explains the selective absorption of ions, since the carriers are specific for a given ion or group of ions. Nitrate and phosphate are both anions, but they require separate carriers. In fact, different carriers are required for $H_2PO_4^-$ and HPO_4^{2-}. Calcium and magnesium are transported by different carriers. The carrier for potassium can also transport cesium, while still another can transport sodium and lithium. Anions and cations are absorbed by the same mechanism, but they use different carriers.

The surfaces of leaves and stems, as well as roots, can absorb nutrients. Any exposed plant surface appears to be able to function in this regard. Carbon dioxide absorption by leaves is the major avenue for obtaining carbon. The processes involved in the transport of a nutrient ion

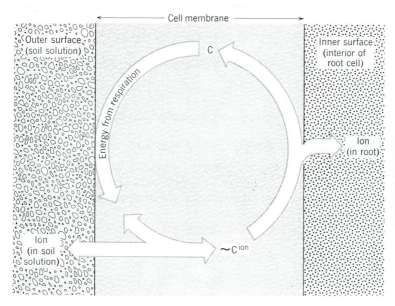

Figure 11-2
A diagrammatic representation of the carrier theory of nutrient uptake. An organic carrier, C, links up with an ion from the soil solution at the exterior surface of the root cell. The ion is carried across the cell's membrane and deposited in the interior of the cell in the roots. (Data from Hanson, 1967. Reprinted from *Plant Food Review*, September 1967, p. 8, publication of The Fertilizer Institute.)

from the soil environment into the root, and its translocation and distribution within the plant, are complex and interrelated.

Factors Affecting Nutrient Uptake

The factors that affect metabolism, and thereby the availability of respiratory energy, will directly affect nutrient uptake. These include the supply of respiratory substrate, temperature, and oxygen supply. The oxygen supply can be significantly altered by management practices. Soil compaction can reduce nutrient uptake through its affect on the availability of oxygen for root respiration. It is interesting that not all nutrients are reduced to the same degree. Lawton established that soil compaction reduced the uptake of potassium more than it did that of phosphorus or nitrogen, and that calcium was least affected in the case of corn. Plants grown in water culture usually obtain maximum growth when air is bubbled through the water.

An increase in the concentration of a nutrient in an environment external to the root will favor an increase in nutrient uptake, when the initial concentration is low.

The moisture content of soil is important, because it influences the rate of movement and diffusion of ions into the outer spaces of root cells. For example, it has been observed that drying the soil reduces phosphorus uptake. This is expected, because phosphorus is so slightly soluble and has such low mobility in soil.

The density and distribution of roots in soil is also important. Roots that enter a zone with a high nitrogen and phosphorus level tend to proliferate. It seems that when phosphate fertilizer is placed with the nitrogen in soil, a greater uptake of phosphorus occurs. The density of roots and the extent of root surfaces become more important as the mobility of a nutrient in soil decreases. Thus, root proliferation is expected to be less important for nitrate uptake than for phosphate uptake. Crops with deeply penetrating root systems generally require less fertilization than those with shallow root systems.

Nitrogen

We live in an ocean of nitrogen, yet the supply of food for human beings and other animals is more limited by nitrogen than any other element. The atmosphere is made up of 79 percent nitrogen (by volume) as inert N_2 gas that resists reacting with other elements to create a form of nitrogen most plants can use. Increasing the soil nitrogen supply for plants consists essentially of increasing the amount of biological fixation or adding fertilizer nitrogen. It is paradoxical that the nutrient absorbed from the soil in greatest quantity by plants is the nutrient most limiting in supply.

The Soil Nitrogen Cycle

During the first billion years of the earth's history, a large amount of reduced nitrogen was released into the atmosphere from the earth's interior. Green plants evolved that produced oxygen and microorganisms oxidized nitrogen to N_2 gas; an atmosphere was formed similar to that existing today. The atmosphere now contains over 99 percent of the nitrogen currently in the earth's nitrogen cycle.

Atmospheric N_2 is characterized by both an extremely strong triple bond between nitrogen atoms and a great resistance to reaction with other elements. Higher plants are incapable of using N_2. The process of converting N_2 into usable forms that vascular plants can use is *nitrogen fixation*. Nitrogen fixation is due to microorganisms (mainly bacteria in soils and algae in water) and to certain atmospheric phenomena, including lightning. Denitrifying bacteria in soils convert available soil nitrogen back into N_2 in a process called *denitrification*. These two processes, fixation and denitrification, are shown as Processes 1 and 5 in Fig 11-3. Fixation and denitrification are approximately equal and responsible for a quasi-equilibrium between nitrogen in the atmosphere and nitrogen in lands and oceans.

A subcycle exists in soil, involving nitrogen in soil organic matter and soil organisms consisting of Processes 2, 3, and 4, as shown in Fig. 11-3. *Mineralization* of organic nitrogen results in available nitrogen as ammonium (NH_4^+). *Nitrification* produces available nitrogen as nitrate (NO_3^-). *Immobilization,* uptake of nitrogen by roots and microorganisms, incorporates the nitrogen back into organic form. The discussion that follows will consider the major processes of the nitrogen cycle in the number sequence presented in Fig. 11-3.

Nitrogen Fixation

There is virtually an inexhaustible supply of nitrogen in the atmosphere, since (at sea level) there are about 77,350 metric tons in the air over 1 hectare (34,500 tons per acre). It takes about 1,000,000 years for nitrogen in the atmosphere to move through one cycle.

Some nitrogen is fixed by electrical discharge (lightning) and other ionizing phenomena of the upper atmosphere. The nitrogen is added to the soil as a compo-

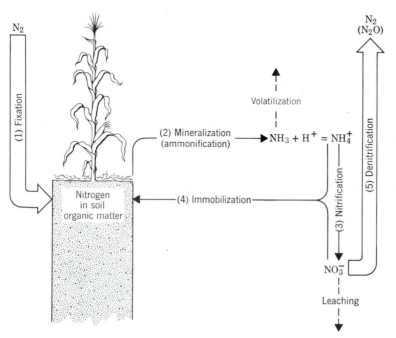

Figure 11-3
The nitrogen cycle is composed of two subcycles. One subcycle consists of the addition of nitrogen to soils by fixation (1) and loss by denitrification (5). The other subcycle consists of cycling nitrogen within the soil involving mineralization (2), nitrification (3), and immobilization (4). Nitrogen can also be lost from the soil by leaching and volatilization.

nent of precipitation. Most nitrogen naturally added to soils is added through biological fixation-symbiotic and nonsymbiotic. Biological nitrogen fixation is a reduction reaction requiring energy supplied by adenosine triphosphate (ATP). Nitrogen-fixing microorganisms contain the enzyme nitrogenase, which combines with a dinitrogen molecule, N_2. Pyruvic acid is the hydrogen donor, and fixation occurs in a series of steps that reduces N_2 to NH_3, as shown in Fig. 11-4. Molybdenum is a part of nitrogenase, and is essential for biological nitrogen fixation. The organisms that fix nitrogen also require cobalt, which is the only known need for cobalt by plants.

Symbiotic Legume Nitrogen Fixation.

Legume plants form a symbiotic relationship with heterotrophic bacteria of the genus *Rhizobium*. The root of the host plant appears to secrete a substance that activates *Rhizobium* bacteria. When the bacteria make contact with a root hair, it curls. An infection thread is formed in the root, through which the bacteria migrate to the center of the root (*see* Fig. 11-5). Once inside the root, bacteria rapidly multiply and are transformed into swollen, irregular-shaped bodies called *bacteroids*. An enlargement of the root occurs and, eventually, a gall or nodule is formed. The bacteroids receive food, nutrients, and probably certain growth compounds from the host plant. The legume host plant is benefited by the N_2 fixed in the nodule. Some of the fixed nitrogen is transported from nodules to various parts of the host plant.

Quantity of Nitrogen Fixed by Legumes.

The quantity of nitrogen added to the soil through the growth of legumes varies

Figure 11-4
Simplified series of steps in nitrogen fixation. (Adapted from Delwiche, in *The Science Teacher,* Vol. 36, No. 3, p. 19, March 1969.)

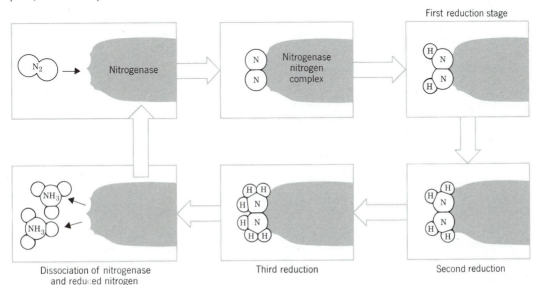

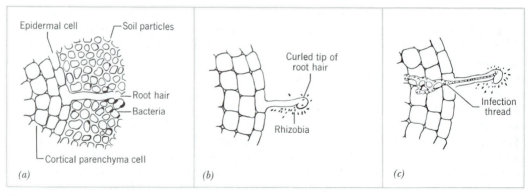

Figure 11-5
Early stages in the formation of a nodule. *(a)* Response of the bacteria to a product of the host plant; organism moves toward root hair. *(b)* Curling of the root hair. *(c)* Early penetration of the infection thread. (From P. W. Wilson, *The Biochemistry of Symbiotic Nitrogen Fixation,* The University of Wisconsin Press, 1940, p. 74. Used by permission of P. W. Wilson and The University of Wisconsin Press.)

greatly, according to conditions such as the kind of legume, the nature of the soil, the effectiveness of the bacteria present, and the seasonal conditions. It appears that the intimate relations existing between nodule bacteria and their host plants are determined mainly by the carbohydrate supply in host plants. Any environmental condition affecting the production of carbohydrates in a plant would automatically affect the quantity of nitrogen fixed by good strains of legume bacteria. It has also been found that symbiotic nitrogen fixation is inhibited by an abundance of available soil nitrogen

One method used to study the quantity of fixed nitrogen is to compare the amount of nitrogen in nodulated and nonnodulated plants. Weber used this method and found that as much as about 160 kilograms of nitrogen per hectare (140 pounds per acre) was fixed when soybeans were grown. This is shown in Fig. 11-6, which also shows that symbiotically fixed nitrogen represented 75 percent of the total nitrogen in the tops of soybean plants. Furthermore, as the

amount of available nitrogen in the soil was increased by the addition of fertilizer, the total amount and percentage of nitrogen fixed by bacteria was markedly decreased. From these and other data, it is reasonable to conclude that 50 to 200 kilograms per hectare (pounds per acre) of nitrogen are commonly fixed per year,

Figure 11-6
Amount of total nitrogen fixed and percent of nitrogen in soybean plants from symbiotic fixation in relation to the application of nitrogen fertilizer. (Data from Weber, 1966.)

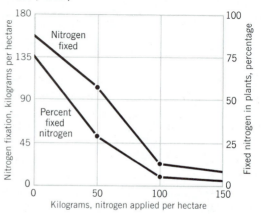

when properly inoculated legume crops are grown. Inoculated alfalfa and clover are capable of fixing so much nitrogen that nitrogen fertilizers do not increase plant growth, but only reduce the amount of nitrogen fixed. Peas and navy beans fix less nitrogen. It is a standard practice to use nitrogen fertilizer to increase yields. Many trees are legumes and fix nitrogen. Black locust is an example of a nitrogen-fixing legume tree.

In actual farm practice, the amount of nitrogen added to a soil by legume bacteria is determined by the methods of disposing of the legume crop. If the crop is turned under as a green manure, the total quantity of nitrogen taken from the air is added. If the crop is cut for hay and sold off the farm, little or no gain is realized; with some legumes, there may even be a net loss of nitrogen. And, if the crop is cut for hay and fed on the farm, about 50 percent of the nitrogen taken from the air by legume bacteria can be returned to the soil, if special care is exercised in handling the manure to prevent loss. It is generally assumed (although not necessarily true for all legumes) that the amount of nitrogen in roots and stubble equals the amount of nitrogen taken from soil. This would mean that the quantity of nitrogen removed in a harvested crop is equal to the nitrogen obtained from the air.

Nonlegume Symbiotic Nitrogen Fixation. Now it is known that many nonlegume species have root nodules and fix nitrogen symbiotically. This means that symbiotic nitrogen fixation is important in the natural range and forest soils, as well as in agroecosystems. Red alder is an example of a nonlegume capable of symbiotic nitrogen fixation. This feature makes red alder a good pioneer species for invading freshly exposed parent ma-

terials and burned-over lands, where soils have little nitrogen-supplying capacity because of low organic matter content. Nonlegumes known to fix nitrogen symbiotically are listed in Table 11-4. The organisms that cause the formation of nodules and the fixation of nitrogen are believed to be actinomycetes. The contribution of nitrogen to the earth's terrestrial ecosystems via symbiotic nonlegume fixation is several times that of herbaceous legumes.

Very recent research in Brazil suggests that some tropical grasses are involved in symbiotic nitrogen fixation. This has raised hopes that the major food crops of the world, such as wheat and corn, may eventually be inoculated and made to fix nitrogen. A large international research effort has been launched to study the problem. If this research leads to the effective inoculation of cereal crops, it will be hailed as one of the greatest scientific advances of recent time. The cost of producing the world's food supply would be reduced, or the same resources would be able to produce more food.

Nonsymbiotic Nitrogen Fixation. There are certain groups of bacteria living in soil, independently of higher plants, that have the ability to use atmospheric nitrogen in the synthesis of their body tissues. Since these bacteria do not grow in association (mutual relationship) with higher plants, they are termed nonsymbiotic. A dozen or more different bacteria have been found that fix N_2 nonsymbiotically. However, the two organisms that have been studied the most belong to the genus *Azotobacter* and the genus *Clostridium*.

Azotobacter are widely distributed in nature. They have been found in soils (of pH 6.0 or above) in practically every locality where examinations have been

Table 11-4

Distribution of Nodulated Nonlegumes

Family	Genus	Species Nodulated	Geographical Distribution
Betulaceae	*Alnus*	15	Cool regions of the northern hemisphere
Eleagnaceae	*Eleagnus*	9	Asia, Europe, North America
	Hippophae	1	Asia and Europe, from the Himalayas to the Arctic Circle
	Shepherdia	2	Confined to North America
Myricaceae	*Myrica*	7	Temperate regions of both hemispheres
Coriariaceae	*Coriaria*	3	Widely separated regions, chiefly Japan, New Zealand, Central and South America, and the Mediterranean region
Rhamnaceae	*Ceanothus*	7	Confined to North America
Casuarinaceae	*Casuarina*	12	Tropics and subtropics, extending from East Africa to the Indian Archipelago, Pacific Islands, and Australia

Reproduced from Stevenson, F. J., "Origin and Distribution of Nitrogen in Soil," *Soil Nitrogen,* Agronomy Monograph #10, 1965, by permission of The American Society of Agronomy.

made. The greatest limiting factor affecting their distribution in soils appears to be soil reaction. These organisms may exist in soils below pH 6.0; however, as a rule, they are not active under such conditions as far as nitrogen fixation is concerned. *Azotobacter* are favored by good aeration, abundant organic matter (particularly of a carbonaceous nature), the presence of ample, available calcium and sufficient quantities of available nutrient elements (especially phosphorus), and proper moisture and temperature relations.

The anaerobic bacteria, *Clostridium,* are much more acid tolerant than most members of the aerobic group; perhaps, for that reason, they are more widespread. It is believed that these organisms can be found in every soil, and that under suitable conditions they fix some nitrogen. It is not necessary that soils be waterlogged for anaerobic bacteria to function. A soil in good tilth may contain considerable areas within the peds favorable for the activities of these anaerobic nitrogen-fixing bacteria.

A question that naturally comes to mind is "How much nitrogen is fixed per acre per year by these nonsymbiotic microorganisms under favorable field conditions?" This question can not be answered definitely because of the many difficulties encountered in making such a measurement under field conditions. From laboratory studies, it is known that the nitrogen fixers use nitrate and ammonium nitrogen, which are normally present in soil. To the extent that soil nitrogen is used, symbiotic fixation of atmospheric nitrogen is inhibited. Furthermore, larger amounts of carbohydrates are used in relation to the amount of nitrogen fixed. It is estimated that only 5 to 20 grams of nitrogen is fixed per 1,000 grams of organic matter decomposed. Considering the amount of decomposing organic matter available in soils, the large number of competing organisms, and the inhibitory effect of the available soil nitrogen, it appears that nonsymbiotic nitrogen fixation is a minor or unimportant factor in crop production. For the natural ecosystem, the small quantity fixed each year over thousands of years is undoubtedly important.

Mineralization

Symbiotically fixed nitrogen is used within plants. It eventually appears as nitrogen in dead tissue of plants and animals, or in animal feces; nitrogen incorporates into soil organic matter and humus. At any one time, over 99 percent of soil nitrogen is in organic matter. About 2 to 4 percent of the total organic nitrogen is mineralized in a single year, which results in one complete turnover of soil nitrogen every 30 to 50 years. Many different kinds of heterotrophic organisms are engaged in organic matter decomposition, with the subsequent mineralization of nitrogen to ammonia (NH_3) (*see* Number 2 of Fig. 11-

3). Mineralization is also called *ammonification*, because the end product is ammonia. Some NH_3 produced at the very surface of the soil escapes by *volatilization*, especially when soil pH is 8 or more.

Most ammonia produced in soil quickly forms ammonium (NH_4^+). There is a strong tendency for ammonium to form because of the presence of hydrogen ions in the soil and the strong bond formed between ammonia and hydrogen from electron sharing (*see* Fig. 11-7). The ammonium ion has a charge of +1 and is available to plants. There is evidence that ammonium is the major form of nitrogen used by plants in forests and rangelands. Ammonium is adsorbed on the cation exchange complex, and it is held against leaching. Some ammonium is fixed in hydrous mica clay minerals, where it occupies space in the crystal lattice normally occupied by potassium.

A hectare furrow slice of soil has about 1,000 kilograms of nitrogen for each percent of organic matter (1,000 pounds per acre furrow slice). This is based on the fact that a furrow slice weighs 2,000,000 kilograms (or 2,000,000 million pounds), and that organic matter is 5 percent nitrogen. If the mineralization rate is 2 to 4 percent per year, there will be approxi-

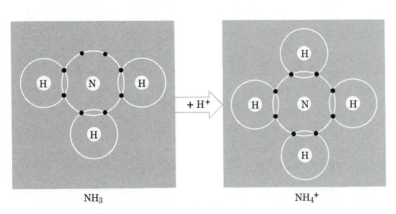

NH₃   + H⁺   NH₄⁺

Figure 11-7
Diagram of formation of ammonium (NH_4^+) showing arrangement of atoms. Valence of nitrogen is −3.

mately 20 to 40 kilograms per hectare of nitrogen mineralized each year for each percent of organic matter in surface soil. Organic matter also exists below the surface plow layer; this organic matter is slowly mineralized as well. It is important to realize that the nitrogen-supplying power of a soil is intimately related to the *organic matter content* and *mineralization rate*. Sandy soils low in organic matter have a poor ability to supply available nitrogen. Desert soils, because of low organic matter content, are likely to supply only 5 percent of the nitrogen needs of a crop produced via irrigation. Organic soils that have been drained, on the other hand, have a maximum potential for supplying nitrogen to plants, since soils are mostly organic matter. Little nitrogen is mineralized in water-saturated organic soils, however, because of oxygen deficiency for aerobic heterotrophic decomposers. In swamps, some plants snare insects to obtain nitrogen.

Nitrification

If ammonium is not absorbed by roots or microorganisms, or is fixed in clay, it is commonly oxidized to nitrate. This process, nitrification, is a two-step process, with nitrite as the intermediate product. Specific autotrophic bacteria are involved. The reactions and organisms are as follows.

1. $NH_4^+ + 1\frac{1}{2}O_2 \xrightarrow{\text{Nitrosomonas}}$
   $NO_2^- + 2H^+ + H_2O + \text{energy}$

2. $NO_2^- + \frac{1}{2}O_2 \xrightarrow{\text{Nitrobacter}}$
   $NO_3^- + \text{energy}$

Nitrification results in the presence of available nitrogen in soils as an anion.

This is of no particular concern, since most plants use nitrate about as readily as ammonium. Paddy rice prefers ammonium. Rice plants in paddies feed mainly on ammonium, because anaerobic conditions inhibit the nitrification of ammonium.

Nitrate is stable in well-aerated soils, and it readily moves with soil water (mass flow) to root surfaces. Nitrate is also readily leached from soils. This has important implications for economy of use and nitrate pollution of ground-water. A product called N-Serve has been developed to inhibit nitrification and to keep nitrogen in its ammonium form in soil to protect it from leaching and denitrification. In well-aerated soils, the ammonium in fertilizers is rapidly converted to nitrate, making the typical plant response to ammonium nitrate fertilizer the same as that of a nitrate fertilizer. Nitrification results in the production of hydrogen ions and is a potential for increasing soil acidity.

Immobilization

Both ammonium and nitrate are available forms of nitrogen for roots and microorganisms. Their use results in the conversion of mineral forms of nitrogen into organic form. The process is *immobilization,* as shown by Number 4 in Fig. 11-3. Immobilized nitrogen is safe in the soil. It is subject to repeated cycling through the nitrogen subcycle in the soil, which involves mineralization, nitrification, and immobilization. Below the root zone or zone of biological activity, nitrate will not be immobilized or denitrified. It is natural for some nitrate to be leached to the water table in well-drained soils of humid regions and to become a component of the ground-water.

Denitrification

Just as it is natural for nitrogen to be added to soils by fixation, it is natural for nitrogen to be lost from soils by denitrification. Denitrification is the reduction of nitrate to gaseous nitrogen and its escape from soil. In fact, in a climax forest or grassland, where the organic matter content of the soil remains about constant from year to year and the amount of nitrogen cycling in the soil remains about constant from year to year, the addition of nitrogen via fixation approaches the losses of nitrogen from the soil by denitrification. Thus, denitrification is one of the most significant processes in the nitrogen cycle, and it accounts for significant losses of nitrogen from soils.

Denitrification is carried out by facultative anaerobic organisms that use nitrate in place of oxygen in respiration, as follows:

$$C_6H_{12}O_6 + 4NO_3^- \rightarrow 6CO_2 + 6H_2O + 2N_2$$
(plus NO, N_2O, and NO_2)

Dentrification occurs under anaerobic conditions in water-saturated soils; it may occur in the interior of moist peds in soils considered to be well drained.

Normally, denitrification is detrimental to agriculture, since nitrogen is lost (as when nitrate fertilizer is applied to poorly aerated soils). Denitrification, however, is an important process that can be used to help prevent excess nitrates from building up in the ground-water of irrigated valleys, where large amounts of nitrogen fertilizer are used. In a large, irrigated valley, excess nitrate is commonly leached to a shallow water table and gradually flows to the lower end of the valley in the ground-water. This takes a considerable period of time. In route, microorganisms gradually denitrify the nitrate, which escapes as nitrogen gas. This enables drainage water from the irrigated valley to have a low value of nitrates, which is not detrimental to other water users downstream. Denitrification has also been used as a means to remove nitrate from manure wastes and sewage effluent.

Human Intrusion in the Nitrogen Cycle

Mineralization of soil organic matter is the major source of available nitrogen for plants. Mineralization of 50 kilograms per hectare (pounds per acre) of nitrogen per year is realistic for many soils. It is obvious that the natural sources of nitrogen in soil are small compared to the needs of a corn crop (*see* Table 11-2). The United States produces over 50 percent of the corn produced throughout the world; this is accomplished by using a large amount of nitrogen fertilizer. The balance sheet for the earth's nitrogen cycle in Fig. 11-8 shows that the amount of industrially fixed nitrogen was equivalent to the terrestrial fixation and to about two times that fixed by legume crops. The biosphere now receives annually about 9,000,000 metric tons more nitrogen per year than is being lost. It is obvious that humans have become important intruders in the earth's nitrogen cycle because of industrial nitrogen fixation. The long-time effects of a buildup of nitrogen in the biosphere are unknown; however, the buildup represents potential danger for nitrate pollution of ground-water and eutrophication of lakes. It is important to realize that adding more nitrogen as fertilizer to the soil does not necessarily result in more leaching of nitrate to the water table. This results from the fact that greatly increased plant growth demands

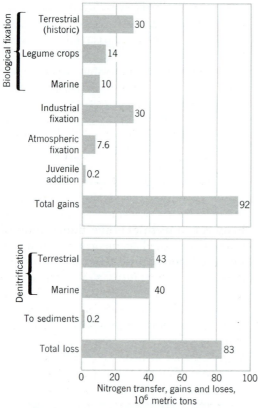

Figure 11-8
Nitrogen balance sheet for the world's biosphere. Nine million metric tons (92-83) more nitrogen is being added than is being removed from denitrification (and loss to sediments). (From C. C. Delwiche, "The Nitrogen Cycle," 1970, *Scientific American*, Inc. All rights reserved.)

more nitrogen uptake. Nitrate losses are increased, however, when the immobilization capacity of the soil is exceeded.

Effect of Nitrogen on Plant Growth

An abundance of nitrogen promotes rapid growth, with a greater development of dark green leaves and stems. Although one of the most striking functions of nitrogen is the encouragement of above-ground vegetative growth, this growth

cannot take place, except in the presence of adequate quantities of available phosphorus, potassium, and other essential elements.

An ample supply of available nitrogen during the early life of a plant may stimulate growth and result in earlier maturity. However, the presence of an excess of nitrogen throughout the growing season frequently prolongs the growth period. This effect is especially significant for certain crops in regions having a short growing season, or in areas where an early fall freeze may do great damage to fruit trees whose season's growth period has been prolonged.

A large supply of available nitrogen encourages the production of soft, succulent tissue that is susceptible to mechanical injury and the attack of disease. Either effect may decrease the quality of the crop. However, the development of softness in the tissues may be desirable or undesirable depending on the kind of crop. For vegetables used for their leaves, pronounced succulence, tenderness, and crispness are desired. Other vegetables and some fruits may have their keeping and shipping qualities impaired when they are grown with an excess of available nitrogen. An excess of nitrogen may encourage lodging in grains, which frequently decreases the quality; a normal amount of nitrogen usually increases plumpness in grains.

Nitrogen Deficiency Symptoms

A nitrogen deficiency is indicated by a light-green to yellow appearance of the leaves. As a rule, older bottom leaves start to turn light-green, then turn yellow at the tip. The entire leaf may turn yellow, even though the tissues are alive and turgid. In a corn plant, the yellowing extends up the mid-rib of the leaf, with the outer edges remaining green the longest (*see* Color Plate 3). A nitrogen-started cucumber may have a small or pointed blossom end; a deficiency of nitrogen may cause the kernels of cereals to become shriveled and light in weight. In fruit trees, the early shedding of leaves, death of lateral buds, poor set of fruit, and development of unusually colored fruit are indications of a lack of nitrogen.

Phosphorus

Phosphorus plays an indispensable role as a universal fuel for all biochemical work in living cells. High energy adenosine triphosphate (ATP) bonds release energy for work, when converted to adenosine diphosphate (ADP). Phosphorus is also an important element in bones and teeth. The relation of phosphorus in soils and plants to animal health, and the extensive occurrence of phosphorus deficiency in grazing animals are well known. Here, the emphasis will be on development of an understanding of the nature of phosphorus in soils and the conditions that control the uptake of phosphorus from soils by plants. Such knowledge helps answer questions like "Is phosphorus likely to be deficient in my garden? If so, what can be done about it?" or "What are the problems of increasing or improving the world food supply that relate to soil phosphorus?"

Phosphorus Cycle in Soils

The earth's crust contains about 0.1 percent phosphorus. On this basis, the phosphorus in a hectare furrow slice of an average soil could produce 50,000 bushels of grain (20,000 bushels per acre). This does not include phosphorus that could be absorbed by roots at depths below the plow layer. Phosphorus, however, commonly limits plant growth. The major problem in phosphorus uptake from soils by plants is the low solubility of most phosphorus compounds and the resulting very low concentration of phosphorus in soil solution at any one time.

Most phosphorus in igneous rocks and soil-parent materials occurs as apatite (*see* Fig. 11-9). Fluorapatite ($Ca_{10}(PO_4)_6F_2$) is the most common apatite mineral. Fluorapatite contains fluorine (F), which contributes to a very stable crystalline structure that resists weathering. The structure is similar to teeth dentine; fluoridation of water is designed to incorporate fluorine into teeth to increase resistance to decay. Apatite weathers slowly and the available phosphorus occurs mostly as $H_2PO_4^-$ in soil solution. The $H_2PO_4^-$ is immobilized by plants and microorganisms; a significant amount of phosphorus in soils is converted into its organic form during soil formation. As with nitrogen, organic phosphorus is mineralized (Process 3) to complete a subcycle of the overall phosphorus cycle.

A major difference between nitrogen and phosphorus cycles in soil is that the available forms of nitrogen (ammonium and nitrate) are relatively stable ions that remain available for plant use. The $H_2PO_4^-$, by contrast, reacts quickly with other ions in the soil solution to become much less soluble or unavailable to plants. Reactions with calcium, iron, and aluminum are the most common. Phosphate is also strongly adsorbed on clay surfaces by replacement of OH from clays (ligand exchange). An equilibrium is established between the concentration of $H_2PO_4^-$ in soil solution and the fixed mineral forms, as represented by Reactions 4 and 5 in Fig. 11-9. The concentration of phosphorus in soil solution is primarily a function of the

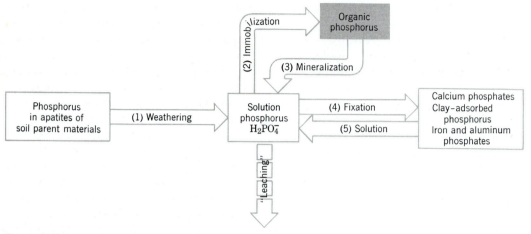

Figure 11-9
Major processes in the soil phosphorus cycle. The availability of phosphorus to plants is determined by the amount of phosphorus in the soil solution.

solubility of the fixed forms of phosphorus. The result is a *very low concentration of phosphorus in soil solution at any one time because of the low solubility of the fixed forms*. In general, there is a decreasing solubility or availability in the order calcium phosphates, clay-adsorbed phosphate, and iron and aluminum phosphates.

Plant Uptake of Phosphorus from Soils

The dominant form of phosphate available to plants is $H_2PO_4^-$. The presence of water is important for phosphorus absorption in soils. Plants absorb about 500 grams of water per gram of growth. The phosphorus in 500 grams of soil water, however, is very inadequate to meet the plant needs, if water and phosphate are absorbed in the ratio that they exist in soils. For analysis, consider:

Transpiration ratio = 500

Phosphorus in plant tissue = 0.3 percent
Phosphorus content of soil solution
$\qquad$ = 0.03 parts per million

It would require 100,000 grams of water to contain the phosphorus needed for 1 gram of plant tissue, calculated as follows:

$$\frac{0.03 \text{ grams phosphorus}}{1,000,000 \text{ grams water}}$$
$$= \frac{0.003 \text{ grams phosphorus}}{X \text{ grams water}}$$

X = 100,000 grams of water

The 100,000 grams of water is 200 times greater than the transpiration ratio. Thus, during the time that plants absorb water, they must in effect deplete the soil solution of its phosphorus and have the phosphorus replaced through 200 cycles. This example shows the great dependence of plants on the diffusion of phosphorus, on the surfaces of soil particles, into and through water films to plant roots, supplying the phosphorus needs of plants. The greater the concentration of phosphorus in soil water, the easier it is for plants to satisfy phosphorus requirements and, in effect, the greater is phosphorus availability in soil.

Effect of pH on Phosphorus Availability

The ions in soil solution are a function of pH. As the pH goes below 5.5, soluble iron and aluminum increase considerably. This causes a fixation of phosphorus as iron and aluminum phosphates. When phosphate reacts with iron and aluminum ions in soil solution, colloidal iron and aluminum phosphates are quickly formed. Over a time of months and years, however, the colloidal forms slowly convert to their crystalline forms, variscite ($AlPO_4.2H_2O$) and strengite ($FePO_4.2H_2O$). Strengite forms more rapidly than variscite as shown in Fig. 11-10.

The conversion of colloidal phosphates to crystalline phosphates is associated with both a marked reduction in the availability or uptake of phosphorus by plants and the effect of phosphorus on plant growth. In an experiment in which Sudan grass was grown in quartz sand, plants receiving crystalline forms of iron and aluminum phosphate grew similarly to plants that

had not received phosphorus. By contrast, the plants receiving colloidal iron or aluminum phosphate grew much better and absorbed more phosphate (*see* Table 11-5). Phosphorus fixation in acidic soils via iron, compared to fixation via aluminum, is more harmful, because strengite forms more quickly than variscite; strengite is less available to plants than variscite.

The best availability of phosphorus is in the range of 6 to 7. Calcium phosphates begin to precipitate at about pH 6.0. Above pH 7.0, the tendency for apatite formation again reduces phosphorus solubility or availability. Increases in pH above 7 also create sufficient OH^- to react with $H_2PO_4^-$ to form $HPO_4^=$ and water, causing the latter form of phosphorus to become the most abundant. Since $HPO_4^=$ is less readily taken up by plants than $H_2PO_4^-$, one can conclude that part of the reduced availability of phosphorus in alkaline soils is due to the presence of hydroxyl ions and the formation of $HPO_4^=$.

Figure 11-10
Transformation of soluble phosphorus into less available forms. Availability decreases with increases in crystallinity. (From Juo and Ellis, 1968. Used by permission of the Soil Science Society of America.)

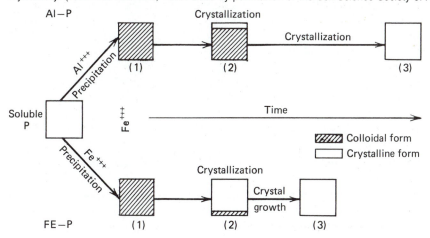

Table 11-5

Yield and Phosphorus Uptake by Sudan grass after Growing 30 Days in Quartz Sand Culture

Form of Phosphorus	Phosphorus Uptake[a]	Yield[a]
Colloidal Al-P	100	100
Colloidal Fe-P	107	96
Crystalline Al-P	22	47
Crystalline Fe-P	13	28
No P	13	27

Adapted from Juo and Ellis, 1968.
[a]Relative to colloidal aluminum phosphate as 100.

Changes in Soil Phosphorus Over Time

The forms in which phosphorus occurs in soils change over time. Some mineral phosphorus is converted to organic phosphorus, as we have already noted. There is a shift over time of the mineral forms to compounds of less solubility. Even the first precipitated calcium phosphates may slowly change from relatively soluble tricalcium phosphate $(Ca_3(PO_4)_2)$ to apatite forms. Freshly precipitated or raw oxides of iron and aluminum phosphate slowly crystallize into strengite and variscite, which expose less surface area and dissolve more slowly. In highly weathered tropical soils, some phosphorus may be encased or coated with iron and aluminum oxides, and protected from solution. This encased or occluded phosphorus is the least soluble of the fixed forms.

Youthful soils, where most phosphorus is in calcium phosphate, have higher available phosphorus than strongly weathered Oxisols. The change in the kinds of phosphorus in soils as a function of soil age is shown in Fig. 11-11. Strongly weathered tropical soils maintain a very low equilibrium concentration of phosphorus in soil solution, because phosphorus is found mainly in minerals of very slow solubility—not because the soils have a low total phosphorus content.

Phosphorus Movement and Loss from Soils

The very low concentration of phosphorus in soil solution at any one time means that leaching removes little phosphorus from soils. The tendency for phosphate ions in soil solution to become fixed makes it difficult for plants to satisfy their phosphorus needs; on the other hand, it decreases leaching of phosphorus from soils, which may produce eutrophication in lakes. Removal of phosphorus from soils into local waters tends to be associated mainly with removal of absorbed

Figure 11-11

Percentage distribution of four forms of phosphorus as related to weathering or soil age. (Adapted from Chang, S. C. and M. L. Jackson, *Journal of Soil Science*, 1958. Used by permission of the Oxford University Press.)

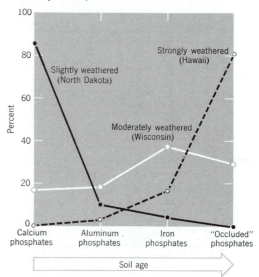

Left is nitrogen-deficient section of corn leaf and cut open section of stem that remains uncolored when treated with diaphenylamine, indicating low nitrate level in plant sap. On right the leaf is dark green and diaphenyl- amine produces a dark blue color indicative of high nitrate level in plant sap.

Nitrogen deficiency on corn. Lower leaves turn yellow along midrib, starting at the leaf tip.

Potassium deficiency on corn. Lower leaves have yellow margins. (Courtesy American Potash Institute.)

Potassium deficiency on alfalfa (and clovers) shows as a series of white dots near leaf margins; in advanced stages entire leaf margin turns white. (Courtesy American Potash Institute.)

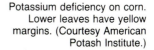

Phosphorus deficiency on corn is indicated by purplish discoloration.

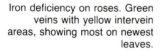

Iron deficiency on pin oak. The leaf veins remain green as the intervein areas lose their green color.

Iron deficiency on roses. Green veins with yellow intervein areas, showing most on newest leaves.

Manganese deficiency on kidney beans. Leaf veins remain green as the intervein areas lose their green color and turn yellow.

Zinc deficiency of navy beans grown on calcareous soil. The small unfertilized zinc deficient plants stand in marked contrast to the taller plants that were fertilized with zinc. A case where zinc fertilization is necessary to produce a crop.

Boron deficiency on sugar beets causes heart rot. Most advanced symptom is on left. (Courtesy American Potash Institute.)

COLOR PLATE 6

Magnesium deficiency on coffee. Veins remain green and intervein areas turn yellow.

Manganese-treated plants in rear showing no deficiency. Plants in foreground are unfertilized and difference in degree of manganese deficiency symptoms is indicative of varietal response to limited soil manganese.

Excess soluble salt symptoms on geranium. The leaf margins turn yellow and become necrotic.

phosphorus on soil particles via erosion than via the leaching of phosphorus in solution to the underground water table.

Phosphorus fertilizers are mainly soluble calcium and ammonium phosphates. They dissolve into soil solution and are subject to fixation. For this reason, phosphorus from fertilizers may move a few centimeters from place of application or, for practical purposes, the fertilizer phosphorus does not move after placement in soil. An obvious consequence is that surface-applied phosphorus is much less effective that phosphorus applied within soil where roots are more abundant, and more water is available for its solution. Another important consequence of phosphorus fixation is that only about 10 to 20 percent of the phosphorus applied to soils is used by the next crop or within the next year. This has caused several times more phosphorus to be added to soils of the United States than is removed in crops.

Anthropic Horizons—Indicators of Human Activity

Anthropic horizons are epipedons that have greatly increased phosphorus content because of human activity. Anthropic horizons are thick and darkly colored like mollic horizons; in addition, however, they have 250 or more parts per million of phosphorus as P_2O_5 that is soluble in dilute acid. Some European farmers have produced anthropic horizons because of the long-continued use of large applications of organic matter, plus materials containing phosphorus. Some of this phosphorus came from bones that were collected on battlefields. No such horizons have been produced in the United States from farming, only an increased phosphorus content.

Human occupancy at a campsite or vil-
lage results in the disposal of refuse containing phosphorus. This is particularly true where the refuse contains many bones. The phosphorus in bones (and teeth) is a calcium phosphate similar to apatite; it accumulates as such in soils or some other insoluble form. Accumulations of phosphorus in soils as a result of human activity makes a determination of the soil phosphorus content an important tool for discovering the locations of ancient campsites, villages, and roads (*see* Fig. 11-12).

Effect of Phosphorus on Plant Growth

The effects of too little or too much phosphorus on plant growth are less striking than those of nitrogen or potassium. It appears to hasten maturity more than most nutrients; an excess stimulates early maturation. Phosphorus deficiency is characterized by stunted plants that have about equally affected root and top growth. Many soils produce forage that is deficient in phosphorus, in terms of the nutritional requirement of animals. Fertilization with sufficient phosphorus to increase the phosphorus content of the forage improves forage quality in these cases.

Phosphorus Deficiency Symptoms

If phosphorus is deficient, cell division in plants is retarded and growth is stunted. A dark-green color associated with a purplish coloration in the seedling stage of growth is a symptom of phosphorus deficiency (*see* Color Plate 4). Later, plants become yellow. Occasionally a pale or yellowish-green color develops, when the lack of phosphorus inhibits the use of nitrogen by a plant. Bronze or purple leaves sometimes are observed at the top of new shoots of phosphorus-starved apple trees.

Figure 11-12
Students removing artifacts from Koster Indian site near the mouth of the Illinois River. During the past 8,500 years, there were 14 periods of occupancy and formation of anthropic horizons. The anthropic horizons are separated by layers of sterile soil representing periods of abandonment and deposition of eroded soil from the slopes above the site.

In the absence of sufficient phosphorus, general maturity of crop and seed formation are usually delayed. Poor pollination is frequently associated with corn that has phosphorus starvation. Perhaps the most characteristic symptom of phosphorus deficiency, among plants in general, is stunted growth.

Potassium

Many soils have an abundance of available potassium, and plants growing on these soils do not respond to potassium fertilizers, even though plants generally use more potassium from soils than any nutrient except nitrogen. This is in stark contrast to what we have just noted, with regard to the general need of nitrogen and phosphorus fertilizers in agroecosystems. Basically, potassium in soils is found in minerals that weather and release potassium ions. The ions are adsorbed on the cation exchange and are readily available for plant uptake. The available potassium accumulates in soils with ustic or

drier soil moisture regimes in the absence of leaching. Such soils are generally neutral or alkaline, do not require lime, and do not need potassium fertilizers, even for high crop yields. Leaching in humid regions removes available potassium and creates a need for potassium fertilizer when moderate or high crop yields are desired. Organic soils are notoriously deficient in potassium, because they contain few minerals that contain potassium. Our discussion of potassium will emphasize the nature of potassium in soils, and factors that affect a soil's ability to meet the potassium needs of plants.

The Potassium Cycle in Soils

The earth's crust has an average potassium content of 2.6 percent. Parent materials and youthful soils could easily contain 40,000 to 50,000 kilograms of potassium per hectare furrow slice (or pounds per acre furrow slice).

The potassium content of soil at depths below the plow layer could be similar.

About 95 to 99 percent of this potassium is in the lattice of the following minerals:

Feldspars
    Microcline $KAlSi_3O_8$
    Orthoclase $KAlSi_3O_8$
Micas
    Muscovite $H_2KAl_3(SiO_4)_3$
    Biotite
        $(H, K)_2(Mg, Fe)_2$
        $Al_2(SiO_4)_3$
Clay
    Hydrous mica $K_2(Si_6Al_2)Al_4O_{20}(OH)_4$

Increasing rate of weathering

Micas weather faster and release their potassium more readily than feldspars, especially biotite. These minerals exist primarily in sand and silt fractions.

During weathering, the potassium ion, K^+, is released into soil solution (*see* Fig. 11-13). Plants absorb potassium as K^+ mainly from soil solution, with a small amount by contact exchange from the cation exchange surfaces. A few kilograms of K^+ exist in soil solution. Up to a few hundred kilograms in the furrow slice of a hectare exist on the cation exchange sites in most mineral soils.

An equilibrium exists between solution potassium and exchangeable potassium, as shown in Fig. 11-13. Consider that weathering is occurring in a soil when plants are dormant. The concentration of potassium in soil solution increases, which forces more potassium onto exchange positions via mass action. During this time, the release of potassium exceeds plant uptake and exchangeable or available potassium increases. During periods of rapid growth, plants may remove potassium from soil faster than it is released by weathering; the equilibrium is shifted to the left. As plants absorb potassium from the soil solution, it dissociates from cation exchange sites in an effort to maintain equilibrium. This sequence of events is typical of the annual changes that occur from winter to summer in available potassium in soils.

An equilibrium also exists between exchangeable and fixed potassium. Fixation occurs by migration of K^+ into vacant positions of the hydrous mica lattice, from which a K^+ had been removed by weathering. Weathering begins at the edges of mineral particles and progresses inward (Fig. 11-14). Along the edges, potassium is

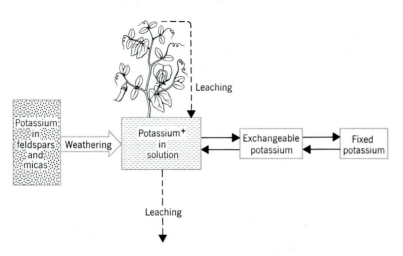

Figure 11-13
The potassium cycle in soils.

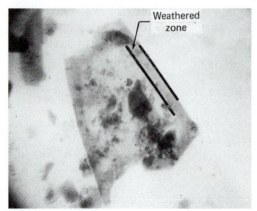

Figure 11-14
Weathered mica silt particle showing loss of potassium from the darkened area around the edges. The loss of potassium results in expansion of the 2:1 (mica) layers. The weathered zone is vermiculite. (Photo courtesy of Dr. M. M. Mortland.)

weathered out, leaving vacant spaces in the lattice; meanwhile, the interior of the particle is still fresh and unweathered. Loss of potassium along the edges removes potassium bridges that hold adjacent layers of the crystals together. Layers separate or expand along the edges. Potassium fixation is the reverse of weathering out of potassium from the lattice (*see* Fig. 11-15).

Fixation and release is a reversible process that is dependent on the concentration of K^+ on the clay surfaces, which in turn is dependent on the concentration of K^+ in soil solution. Complete loss of potassium from between layers causes a complete separation of mineral layers and a loss of potassium-fixing capacity. Potassium fixation in soils conserves potassium that might otherwise be lost by leaching, when potassium release by weathering exceeds plant uptake. Fixation also allows some potassium from fertilization to be temporarily stored in a safe, unavailable position, until plant uptake has reduced

the amount of potassium on the exchange complex. Occasionally, a soil has such high potassium-fixing capacity that most potassium from fertilizer use goes into satisfying the fixing capacity, instead of increasing both the uptake of potassium and the yield. Ammonium ions are similar in size to K^+ and are fixed in the same lattice spaces as potassium.

The discussion thus far has stressed the association between potassium and the mineral components of soil. Apparently, most potassium in plants do not form an integral part of plant and microbial tissue in the way that nitrogen is incorporated into proteins. In fact, a lot of potassium is leached from the leaves of plants during rains. Consequently, organic matter is not a significant source of potassium for plants. Potassium is immobilized, but it does not accumulate in the soil organic fraction. Soils with the lowest available potassium are acid-organic soils. Soils with the highest available potassium tend to be fine-textured soils that are neutral or alkaline.

Forms of Soil Potassium versus Uptake and Plant Growth

Both solution and exchangeable potassium are considered plant-available. Removing the exchangeable potassium also removes the available potassium (solution potassium is also removed), and plants must depend on release of fixed potassium or potassium weathered from minerals. Removing exchangeable potassium from a silt loam soil in Wisconsin reduced the yields of corn (maize) and oats to 62 percent of normal or untreated soil (Table 11-6). The removal of exchangeable and fixed potassium lowered yields to about 20 percent of normal. The addition of 900 kilograms per hectare (800 pounds

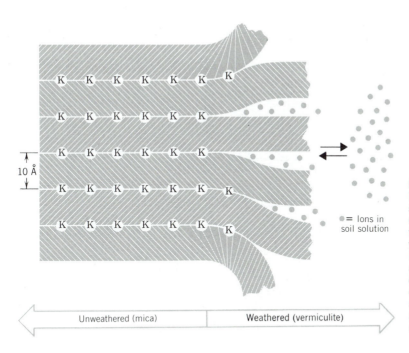

● = Ions in
soil solution

Figure 11-15
Diagram showing the separation and curling of mica layers resulting from the loss of K^+ during the weathering of mica. Fixation of K^+ results in an uncurling and collapse of the layers or a reversal of weathering. Ions in the soil solution diffuse into and out of the expansion space between the layers.

10 Å

Unweathered (mica)   Weathered (vermiculite)

per acre) of potassium in fertilizer to soil, from which exchangeable and fixed potassium had been removed, resulted in corn and oats yields greater than that of the normal soil. The experimental results confirm the validity of the distinctions made in regard to the various forms of potassium in soils and their availability to plants.

The plant uptake data in Table 11-6 show decreases that parallel growth for removal of both solution potassium and exchangeable potassium, plus fixed potassium. Uptake as a result of the potassium fertilizer application resulted in *luxury consumption* of potassium. This is evidence that the fertilizer caused an overabundance of solution and exchangeable potassium, resulting in excessive potassium uptake before most of the fertilizer potassium was dissipated by plant uptake, fixation, or leaching. Luxury consumption of potassium limits the amount of potas-

sium fertilizer that can be applied at one time for the efficient use of fertilizer, and yet maintain a desirable magnesium/potassium ratio in grass forages.

Role of Diffusion in Plant Uptake

As with phosphorus, there is less potassium in soil water than needed for plant growth, if water and potassium are taken up by plants in the same ratio that they normally exist in soils. Diffusion of potassium from cation exchange sites via water films to root surfaces is very important in uptake of potassium from soils by plants. About 90 percent of the uptake was found to be due to diffusion; the remainder was caused by mass flow and root interception in an experiment with soybeans. Root interception can be expected to account for little of the uptake, since only about 1 percent of the surface soil volume is made up of roots, and uptake

Table 11-6
Effect of Different Forms of Potassium on Plant Growth
and Uptake of Potassium

| | Percentage of Untreated Soil | | | |
| | Yield | | Potassium Uptake | |
Soil Treatment	Corn	Oats	Corn	Oats
Normal or untreated soil	100	100	100	100
Exchangeable potassium removed	62	62	41	28
Exchangeable + fixed potassium removed	20	18	23	18
Exchangeable and fixed potassium removed + 900 kilograms of potassium as fertilizer per hectare	149	116	959	452

Adapted from Attoe and Truog, 1945.

from root interception is closely related to root volume.

Effect of Aeration on Potassium Uptake

The supply of oxygen for root respiration can become limited by soil compaction or exclusion of air from soil pore spaces via water. As a consequence, reduced plant growth and reduced nutrient uptake that have been caused by soil oxygen deficiency have frequently been observed. Studies with maize or corn have shown that potassium uptake is more sensitive to poor soil aeration than the other macronutrients. In an experiment where restricted soil aeration was produced by increasing soil moisture content, plant growth was 56 percent of normal; the uptake of potassium was 45 percent of normal. The uptake of nitrogen and phosphorus were 77 and 60 percent, respectively, of normal soil conditions (*see* Table 11-7). When air was forced through

the soils, plant growth was increased and the uptake of potassium was increased more than the uptake of nitrogen and phosphorus.

Potassium Movement and Loss from Soils

Potassium is intermediate between nitrogen and phosphorus, in regard to mobility in soils. Some potassium is leached from soils in humid regions, but the losses do not appear to have any environmental consequences. Many soils have argillic horizons, with a considerable capacity to hold potassium in the exchange and fixed positions. Some potassium leached from surface soil is held in the B horizon and returned to the surface via plant roots. Long-time losses of potassium by leaching result in gradual decreases in the potassium content of soils and the development of soils with a limited supply of available potassium for crops in humid regions. By the time soils become Ultisols and Oxisols,

Table 11-7
Effect of Soil Aeration on Maize (Corn) Growth and Nutrient Uptake

Water content, %	Air Porosity, %[a]	Plant Weight, grams/pot	Nutrient Uptake, % of 15% Soil Water Content		
			N	P	K
15	37	33.7	100	100	100
25	19	30.8	98	76	83
40	0	18.9	77	60	45

Adapted from Lawton, 1945.
[a]Calculated value based on assumed bulk density.

the original potassium minerals have been essentially weathered; deep rooting of trees is important for bringing up potassium from less weathered soil, where the supply of potassium is greater.

Available Potassium in Low Base-Status Soils

There are no hydrous mica potassium-fixing minerals in low base-status soils; sand and silt contain few potassium minerals. These soils are weakly buffered in terms of available potassium. In these soils, the absolute amount of exchangeable potassium and the relative amount are important. Researchers have established a minimal level of 0.10 milliequivalents of exchangeable potassium per 100 grams of soil, and that exchangeable potassium should be equal to 2 to 3 percent of the total exchangeable bases.

Effects of Potassium on Plant Growth

Potassium has a counterbalancing effect on the results of a nitrogen excess. It enhances the synthesis and translocation of carbohydrates, thereby encouraging cell wall thickness and stalk strength. A deficiency is sometimes expressed by stalk breakage or lodging. It also increases the sugar content of sugar beets and sugar canes. The highest dry matter yields of these two crops can be obtained with very high rates of nitrogen fertilization, but the greatest production of sugar results from both moderate nitrogen applications and a sufficient level of available potassium. Root crops such as potatoes also have a high potassium requirement. Less succulent foliage reduces disease and is promoted by good supplies of potassium. There is some evidence indicating that alfalfa is less susceptible to frost injury when it is well fertilized with potassium.

Potassium Deficiency Symptoms

A deficiency of potassium usually shows up as a "leaf scorch" in most plants. Corn indicates a need for potassium by a yellowing of the tips and margins of the lower leaves (see Color Plate 3). This coloration does not move up the mid-rib, as with a nitrogen deficiency, but gradually spreads upward and inward from the leaf tip and edges. When insufficiently supplied with potassium, alfalfa frequently develops a series of white spots near the margin of older leaves (*see* Color Plate 3). Sometimes, this spotting effect is accompanied by a yellowing of the leaf edges; at times, leaf margins turn yellow without the formation of white spots. The edges of leaves finally dry up and curl under. Potato plants indicate a potassium deficiency by a marginal scorch of the lower leaves; frequently, the areas between the veins of potato leaves bulge out, giving a wrinkled appearance. Symptoms for soybeans are the yellowing of leaf margins.

Calcium and Magnesium

There are many similarities between the behavior of calcium, magnesium, and potassium in soils. They are all available as exchangeable cations, and the amount available is importantly related to mineral weathering and degree of leaching. All three are adsorbed as cations, with calcium and magnesium as Ca^{+2} and Mg^{+2}. Some important calcium minerals include feldspars, apatite, calcite, dolomite, gypsum, and amphibole. Important magnesium minerals include biotite, dolomite, augite, serpentine, hornblende, and olivine.

The cations set free in weathering are adsorbed on the cation exchange sites. An equilibrium is established between the exchangeable and solution forms. Diffusion to root surfaces is an important process in uptake from soils. In contrast to potassium, there is no significant fixation into unavailable forms.

The available amount is related to the cation exchange capacity and percentage saturation. Calcium deficiency for plants has occasionally been observed on very acidic soils with low calcium saturation.

Compared to calcium, magnesium is less strongly adsorbed to cation exchange sites, much less exchangeable magnesium exists in soils, and magnesium deficiencies have been observed more frequently. Acidic sandy soils with less than 75 pounds of exchangeable magnesium per acre furrow slice, or 84 kilograms per hectare plow layer, are likely to be magnesium deficient for corn in Michigan. Some soils formed from parent materials rich in serpentine have abundant magnesium for plant growth, a very low calcium saturation, and deficiencies of other nutrients, which produces stunted growth; these soils are referred to as *serpentine barrens*.

A deficiency of calcium is characterized by a malformation and disintegration of the terminal portion of the plant. Calcium is not removed readily from the older tissues that is to be used for new growth when a deficiency occurs. The deficiency symptoms have been established for many plants by the use of greenhouse methods, but they are seldom seen in the field, except in the case of low base-status soils.

Magnesium is a constituent of chlorophyll. As with several other nutrient elements, a deficiency of magnesium results in a characteristic discoloration of leaves. Sometimes, a premature defoliation of the plant results from magnesium deficiency. The chlorosis of tobacco, known as "sand drown," is due to a magnesium deficiency. Cotton plants suffering from a lack of this element produce purplish red leaves with green veins. Leaves of sorghum and corn become striped; the veins remain green, but the areas between the veins become purple in sorghum and yellow in corn (*see* Fig. 11-16). The lower leaves of plants are affected first. In legumes, the deficiency is shown by chlorotic leaves. For coffee, *see* Color Plate 6.

Figure 11-16
Magnesium deficiency symptoms on corn. The veins are green and the intervein areas are yellow.

those of nitrogen, in that both elements undergo changes in oxidation and reduction, are leached from soils as anions, and significant amounts are added to the soil in precipitation.

When coal and oil are burned, sulfur is released into the atmosphere as SO_2. There are many areas where enough sulfur is washed out of the atmosphere by rain to satisfy plant needs. Many studies have shown a relationship between the addition of sulfur to soils in precipitation and the closeness to industrial centers. In fact, so much sulfur is added to some soils that all vegetation nearby has been killed.

Areas in the United States where plant response to sulfur has been obtained are shown in Fig. 11-17. Plants with the greatest need for sulfur include cabbage, turnips, cauliflower, onions, radishes, and asparagus. Intermediate sulfur users are legumes, cotton, tobacco, and alfalfa. Grasses have the lowest sulfur requirements. Sulfur is mobile in plants; its deficiency symptoms are similar to nitrogen. Plants are stunted and light-green to yellow in color.

Sulfur

Sulfur (S) exists in soil minerals, is immobilized into important plant compounds, and eventually accumulates in soil organic matter. Sulfur, similar to phosphorus, is made available in soils via both mineral weathering and mineralization. Plants obtain most of their sulfur from soils as sulfate, (SO_4^{-2}), but some is absorbed through leaves as SO_2. Sulfates are reduced in water-logged soils, to hydrogen sulfide (H_2S gas) and elemental sulfur. Many reactions of sulfur are similar to

Micronutrients

Micronutrients function largely in plant enzyme systems and are required in very small amounts. The factors that determine the amounts available to plants are importantly related to soil conditions and plant species. Iron, for example, is one of the most abundant elements in soils. Some Oxisols are high in iron content, yet need iron fertilization for the profitable production of pineapple. Four of the seven micronutrients are used as cations and three as anions. Cations will be considered first.

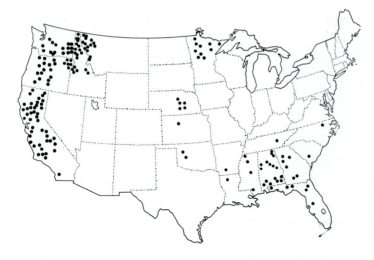

Figure 11-17
Locations where crops
have responded to sulfur
fertilizers. (Data from
Jordan and Reisenauer,
1957.)

Iron and Manganese

Iron (Fe) and manganese (Mn) are weathered from minerals and appear as divalent cations in soil solution; as such, they are available to plants. Reaction with hydroxyl and biological oxidation converts iron and manganese to insoluble forms that are unavailable. Generally, in acidic soils, sufficient Fe^{+2} and Mn^{+2} exist in soil solution to meet the needs of plants. In some very acidic soils, there are toxic concentrations of iron and manganese. Deficiencies are common in alkaline soils, for many plants, because of the insolubility of the hydroxide and oxide forms of iron and manganese. Plants particularly susceptible to iron deficiency include roses, pin oaks, azaleas, rhododendrons; and many fruits and ornamentals. Cereals and grasses, including sugar cane, tend to have a manganese deficiency in alkaline soils.

Deficiency symptoms for iron are striking and are commonly seen on plants growing on calcareous or alkaline soils. Iron-deficient plants have a light-yellow leaf color, which is more evident on younger leaves. The intervein areas are most affected, and the veins retain a darker color. *Iron-chlorosis* is the name given to this condition, as shown in Color Plate 4.

The absence of sufficient manganese dwarfs tomatoes, beans, oats, tobacco, and various other plants. Associated with this dwarfing is a chlorosis of the upper leaves. The veins, however, remain green (*see* Fig. 11-1, and Color Plate 5). The "gray speck" of oats has been attributed to a shortage of this element in some soils.

Copper and Zinc

Copper (Cu) and zinc (Zn) are released in weathering as Cu^{+2} and Zn^{+2}, absorbed by plants, and adsorbed on cation exchange sites. Copper complexes with organic matter, and there is evidence that the complexing can reduce the availability of copper for plants in soils with high organic matter content. Copper availability also decreases with increase in pH. Freshly developed organic soils and leached sandy soils are most likely to be copper-deficient for some plants.

Copper is immobile in plants and deficiency symptoms are highly variable from plant to plant. Gum pockets under the bark and twig dieback occur on fruit trees. Older leaf tips become necrotic in small grains.

Zinc deficiency was first discovered on plants growing on organic soils in Florida. Deficiencies are common on calcareous soils, where high pH reduces zinc availability, and on acidic sandy soils, where zinc has been leached from the soil. Organic soils are also low in zinc reserves and plants are commonly deficient. In some cases, zinc deficiency is caused by high phosphorus fertilization.

Widespread zinc deficiencies occur in the United States on corn, sorghum, citrus and deciduous fruit, beans, vegetable crops, and ornamentals. Pecan rosette, the yellows of walnut trees, the mottle leaf of citrus, the little leaf of the stone fruits and grapes, the white bud of corn, and the bronzing of the leaves of tung trees are all ascribed to zinc deficiencies. In tobacco plants, a zinc deficiency is characterized by a spotting of the lower leaves; in extreme cases, almost total collapse of leaf tissue may occur.

Boron

Most boron (B) in soils is in the mineral tourmaline; it is released in weathering as the borate ion, BO_3^{-3}. The borate ion, or H_3BO_3, is absorbed by plants, and boron accumulates in soil organic matter. Mineral and organic forms of boron are more important in supplying boron to plants. Dry weather that limits decomposition of organic matter in the surface soil has caused a boron deficiency in alfalfa. Fixation at high pH, and leaching of boron from acid soils, results in maximum availability near pH 7.

Many physiological diseases of plants, such as the internal cork of apples, yellows of alfalfa, top rot of tobacco, cracked stem of celery and heart rot and internal black spot of beets, (*see* Color Plate 5), are associated with a deficiency of boron. In sugar beets, boron deficiency appears as a stunting and curling or twisting of the petioles, which is associated with a crinkling of the heart leaves. They have unusually dark-green and thicker leaves, which wilt more rapidly under drought conditions. Older leaves frequently become chlorotic. A rotting of the beets, starting in the crown, may occur. Girdle or canker of table beets is seen as a cracking of the outer skin of beets near the soil surface, followed by a breakdown of the root tissue.

Chlorine

The chlorine requirement of crops is very small, even though chlorine may be one of the most abundant anions in plants. It is unique in that there is little probability that it will ever be needed as a fertilizer. This is due to its addition to the atmosphere from ocean spray and, consequently, the widespread addition of it to soils in precipitation. The presence of chlorine in potassium fertilizer used today is another important manner in which chlorine is added to soils.

Molybdenum

Weathering of minerals releases molybdenum (Mo); it is probably absorbed by plants as the molybdate ion, MoO_4^{-2}. Molybdenum accumulates in soil organic matter, and is adsorbed as an anion, by the clay fraction. In soils with low pH, molybdenum is fixed by its reaction with iron. Liming acidic soils has commonly re-

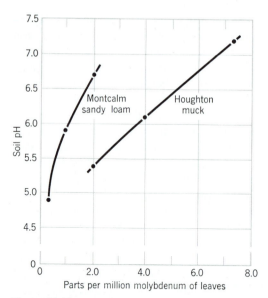

Figure 11-18
Increase in pH due to liming increased the availability of soil molybdenum and increased molybdenum concentration in cauliflower leaves. (Data from Turner and McCall, 1957.)

duced plant molybdenum deficiencies (*see* Fig 11-18). Some calcareous soils in the western United States developed from high molybdenum parent materials, have high levels of available molybdenum, which results in toxic concentrations in forage for grazing animals.

This element is needed for nitrogen fixation in legumes; when it is deficient, legumes show symptoms of nitrogen deficiency. When molybdenum causes other metabolic disturbances in plants, other symptoms are seen. A deficiency of molybdenum in cauliflower causes a cupping of leaves. It is caused by the reduced rate of expansion near the leaf margin, compared to that in the center of the leaf. Leaves also tend to be long and slender, giving rise to the symptom called "whiptail." Interveinal chlorosis, stunting of

plants, and general paleness are also exhibited, depending on the kind of plant.

Bibliography

Allison, F. E., "Nitrogen and Soil Fertility," in *Soil*, USDA Yearbook, Washington, D.C., 1957, pp. 85–94.

Attoe, O. J., and E. Truog, "Exchangeable and Acid-Soluble Potassium as Regards Availability and Reciprocal Relationships," *Soil Sci. Soc. Am. Proc., 10:* 81–86, 1945.

Barber, S. A., "A Diffusion and Mass Flow Concept of Soil Nutrient Availability," *Soil Sci., 93:*39–49, 1962.

Black, C. A., *Soil-Plant Relations*, Wiley, New York, 1968.

Boyer, J., "Soil Potassium," in *Soils of the Humid Tropics*, Nat. Acad. Sci., Washington, 1972, pp. 102–135.

Brill, W. J., "Nitrogen Fixation: Basic to Applied," *Am. Scientist*, July–August 1979, pp. 458–466.

Cook, R. L., and C. E. Millar, "Plant Nutrient Deficiencies," *Mich. Agr. Exp. Sta. Spec. Bull., 353*, 1953.

Delwiche, C. C., "The Nitrogen Cycle," *Sci. Am.*, September 1970, pp. 137–146.

Hanson, J. B., "Roots, Selectors of Plants Nutrients," *Plant Food Rev.*, Spring 1967.

Hodges, T. K., "Ion Absorption by Plant Roots," *Advances in Agron., 25:*163–207, 1973.

Jordan, H. V., and H. M. Reisenauer, "Sulfur and Soil Fertility," in *Soil*, USDA Yearbook, Washington, D.C., 1957, pp. 107–111.

Juo, A. S. R., and B. G. Ellis, "Chemical and Physical Properties of Iron and Aluminum Phosphates and Their Relation to Phosphorus Availability," *Soil Sci. Soc. Am. Proc., 32:*216–221, 1968.

Lawton, K., "The Influence of Soil Aeration on the Growth and Absorption of Nutrients by Corn Plants," *Soil Sci. Soc. Am. Proc., 10:*263–268, 1945.

Lucas, R. E., and B. D. Knezek, "Climatic and

Soil Conditions Promoting Micronutrient Deficiencies in Plants," in *Micronutrients in Agr.,* Soil Sci. Soc. Am., Madison, Wis., 1972, pp. 265–288.

Marx, J. L., "Nitrogen Fixation in Maize," *Science, 189*:368, 1975.

Mortland, M. M., "Kinetics of Potassium Release from Biotite," *Soil Sci. Soc. Am. Proc., 22*:503–508, 1958.

Oliver, S., and S. A. Barber, "An Evaluation of the Mechanisms Governing the Supply of Ca, Mg, K, and Na to Soybean Roots," *Soil Sci. Soc. Am. Proc., 30*:82–86, 1966.

Olsen, S. R., and M. Fried, "Soil Phosphorus and Fertility," in *Soil,* USDA Yearbook, Washington, D.C., 1957, pp. 94–100.

Ray, P. M., *The Living Plant,* Holt, Rinehart and Winston, New York, 1972.

Richards, B. N., *Introduction to the Soil Ecosystem,* Longman, New York, 1974.

Smith, R. L., et al., "Nitrogen Fixation in Grasses Inoculated with *Spirillum* lipoferum," *Science, 193*:1003–1005, 1976.

Stevenson, F. J., "Origin and Distribution of Nitrogen in Soil," in *Soil Nitrogen, Agronomy 10,* Am. Soc. Agron., Madison, Wis., 1965, pp. 1–42.

Struever, S., "The Koster Expedition and the New Archeology," *The Science Teacher, 42*:26–32, 1975.

Subbaroa, Y. V., and R. Ellis, Jr., "Reaction Products of Polyphosphates and Orthophosphates with Soils and Influence on Uptake of Phosphorus by Plants," *Soil Sci. Soc. Am. Proc., 39*:1085–1088, 1975.

Thomas, G. W., and B. W. Hipp, "Soil Factors Affecting Potassium Availability," in *Role of Potassium in Agriculture,* Am. Soc. Agron., Madison, Wis., 1968, pp. 269–291.

Turner, F., and W. W. McCall, "Studies on Crop Response to Molybdenum and Lime in Michigan," *Mich. Agr. Exp. Sta. Quart. Bill., 40*:268–281, 1957.

Viets, F. G., C. E. Nelson, and C. L. Crawford, "The Relationship Among Corn Yields, Leaf Composition and Fertilizer Applied," *Soil Sci. Soc. Am. Proc., 18*:297–301, 1954.

Viets, F. G., "The Plant's Need for and Use of Nitrogen," in *Soil Nitrogen Agronomy, 10,* Am. Soc. Agron., Madison, Wis., 1965, pp. 503–549.

Walker, R. B., "Factors Affecting Plant Growth on Serpentine Soils," *Ecology, 35*:259–266, 1954.

Weber, C. R., "Nodulating and Nonnodulating Isolines," *Agron. Jour., 58*:43–46, 1966.

Youngberg, C. T., and A. G. Wollum II, "Nonleguminous Symbiotic Nitrogen Fixation," in *Tree Growth and Forest Soils,* Proc. of 3rd North Am. Forest Soils Conf., Corvallis, Ore., 1968.

CHAPTER 12

FERTILIZERS AND THEIR USE

For thousands of years people have used lime, marl, ashes, bones, manures, mud, and legumes to add nutrients to soils. In Chapter 6, the use of trees for nutrient accumulation under shifting cultivation was disccussed. Rapid progress in the development of chemical fertilizers occurred after the discovery of the major essential plant elements a little over a century ago. This chapter presents information for gaining an understanding of the nature, manufacture, and efficient use of fertilizers—one of the most important means of increasing the food supply in the world.

Fertilizer Materials

Fertilizers, in a broad sense, include all materials that are added to soils to supply elements essential to the growth of plants. However, the term *fertilizer* usually refers to manufactured fertilizers. Fertilizers do not contain plant nutrients in elemental form as nitrogen, phosphorus, or potassium; however, these elements exist in compounds that provide the ionic forms of nutrients that plants can absorb. The major kinds of nitrogen fertilizers will be discussed first.

Nitrogen Fertilizer Materials

The major source of essentially all industrial nitrogen (including fertilizer nitrogen) results from the fixation of atmospheric nitrogen, according to the following generalized reaction:

$$N_2 + 3H_2 \xrightarrow[\text{pressure and catalysts}]{\text{proper temperature}}$$

$$2NH_3 \text{ (anhydrous ammonia)}$$

The source of hydrogen is usually natural gas (CH_4). A process flow chart of ammonia synthesis is presented in Fig. 12-1.

Ammonia (called ==anhydrous ammonia) is the most concentrated nitrogen fertilizer, being 82 percent nitrogen==. At normal temperature and pressure, anhydrous ammonia is a gas; it is stored and transported as a liquid under pressure. ==After direct application of NH_3 to the soil, NH_3 absorbs a hydrogen ion and is converted to NH_4^+== (*see* Fig. 12-2).

About 98 percent of the nitrogen fertilizer produced in the world is ammonia or one of its derivatives. The manufacture of ammonia and five of its derivatives is shown diagrammatically in Fig. 12-3. A brief summary of some properties of nitrogen fertilizers is given in Table 12-1.

Frequent attempts have been made to determine the relative efficiency of nitrogen fertilizers by applying equal quantities of nitrogren for a given area, in the var-

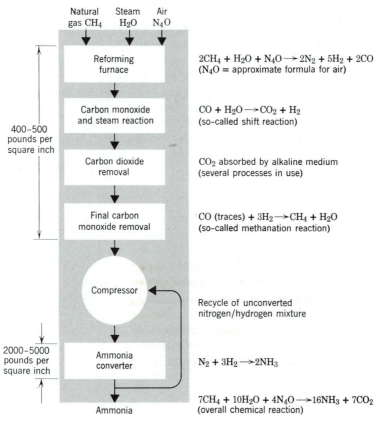

| Natural gas CH_4 | Steam H_2O | Air N_4O |

| Reforming furnace | $2CH_4 + H_2O + N_4O \rightarrow 2N_2 + 5H_2 + 2CO$ (N_4O = approximate formula for air) |

| Carbon monoxide and steam reaction | $CO + H_2O \rightarrow CO_2 + H_2$ (so-called shift reaction) |

| Carbon dioxide removal | CO_2 absorbed by alkaline medium (several processes in use) |

| Final carbon monoxide removal | CO (traces) $+ 3H_2 \rightarrow CH_4 + H_2O$ (so-called methanation reaction) |

400–500 pounds per square inch

Compressor

Recycle of unconverted nitrogen/hydrogen mixture

2000–5000 pounds per square inch

| Ammonia converter | $N_2 + 3H_2 \rightarrow 2NH_3$ |

Ammonia

$7CH_4 + 10H_2O + 4N_4O \rightarrow 16NH_3 + 7CO_2$
(overall chemical reaction)

Figure 12-1
Process flow chart of ammonia synthesis. (From President's Science Advisory Committee, "The World Food Problem," May 1967.)

Figure 12-2
Preplanting application (injection) of anhydrous ammonia about 15 centimeters deep in the soil, where ammonia reacts with hydrogen ions and is converted to exchangeable ammonium.

ious materials, and for a given crop. Since so many factors, such as temperature and moisture conditions, and soil reaction, leaching, kind of crop, and time and method of application, affect the action of any nitrogen fertilizer, relative fertilizing values so obtained may be misleading. Generally, equal amounts of nitrogen from the various carriers, if used in the same manner, produce the same increases in yield.

Urea hydrolyzes rapidly in warm, moist soils to form ammonium carbonate. The ammonium may be used directly by plants, or it may be converted to nitrate and then used as nitrate. Urea-formaldehyde is one of the more recently developed nitrogen fertilizers and is not water soluble. The nitrogen in urea-formaldehyde is released in its available form slowly to provide a continual supply of nitrogen through the growing season. Its high cost greatly limits its use.

Sulfur-coated urea shows good potential for the slow release of nitrogen for crops such as sugar cane and pineapples, which take up to 2 years to mature. Granular urea is sprayed with liquid sulfur and coated with a wax microbiocide. The slow release of nitrogen reduces leaching losses and results in less frequent applications and more efficient use.

Phosphorus Fertilizer Materials

Historically, bones served as a source of phosphorus; yet, as recently as 1840, the value of bones as a fertilizer was found to result largely from their phosphorus content. At about this same time, Liebig suggested that bones be treated with sulfuric acid to increase the solubility or availability of phosphorus. This marked the beginning of the modern fertilizer industry, because it led to the patenting of a process in 1842 for the manufacture of *superphosphate* via the treatment of mineral rock phosphate with sulfuric acid.

Huge deposits of rock phosphate ore

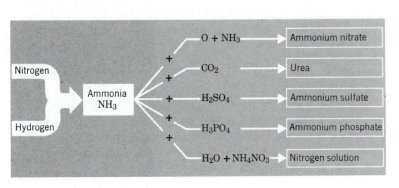

Figure 12-3
Diagrammatic representation of the manufacture of ammonia and some of its derivatives.

Table 12-1
The Principal Fertilizers Supplying Nitrogen

Nitrogen Carrier	Nitrogen %	Remarks
Anhydrous ammonia	82	Used directly, or for ammoniation, nitrogen solutions, etc.
Ammonium nitrate	33	Conditioned to resist absorption of water
Ammonium sulfate	20	Also produced as a by-product of the coking of coal
Ammonia liquor (aqua ammonia)	24	Formed when ammonia is absorbed in water
Nitrogen solutions	Variable	Many kinds formed from solutions of aqua ammonium, ammonium nitrate, urea, etc.
Urea	45	$CO(NH_2)_2$ is hydrolyzed to ammonium in soils
Urea-formaldehyde	35–40	Contains insoluble, slowly available nitrogen
Sodium nitrate	16	Also a naturally occurring salt in Chile
Sulfur-coated urea	39	Slow release, also contains 10 percent sulfur

are located in Florida, east of Tampa, where 73 percent of phosphate ore in the United States is mined. The deposits were formed 10 to 15,000,000 years ago and are buried under a layer of sand (*see* Fig. 12-4). The rock phosphate is in a matrix of sand and clay. Mining consists of removing the overburden and using a hydraulic gun to break up the ore to form a slurry (*see* Fig. 12-5). The slurry is pumped to a recovery plant, where rock phosphate pebbles are separated from the sand and clay by a washing and screening process. The rock phosphate pebbles are ground to form a powder called rock phosphate, which is sometimes used directly as a fertilizer. Rock phosphate (apatite) is very insoluble, so phosphorus has low availability for plant growth.

The only important source of mineral phosphate used to manufacture fertilizers is rock phosphate. The production of ordinary superphosphate, by the acidulation of rock phosphate with sulfuric acid, is shown in the following equation:

$$3[Ca_3(PO_4)_2] \cdot CaF_2 + 7H_2SO_4 \rightarrow$$

(rock phosphate)   (sulfuric acid)

$$3Ca(H_2PO_4)_2 + 7CaSO_4 + 2HF$$

(monocalcium phosphate)   (gypsum)   (hydrogen fluoride)

The ordinary superphosphate produced by the above reaction consists of about one-half monocalcium phosphate and about one-half gypsum. The hydrogen fluoride can be recovered and, in some cases, is used to fluorinate water. Ordinary superphosphate has a phosphorus

Figure 12-4
Location and nature of Florida rock phosphate deposits. (Used by permission of the Florida Phosphate Council, Inc.)

Figure 12-5
In mining rock phosphate, the overburden is removed and the ore matrix is dumped into a "sump" where a hydraulic gun forms a slurry that is pumped to the recovery plant. (Photo courtesy International Minerals and Chemical Corporation.)

content of about 9 percent phosphorus or 20 percent P_2O_5.

Under proper conditions, the reaction of rock phosphate with sulfuric acid will yield phosphoric acid. By treating rock phosphate with phosphoric acid, a more concentrated superphosphate can be produced as follows:

$$\text{Rock phosphate} + 14H_3PO_4 \rightarrow$$

$$10Ca(H_2PO_4)_2 + 2HF$$

The same phosphorus compound is produced with sulfuric acid and phosphoric acid, but without the production of any gypsum when phosphoric acid is used. This more concentrated superphosphate contains about 20 percent phosphorus or the equivalent of 45 percent P_2O_5. Both types of superphosphates are of about equal quality as fertilizers when the same amount of phosphorus is applied.

Ammonium phosphates are produced by neutralizing phosphoric acid with ammonia. The two popular kinds produced are monoammonium phosphate and diammonium phosphate. Ammonium phosphates are good sources of both phosphorus and nitrogen, the phosphorus being water soluble. Some ammonium phosphate is produced as a by-product of the coking industry by using the ammonia produced in the coking of coal to neutralize phosphoric acid.

Basic slag, sometimes referred to as *Thomas phosphate*, is produced as a by-product of the iron industry. Phosphorus is contained in certain iron ores, and steel made from them is brittle if most of the phosphorus is not removed. Slag is produced by oxidizing phosphorus in molten iron via a blast of air blown through it. The molten iron is contained in a converter lined with lime; oxidized phospho-

rus combines with this lime. The slag produced rises to the surface and is drawn off, cooled, and ground so finely that most of it will pass a 100-mesh screen. The phosphorus in slag is soluble in citric acid and is considered available to crops.

Rock phosphate and waste pond or colloidal phosphates are very insoluble forms. These insoluble materials have been used successfully in very acidic low base-status soils of humid tropics. In these soils, soluble phosphorus fertilizers are rapidly fixed; high soil acidity, however, enhances solution of insoluble phosphates. A summary of these and other phosphate materials appears in Table 12-2.

Potassium Fertilizer Materials

Settlement of the eastern seaboard of the United States was stimulated in the early 1600s by a need in England for several important products, including woodashes or potash. To obtain potash, the ashes of trees were leached with water to remove potassium compounds (K_2CO_3), dried, and then calcined to produce potassium oxide. Later, huge mineral deposits of potassium salts were discovered in Germany, and Germany supplied the world market until after World War I. During the 1920s, the Carlsbad, New Mexico potash mines were opened and, more recently, in 1959 the mining of potash was begun in Saskatchewan, Canada (*see* Fig. 12-6).

Common minerals in potash deposits include sylvite (KCl), sylvinite (mixture of KCl and NaCl), kainite ($MgSO_4KCl \cdot 3H_2O$), and langbeinite ($K_2SO_4 \cdot 2MgSO_4$). Processing of ore consists of separating KCl from the other products in ore. The flotation method is commonly used to separate KCl from ore mixture. The ore is ground, suspended in water,

Table 12-2
The Principal Phosphatic Materials

Material	Percentage Available		Remarks
	P_2O_5	P	
Rock phosphate	25–35[a]	11.0–15.4[a]	Effectiveness depends on degree of fineness, soil conditions, and crop grown
Superphosphate (ordinary)	20	8.7	Made by treating ground phosphate rock with H_2SO_4
Triple superphosphate	46	20	Made by treating ground phosphate rock with liquid H_3PO_4
Monoammonium phosphate	48	21	Made by neutralizing H_3PO_4 with ammonia
Diammonium phosphate	53	23	Made by neutralizing H_3PO_4 with ammonia
Basic slag	5–20	2.2–8.8	By-product obtained in the manufacture of steel
Bone meals	17–30[a]	7.5–13.2[a]	Includes raw as well as steamed bone meals
Colloidal phosphate	18–23[a]	7.9–10.1[a]	A finely divided, relatively low-grade rock phosphate or phosphatic clay

[a]Total phosphoric acid instead of amount that is available.

and treated with a flotation agent that adheres to KCl crystals. As air is passed through the suspension, KCl crystals float to the top and are skimmed off (*see* Fig. 12-7). After further purification, the nearly pure KCl is dried and screened for particle size. The fertilizer material is called muriate of potash (KCl) and is about 60 percent K_2O.

Although the great bulk of potassium comes from mines and about 95 percent of potassium fertilizer is KCl, several other sources and materials are worthy of mention. Some potassium is obtained from brine lakes at Searles Lake in California and Salduro Marsh in Utah. If needed, the sea represents an endless supply of potassium. The second most widely used potassium fertilizer, although used to only a minor extent, is K_2SO_4, which is about 40 percent potassium or 48 percent K_2O. A very small amount of KNO_3 is also produced.

All the potassium fertilizer salts are soluble in water and are considered readily available. In general, it can be said that there is very little difference in their effects on crop production, except in rather special cases. Discrimination is sometimes made against carriers having a high percentage of chlorine for special crops such as potatoes and tobacco. The sulfate (or carbonate if available) is usually preferred for tobacco, especially where large

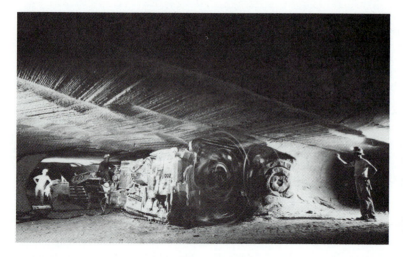

Figure 12-6
A continuous mining machine at work about 1 kilometer underground at Esterhazy, Sask., Canada. (Photo courtesy of American Potash Institute and International Minerals and Chemical Company.)

amounts are to be added, because a crop of superior burning quality is produced. The muriate is just as efficient as the sulfate for potato production, according to most experimental evidence.

Materials Containing Micronutrients

Although the bulk of fertilizer consists of carriers containing nitrogen, phosphorus, and/or potassium, micronutrients are sometimes added to fertilizers to supply one or more of the micronutrients. Some common micronutrient carriers are listed in Table 12-3.

Mixed Fertilizers

Soils vary greatly in their ability to supply crops with available nutrients; also, the mineral requirements of different crops are quite variable. In order to supply nu-

Figure 12-7
A flotation plant for the recovery of potassium minerals. (Photo courtesy of American Potash Institute and the International Minerals and Chemical Company.)

Table 12-3
Some Common Micronutrient Carriers

Carrier	Nutrient Composition
Borax ($Na_2B_4O_7 \cdot 10H_2O$)	11% B
Copper sulfate ($CuSO_4 \cdot 5H_2O$)	25% Cu
Ferrous-sulfate ($FeSO_4 \cdot 7H_2O$)	20% Fe
Manganous oxide (MnO)	48% Mn
Manganese sulfate (variable hydration)	23–25% Mn
Zinc sulfate ($ZnSO_4 \cdot 7H_2O$)	35% Zn
Sodium molybdate ($Na_2MoO_4 \cdot 2H_2O$)	40% Mo

trients for the various requirements of different crops, fertilizers containing two or more essential elements are prepared in many different grades. They are known as *mixed fertilizers* and are made by mixing two or more of the separate fertilizer carriers.

Preparation of Mixed Fertilizers

The preparation of mixed fertilizers can be a relatively simple operation, especially if the mixture is to be of a low grade (i.e., a fertilizer containing a comparatively low percentage of nutrients). It consists essentially of mixing suitable materials in the correct proportion to give the desired grade or analysis. There are many bulk-mixing or blending plants in the United States that follow this simple procedure. On the other hand, many larger fertilizer factories follow more involved processes that require careful chemical and temperature control, particularly in the preparation of some of the carriers of materials used in making the final mix.

After the acidulation of rock phosphate in the manufacture of superphosphate, the product is allowed to stand in a large pile for a considerable time to cure before being ground. If these phosphates are not properly cured, the chemical reactions in-volved are not completed and the fertilizer made from them will harden. Mixtures of superphosphate and potash carriers are often made and stored for a considerable time before being used in the final fertilizer mixture.

Fertilizers in storage tend to assume an unfavorable mechanical condition, chiefly the result of "setting up," which is essentially a cementation, as in Plaster of Paris. It may also be due to the surface tension effects of moisture, which forms films around the particles of fertilizer.

The drilling qualities of certain mixed fertilizers made from fertilizer salts, which take up moisture readily, can be appreciably increased by adding an organic filler, such as muck or diatomaceous earth. Fertilizers are commonly formulated today so that the only material in them classified as filler is what is needed to insure a good physical condition.

The Fertilizer Grade or Guaranteed Analysis

The fertilizer grade or guaranteed analysis expresses the nutrient content. In the United States, the grade is expressed in the order N—P_2O_5—K_2O. The *total* nitrogen is expressed as elemental nitrogen (N). The phosphorus is expressed as *avail-*

able P_2O_5; the potassium is given as *water-soluble* K_2O.

In 1955, the Soil Science Society of America passed a resolution favoring a change from the N—P_2O_5—K_2O basis for expression to the N—P—K or elemental basis. They outlined the following advantages for such a change: First, a greater uniformity would exist in the expression of the nutrient contents of fertilizers, soils, and plants. Plant composition in present-day literature is usually expressed on the elemental basis. The same is true in the fields of animal nutrition and biochemistry, where feeding standards are usually expressed in terms of elemental amounts of phosphorus and calcium. Second, it would result in a greater simplicity. The elemental basis requires fewer symbols for expression. Third, accuracy of expression of the true ratio of the major nutrient elements in a fertilizer requires that the elemental basis be used. Reference to a 10—20—10 fertilizer tends to lead one to believe that the ratio of N—P—K is 1:2:1. The ratio of N—P_2O_5—K_2O is 1:2:1, but the actual N—P—K ratio is 1.0:0.88:0.83. Conversion factors for P and P_2O_5 and for K and K_2O are as follows:

$$P \times 2.29 = P_2O_5$$

$$P_2O_5 \times 0.44 = P$$

$$K \times 1.20 = K_2O$$

$$K_2O \times 0.83 = K$$

Calculation of Fertilizer Formulas

In calculating formulas for mixtures, it is necessary to decide first what percentages of nitrogen, available phosphoric acid, and water-soluble potash are desired in the fertilizer mixture, and then what materials are to be used in making the mixture.

For example, make 1 ton of a 6—12—12 fertilizer using the following ingredients:

Ammonium nitrate:
  33 percent nitrogen
Superphosphate:
  20 percent available phosphoric acid
Muriate of potash:
  60 percent water-soluble potash

The problem is to find out how much of each of these materials is needed. This may be done by use of the equation:

$$X = \frac{A \cdot B}{C}$$

in which X equals pounds or kilograms of carrier required, A equals pounds or kilograms of mixed fertilizer required; with nitrogen, B equals the percentage of nitrogen desired in the mixture, and C equals the percentage of nitrogen in the carrier (ammonium nitrate). By substituting pound values in the above equation, the result is easily determined.

$$X = \frac{2,000 \times 6}{33}$$

$X = 364$ pounds, the amount
    of ammonium nitrate required

Similarly, the required amounts of superphosphate and muriate of potash may be determined.

$$X = \frac{2,000 \times 12}{20}$$

$X = 1,200$ pounds, the weight
    of superphosphate required

$$X = \frac{2,000 \times 12}{60}$$

X = 400 pounds, the weight of muriate of potash required

The total amount of materials used in this fertilizer mixture (364 + 1200 + 400) equals 1,964 pounds. It is necessary to add 36 pounds of filler or physical conditioner to make a ton of the required mixture.

Liquid Fertilizer

A liquid fertilizer is simply a solution containing one or more water-soluble carriers. Materials similar to those used in the manufacture of liquid fertilizers have been added to soils for many years by dissolving them in irrigation water, and as components of conventional dry fertilizers. Advantages of liquid fertilizers over dry fertilizers include: (1) the saving of labor in handling where pumps and pipes can be used, (2) their convenience as foliar sprays, and (3) the convenience of adding pesticides. Disadvantages include: (1) the increased fixation of phosphorus, especially in mixed rather than banded application, (2) corrosion of metal containers and equipment, and (3) the need for special equipment for storage and application.

A popular method used to manufacture liquid fertilizer consists of the neutralization of phosphoric acid with ammonia. Fertilizers with various ratios of nitrogen and phosphorus can be manufactured by varying the degree of neutralization. Potassium can be added to the fertilizer by the addition of KCl. The amount of potassium that can be added, without causing "salting out," is limited and makes the development of high-potash liquid fertilizers impractical.

The time of application and the placement of liquid fertilizer follow closely the principles that apply to the dry forms. When equal amounts of nutrients are applied per unit area, liquid and dry forms produce similar results.

General Nature of Fertilizer Laws

In general, the nature of the laws controlling fertilizer sales in the various states is similar. They all require periodical registration of brands or analyses offered for sale and accurate labeling of the bags or packages. Most states require that the following information be printed on each fertilizer bag, or on an attached tag:

1. Name, brand, or trademark.
2. Analysis (guarantee) or chemical composition.
3. Net weight of fertilizer.
4. Name and address of manufacturer.

Fertilizers offered for sale may be sampled by inspectors any time during the year and at any point in the particular state. The samples are sent to the control laboratory and are analyzed to determine whether the goods meet the guarantee. The results are checked against the guaranteed analysis; by this means, the purchaser is protected from loss through the activities of unreliable companies. The inspection and analysis may be in the hands of the state department of agriculture, of the state agricultural experiment station, or of a state chemist.

Evaluation of Fertilizer Needs

The natural forests and grasslands that existed before the landscape was modified by agricultural pursuits is evidence that essentially all soils throughout the world are capable of supporting plants, provided the climate is favorable. Several important factors account for this situation. There is great diversity in terms of plant needs and tolerances to toxic elements. Nutrient recycling under natural conditions results in repeated use of the nutrients for plant growth. In the natural ecosystem, there may also be abundant vegetation, which is the product of slow growth over many years. When people establish agriculture, the need for additional nutrients is inevitable, if yields are to be maintained above a meager annual yield of 8 to 10 bushels of grain per acre or 500 to 700 kilograms per hectare. Cultivated crops also have different nutrient requirements than natural plants; many nutrients are removed from the land by the harvesting of crops. Therefore, a need is created for evaluating the kind and amount of fertilizer to use, since unnecessary fertilizer is costly and the wrong fertilizer may be harmful.

Deficiency Symptoms and Tissue Tests

Plants, like animals, exhibit peculiar symptoms that are associated with particular nutrient deficiencies. Color photographs of some of these are found in Chapter 11. When the sap of a plant showing a deficiency symptom is tested, a low or deficient amount of the deficient nutrient is usually found. In this way, deficiency symptoms and rapid tissue tests can be used to diagnose plant growth problems. By the time deficiency symptoms appear on a crop, however, the yield potential may be greatly reduced. In addition, tissue tests do not provide information on the quantity of fertilizer to apply. For these reasons, soil tests provide the basis for evaluating the fertilizer needs of most field and vegetable crops.

Soil Tests

Soil testing is based on the concept that a crop's growth response to fertilizer will be related to the amount of *available* nutrients in the soil. For example, the exchangeable potassium is generally a good indication of the available potassium. The more exchangeable potassium there is in a soil, the less plants will respond to potassium fertilizer (*see* Fig. 12-8).

Considerable experimental work is needed to be able to make fertilizer recommendations for specific situations. Fertilizer recommendations must take into account the different nutrient needs of crops, the types of soil, and the climatic conditions. The state agricultural experiment stations have taken responsibility for working on the crops, soil conditions, and the climatic variations that are locally important. The experimental work to develop fertilizer recommendations never

Figure 12-8
Relation of alfalfa yield to amount of exchangeable potassium of several New York soils. Out of 57 soils, 44 gave a yield increase of 20 percent or more when the exchangeable potassium was less than 90 kilograms per hectare. (Data from Chandler, et al., 1945.)

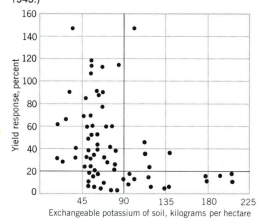

ends, because new crop varieties are developed, soil-testing techniques are continually improved, and crop-yield expectations are generally increasing. Periodically, the latest research results are integrated with previous knowledge and are used to construct reference tables for making fertilizer recommendations.

The fertilizer recommendations in Table 12-4 take into account five important factors. These are soil test levels (in this case, amount of exchangeable potassium), type of crop, yield expectation, type of soil texture, and the climate (since the table is applicable for Michigan). For example, the potassium recommendation for sugar beets expected to yield 24 to 28 tons per acre, growing on soils with loam or finer texture that have 190 pounds of exchangeable potassium per acre furrow slice, is 100 pounds of potassium fertilizer expressed as K_2O or 83 pounds expressed as potassium.

One of the weakest steps in getting good fertilizer recommendations is obtaining a representative sample. A single soil sample of 1 pint (about 1 pound or 500 grams) in size, when used to represent 5 to 10 acres or 2 to 4 hectares, means that 1 pint of soil is used to characterize 5,000 to 10,000 tons of soil. Persons interested in testing soil should consult a testing lab for proper sampling procedures and care of samples after they have been collected.

Plant Analysis for Evaluating Fertilizer Needs

There are several conditions that limit the effectiveness of soil tests in making fertilizer recommendtions for tree crops. The roots of trees are distributed through such a large volume of soil that sampling the upper soil layer or plow layer is of limited usefulness. Furthermore, trees conserve nutrients via the movement of nutrients from leaves into the wood late in the year before the leaves fall. Deep tree roots in unfrozen soil in cold winters may accumulate nutrients all winter long. Since the trees produce over many years, it is possible to evaluate the nutritional status by foliar analysis and to find success by adding fertilizers. A good relationship has been found between potassium content of pine needles, plant height (site index), and potassium deficiency symptoms, as shown in Fig. 12-9 for sandy soils low in available potassium. Deficiency symptoms usually occurred when the potassium content of the needles was less than 0.35 percent.

As with soil tests, foliar analysis must be correlated with yield response before good recommendations can be made. In many cases, this research has been car-

Figure 12-9
Relationship between potassium content of needles and plant height (site index) and deficiency symptoms of Red Pine *(Pinus resinosa)*. (Data from Stone, et al., 1958.)

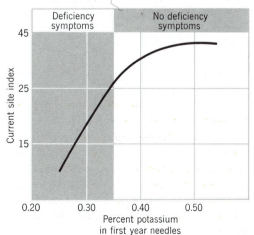

Table 12-4
Potash—Potassium Recommendations for Field Crops on Loams, Clay Loams, and Clays in Michigan

Group 1 (Soil K breakpoints: 0–59 / 60–119 / 120–159 / 160–179 / 180+):
Barley 40–69 bu; Buckwheat 70–100 bu; Clover; Corn 60–89 bu; Corn 90–119 bu; Cover crops; Field beans 15–29 bu (9–18 cwt); Field beans 30–50 bu (19–30 cwt); Corn silage 10–14T; Grass pasture (unimproved); Grasses: Timothy, Orchard, Brome; Kidney beans 15–29 bu (9–18 cwt); Kidney beans 30–50 bu (19–30 cwt); Millet; Oats 50–79 bu; Oats 80–120 bu; Rye; Sorghum; Soybeans 20–40 bu; Soybeans 40+ bu; Sudan grass; Wheat 25–39 bu; Wheat 40–65 bu

Group 2: Alfalfa 3–4T seeding; Alfalfa 3–4T topdressing; Birdsfoot Trefoil; Corn 120–149 bu; Corn silage 15–19T; Sugar beets 18–23T; Wheat 65+ bu

Group 3: Alfalfa 5–6T seeding; Alfalfa 5–6T topdressing; Corn 150+ bu; Corn silage 20–30T; Potatoes 150–299 cwt; Sugar beets 24–28T

Group 4: Alfalfa 7+T topdressing; Potatoes 300–500 cwt

Available Soil Potassium—Pounds of Potassium per Acre[a]				Pounds per Acre Annually Recommended	
Group 1	Group 2	Group 3	Group 4	K₂O	Potassium
0–59	0–59	0–59	0–119	300	249
60–119	60–119	60–119	120–169	200	166
120–159	120–159	120–169	170–209	150	125
160–179	160–199	170–209	210–239	100	83
180+	200–239	210–249	240–279	60	50
	240+	250–269	280–299	30	25
		270+	300+	0	0

[a] bu = bushel; T = ton; cwt = 100 weight.
[b] Multiply by 1.121 to convert to kilograms per hectare.

ried out; foliar analysis has become an important tool for horticulturists and foresters. Researchers have also determined the proper part of the plant to select for testing. Persons interested in use of foliar analysis should obtain instructions for sampling from a testing laboratory.

For annual and perennial field and vegetable crops, plant analysis can be used to determine if a particular element is limiting growth or is in toxic concentration. Caution has to be taken in interpreting results, because a nutrient deficiency reduces plant growth and may result in increased or concentration of other nutrients in the plant. Plant analysis has more value for diagnosis problems than for determining fertilizer needs of annual plants.

Application and Use of Fertilizers

The effectiveness of fertilizers is significantly related to both the type of application and the placement on or in the soil at the time of application. The major factors that affect management decisions concerning the use and application of fertilizers are:

1. Fixation of phosphorus (and to a lesser extent, potassium).
2. Leaching loss of nitrate.
3. Denitrification and loss of nitrogen as N_2.
4. Ammonia volatilization loss.
5. Location of plant roots.
6. Salt effect on seed germination.
7. Soil moisture content and availability for the crop.
8. Time of maximum plant nutrient needs.

9. Management considerations, including labor availability, soil conditions, and so on.

Time of Application

Nutrients in fertilizers are available or water soluble and, therefore, are likely to react with the soil, becoming unavailable or leaching out of the root zone. Nitrate nitrogen is soluble and mobile in soils and is subject to leaching. In fact, excessive use of nitrogen fertilizer can pollute the ground-water. Phosphorus, by contrast, is very immobile in soils. Phosphate ions react with other ions in soil solution to form insoluble, unavailable compounds. Potassium is intermediate in that it is adsorbed to the exchange, which limits its mobility; however, potassium remains available to plants. For these reasons fertilizers are most effective if applied near the time when plant needs are greatest. This is not always practical, since soil conditions, labor supply, and other factors affect the timing of fertilizer applications.

Common Methods of Application

Broadcast. Spreading fertilizer uniformly over the soil surface. Broadcasting is a common method for applying large quantities of bulk fertilizer with truck spreaders before planting. Application by airplane may be advantageous for rough areas, as in the case of forest and pasture fertilization, and for wet areas in the case of rice and fish pond fertilization.

Injection. Placement of liquid fertilizers and anhydrous ammonia into the soil. Injection of anhydrous ammonia is important to prevent volatilization loss (*see* Fig. 12-2).

Banding. Application of fertilizer at planting time in a band near the seed to stimulate early growth of the crop. The fertilizer and seed are separated by a few centimeters distance to prevent high osmotic pressure, from fertilizer salts, near the seed and reduced seed germination. Placement is related to the rooting habit of the crop. Fertilizer is commonly placed to the side and below the seed level for many crops, including corn (maize), and below the seed level for alfalfa and some vegetable crops (*see* Fig. 12-10).

Top Dressing. Application on the soil surface after the crop is growing. This is a common method for fertilizing pastures, winter wheat, rangeland, and lawns. The use of top-dressed phosphorus will be significantly related to moisture conditions at the soil surface.

Side Dressing. Placement of fertilizer along or between rows after the crop is growing. Nitrogen fertilizer is commonly applied as side dressing to reduce the loss of nitrogen before the time when many row crops have a maximum need for nitrogen.

Bedding. Placement of fertilizer in the bottom of a furrow, and then bedding or covering the fertilizer before planting. Used for specific crops as in the case for cotton in the southern United States.

Irrigation. Application of fertilizer in the irrigation water; sometimes called "fertigation." Very common for application of nitrogen. Several opportunities for timing the application are usually available, and there is a low cost for the application. Application either by surface or overhead irrigation.

Foliar. Application of fertilizer to above ground plant parts as a spray. Used for the application of small amounts of micronutrients, especially where applica-

Figure 12-10
Fertilizer placed 5 centimeters to the side and 5 centimeters deeper than the seed will be in the pathway of early developing corn roots. Diffusion of nutrients into the soil surrounding the fertilizer granules has caused root proliferation in this location.

tion to the soil results in rapid fixation of the nutrient.

Application to Shade and Ornamental Trees.

Shade and ornamental trees are commonly fertilized by boring holes about 60 centimeters apart and 30 centimeters deep within the drip line of the tree, and then adding some fertilizer to each hole. Ferric citrate capsules implanted in the trunks of iron-chlorotic pin oak trees will supply the trees with iron for about 3 years.

Other factors are to be considered in many instances. As the overall fertility level of soil is increased, the benefit of localized placement is reduced. Fertilizer purchased in bulk form is less expensive, and lower labor costs result in its use. This makes is possible for many farmers to apply more fertilizer in a less efficient manner and still obtain good results. When only nitrate nitrogen is applied, which is mobile in the soil, almost any placement is satisfactory, so long as the nitrogen permeates the soil and is not leached out of the rooting zone.

Salt Effect of Fertilizers on Germination

When fertilizers are properly placed a couple of centimeters or so from the seed, germination is not affected, but early growth of the plant is stimulated. However, if seed and fertilizer are placed together, germination may be delayed or prevented. The reason is that fertilizers are basically soluble salts and, when they are dissolved in the soil solution, they increase the osmotic pressure of soil solution. The effect on osmotic pressure is related to the solubility of the fertilizer material. Potassium chloride and ammonium nitrate are very soluble and are very likely to prevent or delay germination of seed when placed in contact with seeds. In fact, dropping small amounts of these fertilizers in small areas on the lawn by accident or faulty spreading will kill the grass. Superphosphates have much less solubility by comparison and, when placed with seed, have little or no effect on seed germination, if used at normal rates. The effect of placing some complete fertilizer with wheat seed on the germination of the wheat is shown in Fig. 12-11. Since the increased osmotic pressure of soil solution produced by fertilizer inhibits water absorption by seeds (and roots), one would expect the effects to be less damaging when the soil is kept moist, as is also shown in Fig. 12-11.

Fertilizer Nutrient Interactions

Frequently, the plant response to a fertilizer nutrient is affected by either the use of some other fertilizer nutrient or the level of some other soil nutrient. If nitrogen fertilizer is applied and plant growth is stimulated, this increases the demand for all other plant nutrients. It is not uncommon to find that the response to one nutrient is related to the level of some other nutrient in the soil.

An interesting nitrogen and sulfur interaction exists for some crops in rural areas of the southeastern United States. Several factors are involved that affect the supply of sulfur in the soil and the ability of plants to increase their growth when nitrogen fertilizer is used. In rural areas, only small amounts of sulfur are added to soils in precipitation (near cities rain washes out sulfur from industrial gases). In recent years, ordinary superphosphate, which contains about half calcium sulfate, has been more and more often replaced by triple superphosphate, which does not contain sulfur. Consequently, the effec-

Figure 12-11
Placing fertilizer with
wheat seed was more
damaging on germination
in dry soil than in moist
soil.

tiveness of nitrogen fertilizer is limited in
many cases by an insufficient supply of
soil sulfur.

An interesting phosphorus-zinc interaction is shown in Fig. 12-12. Previous use of phosphorus fertilizer was associated with zinc deficiencies, because the phosphorus reduced a plant's ability to use zinc. Interactions represent a complicat-

ing factor in making fertilizer recommendations and must be considered.

Effect of Fertilizers on Plant Growth

One of the most important factors affecting the growth of plants is the weather. The response that any particular crop will make to the application of fertilizer is

Figure 12-12
Samples of white beans
grown on soil that had
received 0 to 720 pounds
of phosphorus per acre (or
800 kilograms per
hectare) in fertilizer during
the previous 3 years. The
poorer growth and
delayed maturity of plants
on which phosphorus had
been used was due to
zinc deficiency caused by
the use of phosphorus
fertilizer.

largely governed by the weather, particularly the moisture supply. In seasons when it is necessary to delay planting certain crops because of unfavorable weather conditions, the application of fertilizers may speed up the growth processes, thus offsetting somewhat the unfavorable effects of the season. However, looking at the fertilizer-weather relationships from another viewpoint, it is frequently observed that fertilizers in general stimulate early crop growth and, if dry weather prevails about mid-season, fertilizer use may result in decreased yields. This is brought about because soil moisture is more rapidly exhausted by the fertilized crop through increased growth and greater leaf development. On the other hand, fertilized crops may be more drought-resistant because of more deeply penetrating root systems (Fig. 12-13).

The early growth of a crop should not be taken as a measure of the effect of a fertilizer on yield. At times, fertilizers may stimulate early crop growth, but as the season advances this difference disappears; at harvest, no increase is found. Fertilizers may also have little effect on the rate of growth of certain crops, but at harvest a decided increase in yield is noted.

Some of the most striking effects of fertilizers occur on tree crops. About 200 grams of nitrogen applied around the base of Christmas trees in June may change the foliage color from yellow-green to a dark-green by cutting time. The value of the tree may be increased by an amount equal to 10 or more times the cost of the fertilizer.

Effect of Fertilizers on Soil pH

Normally, fertilizers do not produce significant changes in soil pH. However, there are some cases where nitrogen fer-

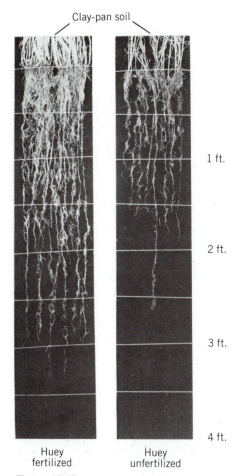

Clay-pan soil

1 ft.

2 ft.

3 ft.

4 ft.

Huey
fertilized

Huey
unfertilized

Figure 12-13
Fertilizer increased the vigor and growth of tops and roots of wheat on the Huey soil in Illinois. Greater rooting depth of fertilized plants increased the supply of available nutrients and water. (Courtesy of Dr. J. Fehrenbacker, University of Illinois.)

tilizers have adversely increased soil acidity. In some cases, proper selection of fertilizers and their use can help to bring about desirable changes in soil pH. A summary of the effects of fertilizers on soil pH follows.

1. The common potassium fertilizers, such as muriate and sulfate of potash, have no permanent effect on soil acidity.

2. Superphosphates, in general, will have no permanent effect on soil reaction. Basic slag, bone meal, and rock phosphate have a tendency to neutralize soil acidity.

3. Fertilizers containing nitrogen in the form of ammonia, or in other forms subject to nitrification (being changed to nitrate), will produce acidity unless sufficient liming material is present in the fertilizer to neutralize the acid that has formed. Some experimental fields that have received applications of sulfate of ammonia fairly regularly for many years, without being treated with lime, have become too acidic to grow corn (*see* Fig. 12-14). This effect is more pronounced on soils such as sands and sandy loams.

4. Nitrogenous fertilizers in which nitrogen is in the nitrate form, and is combined with bases such as sodium or calcium, will result, upon being used by plants, in decreased soil acidity (*see* Fig. 12-14). Some of the fertilizers in this group are nitrate of soda and calcium nitrate. The acidity or basicity of several nitrogen fertilizers is given in Table 12-5.

In general, the systematic use of medium to large amounts of fertilizer at suitable times in the rotation will not appreciably affect soil acidity. Where it is desirable to increase soil acidity (to lower pH), ammonium sulfate can be used as a source of nitrogen.

Effects of Flooding on Fertilizer Use

The transport of oxygen from the shoot to the root is an adaptive mechanism that allows plants to grow in swamps and

Figure 12-14

Manganese concentration in the mid-ribs of corn leaves at tasseling time and soil pH. Differences in soil pH and, consequently, in manganese in plants was caused by applying a total of 1,500 pounds of nitrogen per acre (or 1700 kilograms per hectare) over a 5-year period on the Spinks loamy sand. The use of ammonium sulfate resulted in toxic concentration (over 400 to 500 parts per million) of manganese in plants.

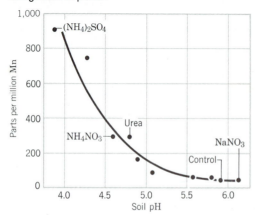

Table 12-5

Equivalent Acidity or Basicity of Several Nitrogen Fertilizers

Fertilizer Material	Nitrogen, %	Equivalent Acidity or Basicity, Kilogram $CaCO_3$ Per 100 Kilogram of Fertilizer
Ammonium nitrate	33	59
Sulfate of ammonia	20	110
Ammo Phos	11	59
Anhydrous ammonia	82	148
Calcium nitrate	15	20[a]
Crude nitrogen solution	44	53
Nitrate of soda	16	29[a]
Urea	45	84

[a]Basicity.

flooded fields. The major crop grown under flooding is paddy or rice. Oxygen diffuses out of the roots of rice plants and creates an oxidized microenvironment in the rhizosphere. This prevents reduced substances from entering the roots and enables aerobic organisms to function in the vicinity of the roots. The major part of the soil, however, remains reduced. This reduced environment affects nutrient availability and fertilizer use.

Nutrients in Irrigation Water

The irrigation water in rice fields is an important source of nutrients. Data compiled in Japan showed that as little as 1 percent of phosphorus and 7 percent of nitrogen to 17 percent of potassium and more than 100 percent of calcium and magnesium that was absorbed by rice was found in the irrigation water (*see* Table 12-6). Silicon is included in the table, even though it is not considered an essential plant element according to Arnon's con-

cept. Silicon is considered to be agronomically essential, since its absence results in poor rice growth and its addition to some low-silicate content soils stimulates rice growth.

Nitrogen Transformations in Flooded Soils

Nitrates normally exist in paddy soils before flooding. After flooding, this nitrate is reduced to N_2 and is lost from soil. The addition of nitrate fertilizer to soil will suffer a similar fate. Organic matter mineralization produces ammonium, which is not oxidized as long as it remains in the reduced soil environment. Ammonium can be adsorbed on the exchange and be readily used by plants. If ammonium diffuses upward into the thin, oxidized surface soil layer, it oxidizes to nitrate. Downward diffusion of nitrate into reduced soil results in denitrification and loss as dinitrogen (*see* Fig. 12-15). Obviously, nitrate fertilizers are not applied to flooded soils. Common carriers

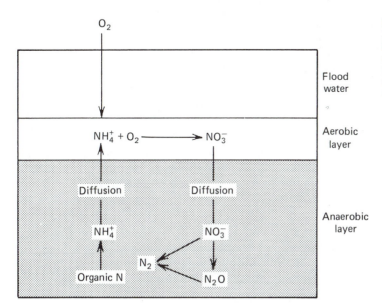

Figure 12-15
Nitrogen transformations in flooded soils. Formation of nitrogen gas (N_2) results in loss by volatilization. (Data from Patrick and Reddy, 1978.)

Flood water

Aerobic layer

$NH_4^+ + O_2 \longrightarrow NO_3^-$

Diffusion   Diffusion

Anaerobic layer

NH_4^+ NO_3^-

N_2

Organic N N_2O

Table 12-6
Sources of Nutrients Absorbed by Rice

Source	Nitrogen	Phosphorus	Potassium	Silicon
Irrigation water	8[a]	1	17	30
Soil and fertilizer	92	99	83	70

Adapted from Yamane and Okazaki, 1982.
[a]Percent of the total.

used are ammonium or urea forms. Deep placement, by hand in paddys in Asia, of large granules of urea or sulfur-coated urea have proved effective in reducing volatilization losses of nitrogen. Timing the application of nitrogen fertilizer via top dressing, when the surface oxidized soil layer is "filled" with roots, encourages uptake of nitrate and reduced nitrogen loss via subsequent volatilization.

Nitrogen Fixation in Flooded Soils

Wetland rice is a unique crop, compared to other agricultural crops; some cases, the same land—year after year without nitrogen fertilizer—will produce without a decline in yields. This situation can be explained on the basis of nitrogen fixation in flooded fields. First, the water in flooded fields or paddys is suitable for the growth of free-living, autotrophic nitrogen-fixing, blue-green algae. Second, anaerobic heterotrophic bacteria (*Clostridium* and *Azotobacter*) are more effective nitrogen fixers in anaerobic soils. Third, the nitrogen fixing blue-green algae *Anabaena*, forms a symbiotic relationship with the small water-floating fern, *Azolla*. These algae live in the pores of the fern fronds. The fern has adventitious roots that hang down in the water and can penetrate mud.

The *Anabaena-Azolla* system has been used in China for centuries and is being studied intensively in many countries today. By inoculation with *Anabaena-Azolla* (referred to as *Azolla*), wet soils can be used for production of *Azolla*. However, *Azolla* dies if the soil dries out. Where land remains wet until it is prepared for transplanting rice, the *Azolla* can be grown and incorporated into the soil as green-manure when the land is puddled. *Azolla* growing on open bodies of water forms a floating mat that can be skimmed off and used as green-manure (or cattle feed). A mat 2 centimeters thick contains about 50 kilograms of nitrogen per hectare or 50 pounds per acre. If rice paddy is inoculated at transplanting time, *Azolla* grows until shade kills it; the nitrogen is mineralized and then made available to the rice crop.

Managing *Azolla* requires much labor, because it is essentially another crop. The inoculum requires management so that a supply is available when needed, and the *Azolla* is sensitive to many environmental factors. Two important factors that are limiting the widespread use of *Azolla* are its high phosphorus requirement and sensitivity to high temperature.

Phosphorus Availability in Flooded Soils

Although many rice fields are deficient in phosphorus, flooding increases phosphorus availability in soils. The major form of phosphorus in many rice soils is iron phosphate. Flooding the soil reduces iron in ferric phosphates and releases the phosphorus to soil solution. There is a greater volume of water for the solution of phosphorus and a greater opportunity for phosphorus diffusion. In addition, the change in pH due to flooding may be an important factor in many soils. The effects of flooding Ultisols and Oxisols are not as great as on other soils. This may be related to the greater amounts of occluded phosphorus. Good results from broadcasting phosphorus fertilizer on wetland or flooded soils for rice are likely due to the great concentration of rice roots in a thin layer of topsoil.

Iron and Manganese Availability and Toxicity

The chemistry of iron and manganese in flooded soils is similar; flooding generally increases the availability of both. Iron deficiencies in rice are due to the high iron requirement of rice and the oxidizing nature of root surfaces that oxidize iron and reduce its solubility. High soil pH also favors low iron availability. Conversely, reducing conditions favor the reduction and solubility of iron, and iron toxicity may occur. The toxicity problem is most important in acidic Ultisols and Oxisols that are high in iron content. Common iron carriers added to both flooded and nonflooded soils are quickly fixed. Foliar sprays are the best alternative for treating iron-deficient paddy rice.

Manganese is seldom deficient in rice grown on flooded soil. Rice soils are generally higher in manganese than iron, and flooding increases solution of manganese due to reduction. The high levels of available iron generally tend to retard manganese uptake and development of manganese toxicity.

Other Micronutrient Conditions

Zinc deficiency is a common problem in rice production. Zinc availability is decreased by flooding, and it has a low availability in soils with a high pH. High levels of phosphorus fertilization induce zinc deficiency. Copper availability is also reduced due to flooding, but deficiencies of copper are rarely observed in crop production on flooded soils.

Flooding increases the availability of molybdenum and scarcely affects the availability of boron. Deficiencies for crop production on flooded soils are rare. Chlorine has not been found deficient; this is likely due to the chlorine in rainfall and irrigation water. In summary, the flooding of soils for rice production is beneficial, based on its effects on nutrient availability. In addition, flooding aids weed control, prevents drought, and facilitates many tillage and transplanting operations.

Manure as a Fertilizer

After centuries during which animal manures were highly valued for their fertilizing quality, manure on many American farms has become a liability. Farms have become more specialized and bigger so that most manure is produced on fewer larger farms, instead of on a large number of small farms. A farm with 10,000 head of cattle produces about 260 tons (236 metric tons) of manure a day. The magnitude of the manure-disposal problem is dramatized by the fact that domestic animals in the United States produce body waste equivalent to 1,900,000,000 people. The emphasis has changed from application of manure on soils to increase yields to the use of soil for animal waste disposal. In many developing nations, however, cattle manure is a major source of fuel (*see* Fig. 12-16). This section considers the production, composition, storage, and handling of animal manures.

Components of Manure

Farm manure consists of two components, solid and liquid. Solid excrement, on the average, contains one-half or more of nitrogen, about one-third potassium, and nearly all the phosphorus that is excreted by any animal. Nitrogen in feces exists largely in two forms: first, as residual proteins that have resisted decomposition in the digestive processes; and, second, as proteins that have been synthesized in the cells of bacteria. Over one-half of the nitrogen may be present as synthesized protein. This form is readily broken down when added to soils, so that nitrogen is available to plants. Solid excrement also contains large quantities of lignin. In other words, a large share of the organic matter in feces is humified; a compound is formed similar to humus that is formed

Figure 12-16
Cow dung being shaped into cakes, plastered on a wall, and allowed to dry. The dried chips are used for cooking food in many developing nations due to a fuel wood shortage.

in soils. As much as 50 percent of the organic matter in solid excrement may be in the humified state. The nitrogen contained therein is only slowly available to plants when added to soils.

The liquid fraction, or urine, contains plant nutrients that have been digested and used by the animal body and later excreted. All plant nutrients in this fraction are soluble and either directly available to plants or readily become so. Liquid portions of manure differ from solids not only in regard to the availability of nutrients, but also in its low content of phosphorus and high content of potassium and nitrogen. The distribution of plant nutrients between liquid and solid portions of manure is shown in Fig. 12-17. When voided, nitrogen in urine exists largely as urea and hippuric and uric acids. These compounds are not volatile at ordinary temperatures; however, manure contains organisms that are capable of rapidly breaking these compounds down via the formation of ammonia, which combines with water and carbon dioxide to make ammonium carbonate. This compound is unstable; even in solution, it tends to decompose and lose ammonia, especially at higher temperatures $[(NH_4)_2CO_3 \rightarrow 2NH_3 + H_2O + CO_2]$. This compound may lose all its ammonia on drying. The unstable nature of nitrogen in urine presents a major problem in the handling of manure.

Quantity and Composition of Excrements

Many factors influence the quantity of manure produced and its composition, such as: (1) the kind and age of the animal, (2) the kind and amount of feed consumed, (3) the condition of the animal, and (4) the milk produced or work performed by the animal. Wide varia-

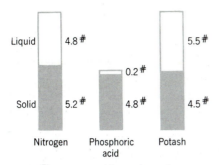

Liquid 4.8#   5.5#

0.2#

Solid 5.2#   4.8#   4.5#

Nitrogen   Phosphoric   Potash
        acid

Figures represent pounds per ton

Figure 12-17
Distribution of plant nutrients between liquid and solid
portions of a ton of average farm manure. (Multiply
values by 0.5 to convert to kilograms per metric ton.)

tions are often found in the manure of
animals, even of a given class. Animals
of different ages and doing different
kinds of work require different amounts
and proportions of nutrients to maintain

them. A young animal, for example, that
is building muscle and bone needs considerable phosphorus, nitrogen, calcium,
and other elements; the manure produced by such animals will contain much
less of these elements. Since the composition of manure is so variable, data as
presented in Table 12-7 can only be approximate.

Urine makes up only 20 percent of the
total weight of the excrement of horses,
but 40 percent of that from hogs. These
represent two extremes. Since urine
makes up only 20 to 40 percent of the total weight of manure from any animal,
and yet contains approximately two thirds
of the total potash and somewhat less than
one-half of nitrogen, it is evident that,
pound for pound, urine is more concentrated; hence, it is more valuable than the
solid portion.

Table 12-7

Quantity and Composition of Fresh Manure Excreted by Various Kinds
of Farm Animals

Animal	Excrement	Pounds per Ton[a]	Water, %	Nitrogen, lb	P_2O_5, lb	K_2O, lb	Tons Excreted[b] per Year
Horse	Liquid	400	—	5.4	Trace	6.0	—
	Solid	1,600	—	8.8	4.8	6.4	—
	Total	2,000	78	14.2	4.8	12.4	9.0
Cow	Liquid	600	—	4.8	Trace	8.1	—
	Solid	1,400	—	4.9	2.8	1.4	—
	Total	2,000	86	9.7	2.8	9.5	13.5
Swine	Liquid	800	—	4.0	0.8	3.6	—
	Solid	1,200	—	3.6	6.0	4.8	—
	Total	2,000	87	7.6	6.8	8.4	15.3
Sheep	Liquid	660	—	9.9	0.3	13.8	—
	Solid	1,340	—	10.7	6.7	6.0	—
	Total	2,000	68	20.6	7.0	19.8	6.3
Poultry	Total	2,000	55	20.0	16.0	8.0	4.3

Compiled from Van Slyke, 1932.
[a]Multiply by 0.5 to convert to kilograms per metric ton.
[b]Clear manure without bedding; tons excreted by 1,000 lbs of live weight of various animals.

Losses by Volatilization

Losses incurred by volatilization fall principally on nitrogen and organic matter. Large quantities of ammonia are produced in manure from urea and other nitrogenous compounds. In the earlier stages of manure decomposition, ammonia is combined largely with carbonic acid as ammonium carbonate and bicarbonate. These ammonium compounds are rather unstable, and gaseous ammonia may be readily liberated (*see* Fig. 12-18). The tendency to lose ammonia nitrogen increases with the increase in concentration of ammonium carbonate and the increase in temperature.

At ordinary temperatures, little or no loss of ammonia from manure occurs at pH 7.0 or below. High temperatures, produced by aerobic decomposition in a loose manure pile, are conductive to a very rapid loss of ammonia.

Freezing also tends to increase the loss of ammonia by increasing the concentration of solution via crystallization of water. This loss may be considerable when manure is spread and becomes frozen.

Air movement greatly affects the loss of ammonia. Wind movements hasten evaporation of water, which decreases the capacity of water to hold ammonia. Thus, manure that is permitted to dry out may lose appreciable quantities of ammonia. This fact emphasizes the importance of ammonia loss due to air circulation in loosely piled manure heaps and in over-fermented manure that is forked frequently. Losses also may be considerable if manure is spread and permitted to dry before plowing under.

It has been pointed out that when manure decomposes, it suffers important losses in organic matter. These losses occur mainly in carbohydrate constituents. One of the important end-products of carbohydrate decomposition is carbon dioxide, most of which is lost from manure by volatilization. Shrinkage that accompanies the partial decomposition of manure is evidence of organic matter loss.

Application of Manure

Prompt spreading of manure is generally considered most effective; however, when in good storage, it is likely to lose less value than if spread on a field without being plowed under or worked into the soil immediately. Losses of applied manure may occur in three ways: (1) volatilization of ammonia nitrogen as a result of drying or freezing; (2) surface runoff water carrying soluble portions of all three nutrients; or (3) leaching of nutrients.

Much experimental work with commercial fertilizers indicates that their effectiveness is decreased if they are applied a considerable time before seeding the crop. This effect is attributed to leaching losses and to the fixation in less soluble forms of plant nutrients by soil. Fresh and properly stored manures contain large amounts of soluble nutrients, and the

Figure 12-18
Volatilization loss of ammonia and total nitrogen from fermented cow manure exposed to drying. (Based on data of Heck, 1931.)

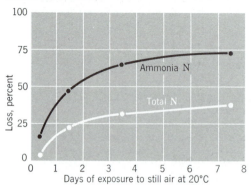

same principles apply. Manure applied to corn land in the spring is likely to give greater returns than the same amount of manure applied in the fall. This effect will probably apply only to the crop of the first year and not to later crops. If manure has already lost its readily soluble nutrients, the time of application is less important.

Little loss of nutrients from manure may be expected when applied on level, medium-to-fine-textured soils. These soils are safe for fall and winter spreading of manure, although they fix considerable quantities of nutrients in less available forms. It is not advisable to place manure on sandy soils or on hilly fields much ahead of plowing time because of leaching and erosion losses. It must be remembered, however, that manure applied on sloping fields decreases soil erosion losses, thereby counterbalancing to some extent the loss of nutrients from manure. Illinois law, however, restricts application of manure (and fertilizer) on slopes over 5 percent, when soils are frozen, to protect water quality in streams and lakes.

The rate at which manure should be spread will depend on the amount produced on farms. As much cultivated land should be covered as possible. It is usually better to cover an entire field with a light application than to give only a part of the field a heavy coating.

Residual Effects of Manure

An application of manure usually shows a favorable influence on crop yields for several years. These beneficial effects are distributed over a longer time than those of chemical fertilizers. Striking results in showing the long-continued effects have been obtained (*see* Fig. 12-19) by making heavy applications of manure for several years in succession, and then discontinuing application.

Using Manure for Biogas Production

Biogas production is based on the anaerobic decomposition of manure and other organic waste materials. During anaerobic decomposition, various gases are produced, the major one being methane (CH_4). The remaining sludge and effluent

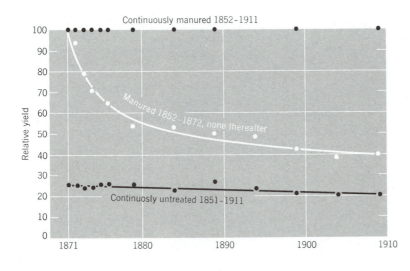

Figure 12-19
Residual effects of heavy applications of manure (14 tons per acre annually) on yield of barley grown continuously on the soil of the Rothamsted Experiment Station in England. (Data from Salter, 1939.)

contains the original plant-essential elements removed from the soil. Since the loss during digestion is mainly carbonaceous material, the resulting sludge and effluent are enriched in elements such as nitrogen and phosphorus, compared to the original material used for biogas production. Basically, a slurry of manure and other organic wastes are fed into a hopper that is connected to an anaerobic digestion chamber. Production of gas creates a pressure in the digestion chamber, and the gas is piped off to light lamps, cook food, heat rooms, or to run small engines. Slowly, material passes through the digestion chamber and exits as sludge and effluent. The sludge and effluent are excellent for use as fertilizer. A biogas generator with a metal, floating digestion chamber is shown in Fig. 12-20.

Several million biogas generators are in use in China, where the primary interest is in the nutrient value of the composted manure that is produced; gas production is considered the by-product. Since most harmful organisms are killed in digestion, biogas generation is a good means to dispose of human wastes and to help keep the environment more sanitary. The gas is combustible and safety precautions are needed to prevent explosions due to exposure of gas to a naked flame.

Fertilizer Use and Environmental Quality

Algae are the most abundant plants growing in surface waters; their growth is commonly limited by a lack of nitrogen and/or phosphorus. Consequently, any activity that enriches surface waters with nitrogen or phosphorus might contribute to excessive algal growth or an algal bloom. Furthermore, nitrogen in nitrate form can be harmful in surface or ground-waters used for human consumption. This section will consider some important factors related to the pollution of waters resulting from use of nitrogen and phosphorus fertilizers.

Figure 12-20
Biogas generator with intake hopper on the left, floating metal gas holder, and the anaerobic digestion chamber in the center, and the sludge discharge at the right. Note the gas line coming off the top of the gas holder.

Natural Nitrate Content of Soil Percolate

Nitrogen fertilizers are mostly water soluble, and the nitrate form of nitrogen is mobile in soils. During rains, nitrogen fertilizer is readily carried into the soil. For this reason, nitrogen fertilizers are not likely to be carried into surface waters by runoff water. The nitrate from nitrogen fertilizer, however, might migrate downward through the soil with percolating water and be carried to underground water reservoirs or ground-waters. In fact, it is natural for some nitrate ions to move to the ground-water in natural ecosystems. This is due to nitrate being a common form of available nitrogen in soils for plants. Also, the nitrate may be carried to the water table in wet seasons, when plants are relatively inactive. A look at some experimental data should be helpful.

In Chapter 3 (*see* Table 3-2), we discussed some lysimeter data collected at Coshocton, Ohio, which showed nutrient losses in soil percolate. One lysimeter, in 1948, grew a grass crop and no fertilizer or manure was applied. This situation can be considered comparable to a natural grassland. The loss of nitrate nitrogen per hectare was 5.1 kilograms per year, and the percolate was 15.7 centimeters. Since there are 10 hectare centimeters of water in 1,000,000 kilograms, the parts per million (ppm) of nitrate nitrogen (N) can be calculated from the following equation:

$$\text{ppm N} = \frac{\text{kg N leached per hectare} \times 10}{\text{cm of percolate}}$$

Substituting in the above equation shows that the natural soil percolate had 3.2 parts per million of nitrate nitrogen. This value is very reasonable since 2 to 3 parts per million of nitrogen are considered adequate for algal growth, and this is also the nitrogen content commonly found in springs and shallow wells. The United States Health Service considers 10 parts per million of nitrate nitrogen the maximum allowed in water used to make baby formula.

There has always been a natural amount of nitrate in ground-water which results from the natural reactions of the nitrogen cycle. Since ancient times, the manner in which human and other animal wastes were disposed created varying degrees of nitrate pollution of water supplies. Today, however, the problem is different because of the magnitude of nitrogen used as fertilizer and the great concentrations of cattle on large dairy and beef farms.

Factors Affecting Nitrate Pollution of Ground-water

The factors that affect the amount of nitrogen in drainage water or soil percolate can be represented in equation form as follows:

nitrogen in soil percolate

= amount of nitrogen available as nitrate

− amount of nitrogen immobilized

Factors Affecting Available Nitrate	*Factors Affecting Nitrogen Immobilized*
1. Organic matter mineralization	1. Kind of crop grown
2. Fixation from atmosphere	2. Yield or growth of crop
3. Nitrogen added in precipitation	3. Amount and distribution of rainfall
4. Nitrogen added as manure	
5. Nitrogen added in fertilizer	
6. Denitrification	

A modest amount of nitrogen fertilizer could have no effect, increase, or decrease the amount of nitrogen in the percolate, depending on the particular situation. If, however, an excessive amount of nitrogen fertilizer was applied, and the soil's immobilization capacity was greatly exceeded, nitrate pollution of the ground-water would occur. The continued use of nitrogen at rates of 150 pounds or more per acre (170 kilograms per hectare) per year caused an accumulation of nitrate nitrogen in the Marshall silt loam in Missouri, when the land was used continuously for corn production (see Fig. 12-21). The 0- or 100-pound rate produced similar amounts of nitrate nitrogen in the soil profile. Thus, we see that farmers should avoid using too large an amount of nitrogen fertilizer, particularly when plant growth is inactive. The danger of nitrate pollution is also greatest on sandy soils with high percolation capacity. In these cases, it is important to use modest amounts of nitrogen fertilizer and to take care in applying fertilizer only when crops are actively growing.

Studies in Missouri and Colorado show that nitrate pollution of ground-water is related to numbers of livestock. Nitrate nitrogen is produced during the decomposition of manure in the same ways that organic matter decomposition in soils produces nitrate. Where animals congregate in large numbers, as in feedlots, growing plants are absent and the nitrate produced from manure decomposition is not immobilized. Under these conditions, rainwater can leach nitrate out of the feedlot and transport nitrates to the ground-water table. The same situation applies to corrals.

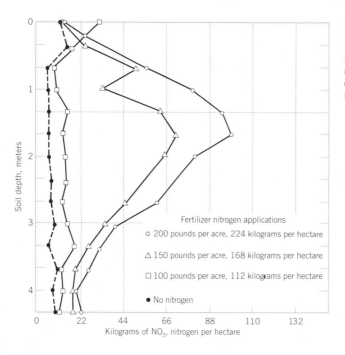

Figure 12-21
Nitrate nitrogen in the Marshall silt loam after seven annual applications of nitrogen fertilizers in a continuous corn program. (Data from Linville, 1968.)

Fertilizer nitrogen applications

○ 200 pounds per acre, 224 kilograms per hectare

△ 150 pounds per acre, 168 kilograms per hectare

□ 100 pounds per acre, 112 kilograms per hectare

● No nitrogen

Soil depth, meters

Kilograms of NO_3, nitrogen per hectare

Phosphate Pollution of Surface Waters

Phosphate from fertilizer, in contrast to nitrate, reacts with soil constituents to form insoluble compounds that are immobile in soils. For this reason, there is little possiblity that ground-waters would become polluted from the use of phosphorus fertilizers. Erosion of soil particles can, however, carry phosphate adsorbed on soil particles into surface waters. This type of contamination occurs naturally, as well as resulting from erosion of agricultural land. In fact, considerable concern exists when unprotected sloping land lies exposed for long periods of time, where highways or subdivisions are constructed. The real problem here is not so much the use of phosphorus fertilizer but, instead, erosion.

Erosion is a selective process. The finer particles are removed, and these finer particles are richer in plant nutrients. Research at the University of Wisconsin showed that the fine, eroded soil material had 3.4 times more phosphorus than soil itself. Where fertilizers promote a more vigorous plant cover and reduce soil erosion, fertilizers reduce the danger of enriching surface waters with phosphate.

Bibliography

Azevedo, J., and P. R. Stout, "Farm Animal Manures: An Overview of Their Role in the Agricultural Environment," *Manual 44,* Cal. Agr. Exp. Sta., Davis, Calif., 1974.

Bear, F. E., "The Effect of Increasing Fertilizer Concentrtion on Exchangeable Cation Status of Soils," *Soil Sci. Soc. Am. Proc. 16:*327–330, 1952

Brill, W. J., "Nitrogen Fixation: Basic to Applied," *Am. Sci., 67:*458–466, July–August 1979.

Brown, L. R., "Human Food Production as a Process in the Biosphere," *The Biosphere,* Sci. Am., San Francisco, 1970, pp. 93–103.

Chandler, R. F., Jr., M. Peech, and C. W. Chang, "The Release of Exchangeable and Nonexchangeable Potassium from Different Soils upon Cropping," *Soil Sci. Soc. Am. Proc., 10:*141–146, 1945.

Committee on Elemental Guarantees of Soil Sci. Soc. Am., "N. P. K. Simpler Terms for Fertilizer," *Crops and Soils, 14* (6), 1962.

Craswell, E. T., and P. L. G. Vlek, "Nitrogen Management for Submerged Rice Soils," *Symposia Papers 2,* 12th Int. Congress Soil Science, Delhi, 1982, pp. 158–181.

FAO, "China: Azolla Propagation and Small-scale Biogas Technology," *FAO Soils Bull. No. 41,* Rome, 1978.

Heck, A. F., "Conservation and Availability of the Nitrogen in Farm Manure," *Soil Sci., 31:*467–479, 1931

Jones, U. S., *Fertilizers and Soil Fertility,* Reston, Reston, Va, 1982.

Linville, K. W., *Residual Nitrate in Missouri Soils,* Master of Science Thesis, University of Missouri, 1968.

Massey, H. F., and M. L. Jackson, "Selective Erosion of Soil Fertility Constituents," *Soil Sci. Soc. Am. Proc., 16:*353–356, 1952.

Neely, D., "Pin Oak Chlorosis-Trunk Implantations Correct Iron Deficiency," *Jour. Forestry, 71:*340–342, 1973.

Patrick, W. H. Jr., and C. N. Reddy, "Chemical Changes in Rice Soils," in *Soils and Rice,* IRRI, Manila, 1978, pp. 361–380.

President's Science Advisory Committee, *The World Food Problem,* Vol. 2, The White House, Washington, D.C., May 1967.

Randhawa, N. S., and J. C. Katyal, "Micronutrients Management for Submerged Rice Soils," in *Symposia Papers No. 2,* 12th Int. Congress Soil Science, Delhi, 1982, pp. 192–211.

RuKun, L. J. Bai-Fan, and L. C. Kwei, "Phosphorus Management for Submerged Rice Soils," in *Symposia Papers No. 2,* 12th Int. Congress Soil Science, Delhi, 1982, pp. 182–191.

Salter R. M., and C. J. Schollenberger, "Farm Manure," *Ohio Agr. Exp. Sta. Bull., 605,* 1939.

Stoeckeler, J. H., and H. F. Arneman, "Fertilizers in Forestry," *Adv. Agron., 12:*127–195, 1960.

Stone, E. L., G. Taylor, and J. DeMent, "Soil and Species Adaptation: Red Pine Plantations in New York," *First Nor. Am. For. Soils Conf. Proc.,* Michigan State University, East Lansing, Mich., 1958, pp. 181–184.

Tisdale, S. L., and W. L. Nelson, *Soil Fertility and Fertilizers,* 3rd Ed., Macmillan, New York, 1975.

Van Slyke, L. L., *Fertilizers and Crop Production,* Orange Judd, New York, 1932.

Wantanabe, I., "Biological Nitrogen Fixation in Rice Soils," in *Soils and Rice,* IRRI, Manila, 1978, pp. 465–478.

Warncke, D. D., D. R., Christenson, and R. E. Lucas, "Fertilizer Recommendations for Vegetable and Field Crops in Michigan". *Ext. Bull., E.*-550, Michigan State University, East Lansing, Mich., June 1976.

Wittwer, S. H., and G. F. G. Teubner, "Foliar Absorption of Mineral Nutrients," *An. Rev. Pl. Phy., 10:*13–32, 1959.

Wolcott, A. R., "The Acidifying Effects of Nitrogen Carriers," *Agr. Ammonia News,* Agr. Amm. Institute, Memphis, July–August 1964.

Yamane, I., and M. Okazaki, "Chemical Properties of Submerged Rice Soils," in *Symposia Papers No. 2,* 12th Int. Congress Soil Science, New Delhi, 1982, pp. 143–157.

CHAPTER 13

SOILS, PLANT COMPOSITION, AND ANIMAL HEALTH

Considerable interest exists today in health foods and in living in more natural ways. One important concern is whether the use of fertilizers affects food quality and health. "Do fertile soils produce better or more nutritious food than infertile soils?" "If a plant grows well and is healthy, will it be a satisfactory source of food for man or animals?" This chapter provides some insight into the role of soils and fertilizers in plant composition and into some relationships between plant composition and animal health.

Factors Affecting Plant Composition

Plant composition is affected by soil, environmental conditions, and genetic factors, as summarized in Fig. 13-1. The genetic factor is probably the most important. The genetic factor not only controls the uptake and mineral composition, but it also controls the amounts and kinds of proteins, vitamins, carbohydrates, and so on. It is interesting, however, that the vitamin C content of tomatoes is determined by the amount of sunlight striking the tomato. This discussion will emphasize the soil and genetic factors affecting the mineral composition of plants.

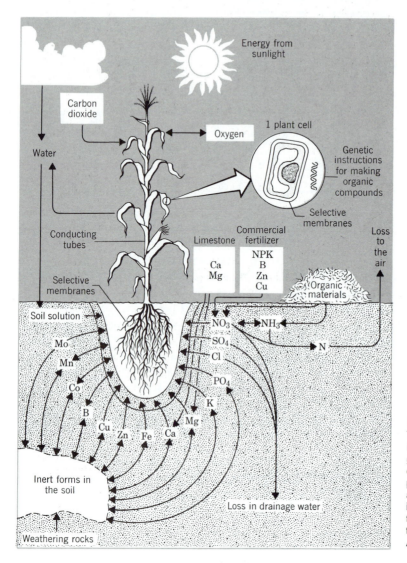

Figure 13-1
The composition of plants is controlled by soil, genetic, and environmental factors. The plant processes operate according to a set of instructions determined by the genetic inheritance of the plant. (Data from Allaway, 1945.)

Plant Species and Plant Composition

Perhaps one of the most notable differences in plant composition exists between grasses and legumes (Table 13-1). Legumes contain a higher percentage of nitrogen and calcium and a lower percentage of potassium than grasses grown under comparable conditions.

Tobacco is known for its high potassium content, which commonly exceeds 4 percent in leaves. Certain varieties of cereals, having a high resistance to disease and insects, have a silica-encrusting layer in the epidermis and may contain up to 15 percent silica.

The higher base content of the foliage of hardwoods, as compared to spruce, was used to explain the higher pH of soils de-

Table 13-1

Average Composition of Legumes and Grasses

Plant	Percentage Composition			
	Nitrogen	Potassium	Calcium	Magnesium
Grasses	0.99	1.54	0.33	0.21
Legumes	2.38	1.13	1.47	0.38

Data of Snider, 1946.

veloped under the hardwood forest (*see* Chap. 9). These examples support the fact that the composition of plants is influenced by genetic factors.

Level of an Individual Nutrient

An increase in the amount of a nutrient in soil that is available or readily soluble may or may not cause an increase in the percentage of that nutrient in a plant. It depends on the extent to which the total growth of the plant is increased. Several possibilities exist. If the nutrient is in short supply, and the growth of the plant is limited by it, addition of the nutrient would probably result in a great increase in the amount absorbed. A correspondingly large increase in plant growth could occur so that the percentage nutrient concentration of the plant remained the same, as shown on the far left side of Fig. 13-2. An increase in growth and nutrient concentration or percentage composition can occur simultaneously in the zone of poverty adjustment (*see* Fig. 13-2). A point

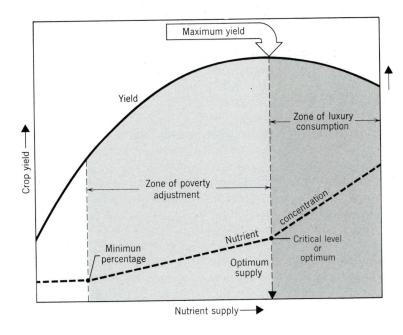

Figure 13-2
Schematic graph of the manner in which nutrient concentration and crop yield vary with the supply of nutrients. (Data from Brown, 1970.)

is finally reached, with increasing nutrient supply, where the maximum yield is obtained; further increases in the nutrient supply result in decreased yields. Plants may, however, continue to take in nutrients. A sharp increase in nutrient concentration in the plant results. This is the luxury consumption range shown in Fig. 13-2.

Cation Content of Plants a Constant

Potassium, calcium, and magnesium all play some of the same roles in a plant; for instance, their role in the buffer system of plant cells. In this regard, they can partially substitute for each other. This view is consistent with the common observation that the milliequivalents of these three cations in plant tissue tends to be a constant. Further support of cation constancy in plants is that a large application of potassium sometimes produces magnesium deficiency. Note in Fig. 13-3 that increasing the amount of potassium in the nutrient solution caused large and comparable decreases in the amount of calcium and magnesium in alfalfa plants; the total amounts of potassium, calcium, and magnesium in the tissue remained the same.

Corn grown on highly calcareous Harpster soils of north-central Iowa commonly has a severe potassium deficiency symptom, so that during the summer these areas of soil can be readily located by observing the growth of corn. Potassium-deficient plants were found to contain as much as 10 times more calcium and magnesium than potassium. Contrast this with the fact that grasses tend to contain as much or more potassium as the sum of calcium and magnesium (Table 13-1).

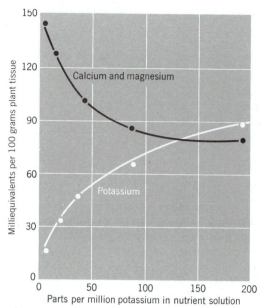

Figure 13-3
The calcium, magnesium, and potassium content of alfalfa plants as a function of the potassium concentration in the nutrient solution. (Adapted from Wallace, 1948.)

Plant Composition Changes with Age

Nutrient accumulation occurs at a faster rate than plant weight when the plant is young, whereas the reverse is true when the plant approaches maturity. This causes a declining concentration of nutrients in plants with increasing age. It is a common practice for dairy farmers to cut forages when they are relatively immature in order to provide a more nutritious feed.

The accumulation of dry matter in various parts of a corn plant and the accumulation of nitrogen, phosphorus, and potassium through the season are given in Fig. 13-4. In this instance, corn (maize) plants had absorbed 40 percent of their nitrogen, 30 percent of their phosphorus,

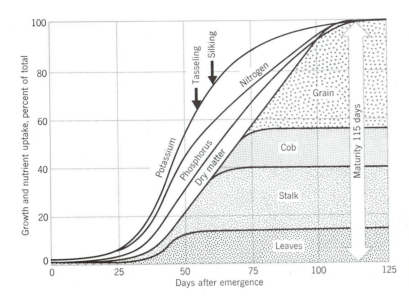

Figure 13-4
Uptake of nutrients in relation to the accumulation of dry matter by corn plants. (Data from Hanway, 1960.)

and over 50 percent of their potassium by the time they had made 20 percent of their growth.

Composition Varies with Plant Part

The pattern of organ development for corn is also shown in Fig. 13-5. The grain accounted for nearly one-half of the weight of mature plants, and it was largely produced during the last one-third of the growing season. During this time, the plant accumulated about 40 percent of its dry matter and only 25 percent of its nitrogen. Thus, the production of the grain occurred at the expense or loss of some nitrogen from the earlier formed organs—stems, leaves, and roots (*see* Fig. 13-5). The translocation of many nutrients and food is a continuing process in plant growth, and different organs have a different priority for the materials.

Fruit or seed production has the highest priority. It is natural for nutrients accumulated in the vegetative parts to be translocated and used later for seed production. Nutrients in excess of fruit growth remain in other plant organs. This causes seeds of plants to have a similar composition when grown under widely different conditions, whereas the composition of vegetative parts may vary greatly.

Soils, Fertilizers, and Animal Health

We have noted that plants are commonly deficient in mineral elements and exhibit deficiency (disease) symptoms. It should be no surprise that plants may contain inadequate amounts of certain mineral elements for optimum animal health. One of the most striking cases is the phosphorus deficiency seen in cattle grazing on Oxisols low in available phosphorus in the tropics. In some cases, plants contain toxic amounts of elements that create animal diseases. We must be cautious, however, in making a sweeping generalization about soil fertility and *human* health.

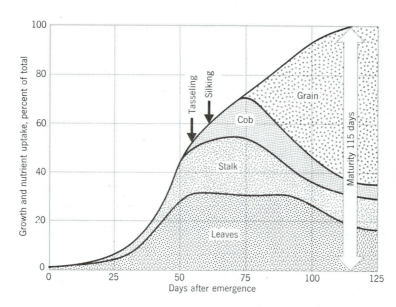

Figure 13-5
Part of the nitrogen, as well as phosphorus, potassium, and some other nutrients, are translocated from vegetative tissue to the grain during the later part of the growing season. An amount of nitrogen equal to about one-half the nitrogen in the corn grain was absorbed during the time the grain was produced. A similar amount of nitrogen was translocated from the other organs to the grain during the same period. (Data from Hanway, 1960.)

There is a great variation in plant composition due to genetic differences. Human diets commonly consist of foods from many locations with different soils and environments. Some major soil-plant-animal health relationships will be discussed in the following sections.

Plant and Animal Requirements Compared

Animals require over 20 different mineral elements, most of them the same as those required for plants. Boron, however, is needed by plants but not by animals. The oldest and most common dietary supplement used by people is salt (sodium chloride) to supply sodium, which is not essential for most plants. Animals, however, need chromium, iodine, and selenium in addition to the elements needed by plants. These elements are in soils and are absorbed by plants even though they are not required. Thus plants can be perfectly

healthy and unable to satisfy the dietary needs of animals.

Soil—Plant—Animal Health Relationships

Locations of mineral nutritional diseases in animals are shown in Fig 13-6. The large iodine-deficient or goiter belt in the northern states should be noted. Phosphorus deficiencies are widespread throughout the states, and selenium toxicity is common in the western states. Almost every state is represented by either a deficiency or toxicity of one or more nutrients for animals.

Phosphorus.   Low soil phosphorus commonly limits plant growth, and it is, perhaps, the most critical mineral element deficiency for grazing livestock. Weak bones and phosphorus deficiency of cattle that graze on grass is quite common. Some of the complex relationships be-

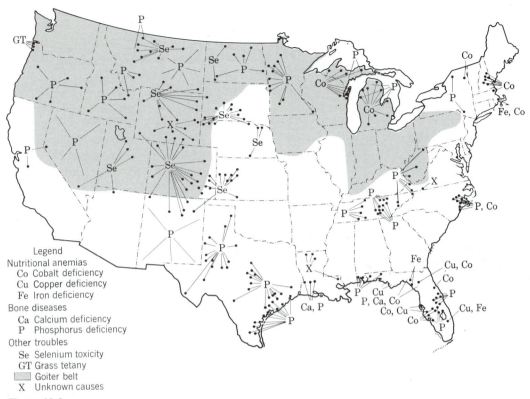

Legend
Nutritional anemias
Co Cobalt deficiency
Cu Copper deficiency
Fe Iron deficiency
Bone diseases
Ca Calcium deficiency
P Phosphorus deficiency
Other troubles
Se Selenium toxicity
GT Grass tetany
Goiter belt
X Unknown causes

Figure 13-6
Occurrence of mineral nutritional diseases in animals. The dots show the approximate locations of the observed deficiency. The lines not terminating in dots indicate a generalized area where specific locations have not been reported. The goiter region is also a generalized area.

tween phosphorus levels in soils and phosphorus deficiency in animals are shown in Fig. 13-7. As the amount of phosphorus fertilizer increased (or available soil phosphorus increased), the phosphorus content of oat grain and alfalfa hay were markedly increased. Since cattle require about 0.3 percent phosphorus in their diet for optimum growth, oat grain contained sufficient phosphorus without fertilization. Oat straw remained phosphorus-deficient for cattle, even with fertilization. Cattle or livestock with diets consisting mainly of grasses usually need

a phosphorus supplement to ensure adequate dietary phosphorus. The use of phosphorus fertilizer changed alfalfa hay from an inadequate source of phosphorus to an adequate source. By contrast, the racehorse industry has centered on the high phosphate soils of Kentucky and Florida.

Phosphorus is not a serious problem in human nutrition. Humans eat large amounts of cereal grain and meat, which are good phosphorus sources. For humans, the use of phosphorus fertilizer is more important for increasing food sup-

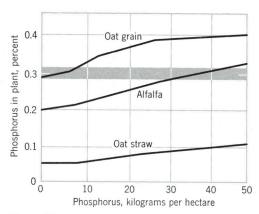

Figure 13-7
Effect of rate of phosphorus fertilizer application on concentration of phosphorus in oats and alfalfa grown on Clarion loam. About 0.3 percent of phosphorus (dashed line) is required by dairy cattle for normal growth. (Data from Allaway, 1975.)

ply than increasing the phosphorus content of plants.

Calcium.   Plants and animals need relatively large amounts of calcium. Calcium is important for formation of bones, teeth, and egg shells. Rarely, or only occasionally, are plants calcium-deficient because some other soil factor becomes limiting before low soil calcium limits growth. Adequate calcium supply in animal diets appears importantly related to differences in plant species. Red clover grown on acidic soils of the northeastern United States contains more calcium than grass grown on calcareous soils of arid and semiarid regions of the western United States. Snap beans and peas usually contain three to five times more calcium than corn or tomatoes. A diet for humans that is adequate in calcium depends on food selection. Liming acid soils enables people to grow more legume crops (beans, peas) that normally have a higher calcium content. Lime increases the soil calcium sup-

ply, however; it also tends to increase the calcium content of all crops.

Livestock diets commonly consist of large amounts of low calcium grasses and grains. Adding calcium to the diet is a common practice. Cows deficient in calcium produce less milk than cows on adequate calcium diets, however, the concentration of calcium in the milk is similar. This means that the milk of cows remains similiar in calcium regardless of its level in a cow's diet. Remedying a calcium deficiency in cattle diets results in more milk, not better milk.

Magnesium.   The similar behavior of calcium and magnesium in soils means that acidic soils low in calcium are usually low in magnesium. Magnesium deficiency is fairly common for plants growing on acidic, sandy soils. High levels of available soil potassium inhibit magnesium uptake by plants, sometimes causing magnesium deficiencies. As with calcium, the accumulation of magnesium in plants is strongly affected by species; leguminous plants are high in magnesium and while grasses, tomatoes, corn, and other nonleguminous crops are low in magnesium regardless of the soil magnesium level.

Grass *tetany,* or grass *staggers,* is the major animal disease caused by low magnesium in the forage of grazing cattle and sheep. Pregnant and lactating animals are the most susceptible. Animals become nervous, stagger, and fall. Injection of magnesium soon after the symptoms appear will result in recovery in a few minutes. Untreated cases severe enough to cause falling result in death if untreated. Magnesium is commonly added to diets of cattle grazing on green pastures.

High potassium fertilization has been associated with an incidence of grass tetany resulting from the effect of high soil potassium on reduced uptake of magnesium (cation-constancy effect). Magnesium fertilizers can be used to increase plant magnesium and to prevent tetany. Grass tetany tends to be more prevalent in the early spring and late fall when weather is cool, because more forage is made up of grasses at that time.

Cobalt. Cobalt is required by bacteria that fix nitrogen in the nodules of legumes. This is the only known need for cobalt by plants. Microorganisms in ruminants incorporate cobalt into vitamin B_{12}. Nonruminants, including humans, can not synthesize vitamin B_{12} and must obtain the vitamin in their diet.

Cattle and sheep that *are not* fed legumes almost always need supplementary cobalt. The relationship of soil cobalt and health of ruminants is one of the most striking examples of soil-plant relationship to animal health. People get their vitamin B_{12} from animal products such as milk, cheese, meat, and eggs. Only persons on a strictly vegetation diet are likely to have a diet deficient in vitamin B_{12}.

Cobalt deficiency was observed by the earliest settlers in parts of New England, but the cause of the disease was not recognized. In New Hampshire, it was called "Chocorua's Curse" and "Albany Ail" from local place names. In southern Massachusetts, it was known as "neck's ail," not because the disease affected the neck of the animal, but because it was most common on necks of land that projected into bays.

It is now known that cobalt deficiency in New England has its origin in the geologic history of soils. The low-cobalt soils of New Hampshire, for example, were formed in glacial deposits derived from

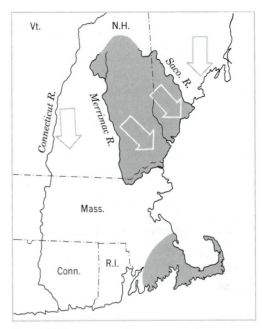

Figure 13-8
Low cobalt areas (shaded) in New England. The arrows indicate direction of glacial ice movement. (Data from Kubota, 1965.)

granite of the White Mountains, which contained very little cobalt (*see* Fig. 13-8).

Moreover, Spodosols in the area, formed from sandy glacial drift, tend to lose their cobalt via leaching. Cobalt moves downward at a more rapid rate than iron, especially in Spodosols of poor drainage.

Cobalt deficiency among grazing animals also is a problem on wet sandy soils of the lower coastal plain in the southeastern United States. The low-cobalt soils there are Humaquods—poorly drained Spodosols soils with organic pans. The original sandy deposits had little cobalt to begin with, and that was leached from the upper horizons as soil development proceeded. Consequently, there is little cobalt available to forage plants growing on these soils.

Copper. Plant copper deficiencies are most common on organic soils and very sandy soils (in Florida—*see* Fig. 13-6). Ruminants are sensitive to the copper deficiency that results from a diet of low copper feed. Copper fertilization of pastures will increase plant copper and prevent copper deficiency in animals. In parts of Australia, livestock production was impossible until copper fertilizers were used on the pastures. Care must be used in applying copper to soils to prevent toxic copper levels in both plants and animals. Monogastric animals, including people, are less sensitive to low copper levels in feed; no link has been found between copper in the soil and human health.

Iodine. Iodine is an element needed by animals, yet is not essential to plants. The relationship between iodine levels in soils and plants and the incidence of goiter has been one of the most striking soil-plant animal health relationships. The goiter belt where soils are low in iodine is shown in Fig. 13-6.

Iodine is taken up by plants and is passed along the food chain to animals. Iodine is volatilized over oceans, carried in the atmosphere, and carried down to the land in precipitation. The centers of continents are usually areas low in iodine, because they receive very little iodine in precipitation. Coastal areas receiving ocean spray and ocean-grown plants have abundant iodine. Iodine in plants can be increased by soil fertilization with iodine; however, the addition of iodine to salt has proved to be a simple and inexpensive solution to goiter prevention. The development of iodized salt is considered by some to be one of the greatest scientific contributions to human health.

Iron. Iron deficiencies are common in both plants and humans. Some nutritionists consider iron deficiency or anemia to be the most frequently observed dietary deficiency in people. Animals generally make efficient use of body iron. Iron deficiency in people is closely related to a loss of blood and inefficient body use of dietary iron. A major source of iron for humans is meat. The result is little or no relationship between soil-iron, plant-iron, and animal health (*see* Fig. 13-9).

Molybdenum. Plants and animals need very small amounts of molybdenum. Plants deficient in molybdenum needed for good animal nutrition have been found growing on certain acidic soils. The major nutritional problem is molybdenum toxicity (*molybdenosis*), which develops in grazing animals when forage has over 10 to 20 parts per million of molybdenum. The toxicity problem arises from the fact that forage plants have a wide range of tolerance for molybdenum, while animals do not. Legumes accumulate more molybdenum than common grasses, and they take up more molybdenum in wet soils than in dry soils. There is no effective method for reducing molybdenum uptake from soils by plants. The data in Fig. 13-9 show that large increases in molybdenum can occur in plants with increases of molybdenum in the growing medium.

Most problem areas of molybdenum toxicity are related to the geologic origin and wetness of soils. Common sources of molybdenum are granite, shales, and fine-grained sandstones.

Soils producing forages with high levels of molybdenum are generally confined to valleys of small mountain streams in the western United States. Only a very small part of any valley actually produces high

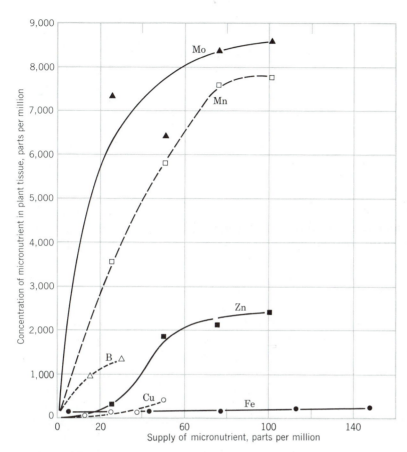

Figure 13-9
The iron (and copper) content of tomato leaflets was little affected by a large increase in supply of nutrient in the growing medium. (Data from Beeson, 1955.)

molybdenum forages. These soils are wet or poorly drained, alkaline, and high in organic matter. Also the alluvium from which they were formed was originally derived from granites or high-molybdenum shales. Molybdenosis in Nevada and California is associated with soils formed from the high-molybdenum-content granite of the Sierra Nevada Mountains.

A typical area where high-molybdenum forages grow and the effect of these forages on cattle, are shown in Fig. 13-10. Molybdenum toxicity in cattle has also been found where dusts from molybdenum-processing industries, or waste water from the tailings of uranium mines, has

contaminated pastures. Molybdenum toxicity has also occurred in cattle-grazing pastures on organic soils in Florida. Molybdenum toxicity has been found in Hawaii on volcanic ash soils, at high elevations where soil is wet most of the time.

Selenium. Selenium is not needed by plants but is needed in small amounts by warm-blooded animals and, probably, people. Large areas of the United States have soil so low in selenium that forages for livestock are deficient in this element. Selenium deficiency is a serious livestock problem. Acid soils formed from low-selenium parent materials are areas of likely

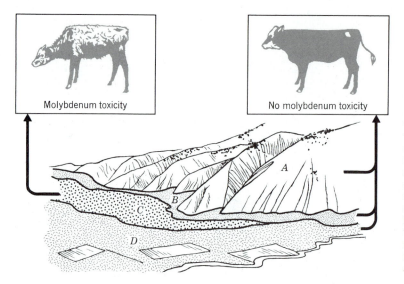

Figure 13-10
Plants containing toxic levels of molybdenum are found only on wet soils formed from high-molybdenum parent materials, area C. Areas A and B have well-drained soils in which the molybdenum is not readily available to plants. Area D is wet, but has soils formed from low molybdenum parent materials. (Data from Allaway, 1975.)

selenium deficiency in livestock. These conditions are prevalent in the northwestern, northeastern, and southeastern states. Much of the wheat in the United States is produced in areas where soils have adequate selenium, thus making bread a good source of selenium for people. People in the United States appear to have neither deficiencies or toxicities of selenium. Areas where forages and feed grains are likely to be deficient, adequate, or toxic for livestock, are shown in Fig. 13-11.

In 1934, the mysterious livestock maladies on certain farms and ranches of the Plains and Rocky Mountain states were discovered to be due to plants with so much selenium that they were poisonous to animals grazing there. Affected animals had sore feet, lost some of their hair, and many died. Over the next 20 years, scientists found that the high levels of selenium occurred only in soils derived from certain geologic formations of high selenium content. Another important discovery was that a certain group of plants, called *sele-*

nium accumulators, had an extraordinary ability to extract selenium from soil. These selenium accumulators were primarily shrubs or weeds native to semiarid and desert rangelands. They usually contained 50 parts per million or more of selenium, whereas range grasses and field crops growing nearby contained less than 5 parts per million.

An interesting story relating selenium toxicity to George Custer's fateful day at the Little Big Horn River in Montana on June 25, 1876 was written by E.V. Wilcox. He learned from an old horse packer that the horses used in the battle had been overwintered where soils were high in selenium and that the horses were allowed to forage native range plants because of a shortage of army hay. Enroute to the battle site, a legume selenium accumulator, *Astragalus bisulcatus,* was in its succulent and palatable stage. The horses likely grazed on it enroute. The horse packer also related that the horses had peculiar symptoms similar to those now known to be selenium toxicity. Studies of soils in the

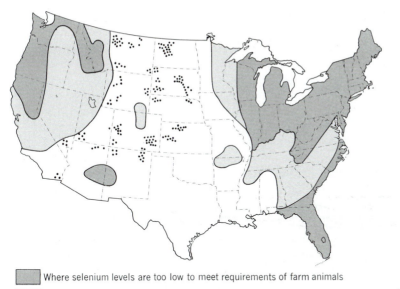

◼ Where selenium levels are too low to meet requirements of farm animals

☐ Where selenium is adequate to meet requirements of farm animals

▨ Where selenium is both adequate and inadequate in same locality

• Where selenium toxicity may be a problem

Figure 13-11
Areas where forages and feed crops contain various levels of selenium. (Data from Allaway, 1975.)

area made in the 1930s showed that Astragalus bisulcatus had 140 times more selenium content than nonaccumulator plants growing alongside. Sitting Bull's horses came from an area outside the selenium toxicity belt. Wilcox suggests that one factor in the defeat was the selenium toxicity and poor performance of Custer's horses.

Zinc. Both plants and animals develop zinc deficiencies. Zinc deficiencies in crops have frequently been observed on fields where soils have been graded smooth, so that irrigation water can be applied more uniformly. The deficiency occurs where the topsoil has been removed and a calcareous, low organic content soil is exposed. Research indicates that zinc fertilization of food and feed crops may be potentially useful in improving plants as

sources of dietary zinc. No close relationship, however, has been found between available zinc in soils and the occurrence of zinc deficiencies in people.

Soil Fertility Depletion and Nutritive Quality of Plants

We have noted that cattle of the early settlers in New England suffered cobalt deficiency on soils of low, but natural fertility. Shakespeare makes reference to goiter in mountainous regions now known to have soils low in iodine. It has been well documented that animal and human health have been adversely affected by mineral nutrient deficiencies in food grown on natural soils or on soils little affected by humans. Some people are concerned that soil depletion results in changes in plant composition, causing a decline in human health. Some of the most dramatic cases in humans or animals of nutritional deficiencies associated with soils date back a long time. They are due to naturally occurring deficiencies, instead of to those of soil depletion.

Scientists at Michigan State University established an experiment on a farm where the land had been neglected; soils were very acidic and depleted of nutrients from many years of cropping. Part of the land was used for crop production with no attempt made to restore the depleted nutrients or to correct soil acidity by liming; part of the land was limed and heavily fertilized. Two similar dairy herds were given feed only from these two areas to test for differences in nutritional quality of the crops. At the end of 5 years, no differences were observed in the nutritional quality of the milk, health, or reproduction of these herds.

The average concentration of phosphorus in food crops grown on commercial farms is probably as high or higher than that in similar crops grown 50 years ago, because phosphorus fertilizers have been used extensively. The concentration of iron and manganese in plants is controlled most often by factors that affect the ability of the plant to use these elements and seldom by the total amount of the element in the soil. The sodium, chlorine, and iodine required by people have been supplied by direct supplementation of diets. Also, changes in the levels of these elements in soil or in food crops would have no effect on human nutrition.

Of all the mineral elements required by people and animals, downward trends in concentration of zinc, magnesium, and possibly sulfur in food and feed crops, would appear to be more likely than for any other minerals. Even with these elements, any evidence of a decline must rest on circumstantial evidence such as increasing reports of magnesium deficiency in cattle or the need for zinc or sulfur fertilizers to obtain optimum crop yields. There is no evidence that a decline (if there has really been a decline) in the concentration of zinc, magnesium, or sulfur in food crops has any effect on the nutritionl status of people; this status is strongly dependent on food selection practices, dietary habits, and use of these elements in diet.

In considering the effects of human use of soils for agriculture, it is necessary to distinguish between depletion of the soil's supply of essential elements for plants and animals and deterioration of the soil due to the washing or blowing away of the surface soil to expose hard or rocky subsoil material. Depletion of the soil supply of essential nutrients has generally been recognized by agricultural research workers, and it has usually been corrected by proper use of fertilizers before there is any decline in the nutritional quality of the crops produced for people or animals. It is often more difficult to correct or reclaim areas that have been damaged by excessive erosion or soil blowing. Some historical records of failure of settlements in certain parts of the world can be attributed to a failure of crop production from the destruction of soil by erosion, or by the salting out of irrigated lands. There are no records where failure of settlements can be attributed to crops of poor nutritional quality that result from depletion of the soil supply of essential elements.

Other Soil—Animal Health Concerns

There are several health concerns that are not closely related to the mineral nutrient composition of plants, but are still related to soils. These include nitrate accumulation in plants, relation of soils to the vitamin content of plants, and potentially dangerous elements resulting from environmental pollution.

The Nitrate Problem

Nitrate by itself is not very toxic to animals, when present in food and feed crops. Under certain conditions, however, nitrates in the digestive tracts of animals or in stored foods may be converted to nitrites. Nitrites are very toxic to animals; once they are absorbed into the blood, they react with hemoglobin in a way that interferes with the transport of oxygen in the bloodstream.

High levels of nitrate in plants usually result from high levels of nitrate in soils, plus the impact of some environmental factor that interferes with the metabolism of nitrates into amino acids and protein. Nitrate accumulation in plants is favored by drought and cloudy weather. Extremely high levels of nitrate in soils, however, may cause high levels in plants, even though environmental conditions are favorable for plant growth and use of nitrates in plant to form amino acids and proteins.

Different plants show different tendencies to accumulate nitrates. Leafy vegetables, cereals, and grasses cut at the hay stage, and some annual weeds, are likely to contain high levels of nitrate. Pigweed is known as a nitrate accumulator. Perennial grasses and legumes are much less likely to be high in nitrate. Nitrate tends to remain high in leaves and low in seeds. Nitrates have not been a problem in the consumption of grains used as food or feed crops.

Forages high in nitrogen are especially dangerous to ruminants such as cattle and sheep. The rumen of these animals provides an environment conducive to reducing nitrate to toxic nitrite. Losses of cattle and sheep because of nitrite poisoning have been a serious problem in livestock production for many years. The problem is particularly acute during seasons of cloudy weather or drought, which interrupts plant growth.

High levels of nitrate in soils may result from excessive use of nitrogen fertilizers, excessive use of readily decomposable composts and manures, accumulation of nitrate from organic matter during fallow or drought periods, or breakdown of organic matter and green-legume manure crops. However, plants containing high levels of nitrate often have been found growing on soils that have not received any fertilizer, manure, or compost. These plants are likely to be nitrate accumulators. A major step in an attempt to control nitrate accumulation in plants is to use manures, fertilizers, and crop residues that will provide ample, but not excessive supplies of available nitrates for plants.

Soil Fertility and Vitamins in Plants

Vitamins are organic compounds synthesized in plants; animals and people need at least 14 vitamins to remain healthy. About 30 years ago, there was a growing concern about the effect of soils on the vitamin content of plants. Studies relating to carotene and vitamin C were numerous because of our great dependence on plants for these vitamins in our diet. It was found that yellow plants having an iron deficiency were low in carotene. When plants were green and grew normally, carotene content was a function of species and variety and not soil. Since it is uneconomic to grow deficient plants low in carotene, there is no need for concern.

The vitamin C level in plants is dependent on the amount of sunlight striking a plant. Tomatoes heavily fertilized and shaded by luxuriant vegetative growth have less vitamin C than unfertilized plants with less foliage and more exposure to the sun. The other vitamins appeared to be controlled mainly by plant species or variety. Human vitamin deficiencies have been critical over the years, but no vitamin deficiencies have been known to be caused by a soil deficiency where the food is grown.

Potentially Toxic Elements in Plants from Environmental Pollution

Potential poisoning of humans by arsenic, cadmium, lead, and mercury are real concerns today. Arsenic has accumulated in soils from sprays used to control insects and weeds and to defoliate crops in facilitating a harvest. Arsenic accumulates in soils and has injured crops, but has not created a hazard for humans or animals.

Cadmium poisoning has occurred in Japan from the dumping of mine waste into rivers, where fish accumulated cadmium. Cadmium is an industrial waste and appears in sewage sludge. Using soils for sewage disposals represents a potential danger. No natural soils have harmful levels of cadmium.

Lead is discharged into the air from automobile exhausts and other sources, and it eventually reaches the soil. In soil, lead is converted to forms unavailable to plants. The lead taken up tends to remain in roots. Soils must become very polluted with lead before significant amounts move into tops of plants. Mercury is discharged into the air and water from use in pesticides and industrial activities. Inorganic mercury is not highly toxic, but methyl mercury is. Under conditions of poor aeration, inorganic mercury is converted to methyl mercury. Plants do not take up mercury readily from soils; however soils should not be used to dispose of mercury because of the highly toxic nature of methyl mercury.

The soil is being used more and more for sewage disposal. The application of high levels of heavy metals to soils, and their potential uptake by plants, represents an important limitation in the use of soils for sewage disposal and industrial waste. Sewage known to be lacking in heavy metals, however, can be added to soils without any danger of heavy metal poisoning to humans or other animals.

Bibliography

Allaway, W. H., "The Effect of Soils and Fertilizers on Human and Animal Health," *USDA Agr. Information Bull., 378,* 1975.

Beeson, K. C., "The Effect of Fertilizers on the Nutritional Quality of Crops," in *Nutrition of Plants, Animals, and Man,* Centennial Symposium Proc., Michigan State University, E. Lansing, Mich., 1955.

Brown, J. R., "Plant Analysis," *Missouri Agr. Exp. Sta. Bull.,* SB881, 1970.

Dexter, S. T., et al., "Nutritive Values of Crop and Cow's Milk as Affected by Soil Fertility," *Mich. Agr. Exp. Sta. Quart. Bul., 32:* 352–359, 1950.

Grunes, D. L., P. R. Stout, and J. R. Brownell, "Grass Tetany of Ruminants," in *Advances in Agronomy,* Vol. 22, Academic, New York, 1970, pp. 331–374.

Hanway, J., "Growth and Nutrient Uptake by Corn," *Iowa State University Extension Pamphlet, 277,* Ames, Iowa, 1960.

Jackson, J. E., and G. W. Burton, "An Evaluation of Granite Meal as a Source of Potassium for Coastal Bermudagrass," *Agron. Journ., 50:*307–308, 1958.

Kubota, J.,"Soils and Animal Nutrition," *Soil Cons.,* November 1965, pp. 77–78.

Kubota, J., "Areas of Molybdenum Toxicity to Grazing Animals in the Western States," *Jour. Range Manag., 28:*252–256, 1975.

Kubota, J., "The Poisoned Cattle of Willow Creek," *Soil Cons.*, April 1975, pp. 18–21.

Snider, H. J., "Chemical Composition of Hay and Forage Crops as Affected by Various Soil Treatments," *Ill. Agr. Exp. Sta. Bull., 518,* 1946.

Stanford, G., and W. H. Pierre, "The Relation of Potassium Fixation to Ammonium Fixation," *Soil Sci. Soc. Alm. Proc., 11:*155–160, 1946.

The Fertilizer Institute, *"Our Land and Its Care,"* Washington, D.C., 1962.

Trelease, S. F., "Bad Earth," *Scientific Monthly, 54:*12–28, 1942.

Viets, F. G., and R. H. Hageman, "Factors Affecting the Accumulation of Nitrate in Soil, Water, and Plants," *USDA Agr. Handbook*, No. 413, Washington, D.C., 1971.

Wallace, A., et al., "Further Evidence Supporting Cation-Equivalent Constancy in Alfalfa," *Jour. Am. Soc. Agron., 40:*80–87, 1948.

Wilcox, E. V., "Selenium Versus General Custer," *Agr. Hist., 18:*105–107, 1944.

CHAPTER 14

SOIL EROSION AND ITS CONTROL

In the development of a new country, little attention is given to conservation. Usually, the natural resources that the country affords are present in such quantity that they appear inexhaustible. The limitation of supply is in manpower and the essentials of life that the country does not produce. Later, when population density has become great, the necessity for conserving resources becomes apparent and frequently acute. Often, as in the United States, the need for conserving soil is one of the last to be recognized. Our surplus production of a few crops, such as cotton, wheat, and corn, tends to obscure the waste in soil productivity that is being sustained. The problem as a whole attracted no widespread attention until about 1933. Since that time, public interest in the work has developed rapidly.

Soil Erosion Defined

The basic definition of the word *erosion* is to wear away. Since the earth was first formed, there has been a continual wearing away of the surface. Many agents are responsible, but the discussion here will be limited to cultivated fields. It will be restricted to water and wind erosion.

Types of Water Erosion

Erosion by water may be divided into four categories: (1) splash, (2) sheet, (3) rill, and (4) gully. Strictly speaking, sheet erosion refers to the uniform removal of soil from the surface of an area in thin layers. For sheet erosion alone to occur, it is necessary that there be a smooth soil surface, which is seldom the case. Usually a soil surface that is designated "smooth" contains small depressions in which water will accumulate. Overflowing from these at the lowest point, the water cuts a tiny channel as it moves down the slope. Duplicated at innumerable points, this process presently creates a surface that is cut by a multitude of very shallow trenches called *rills*. None of these may grow to appreciable size or depth, so the surface soil is rather uniformly removed from the field. Accordingly, sheet erosion and rill erosion work hand in hand; the combined process is usually called *sheet erosion,* as distinguished from gully formation.

Although sheet erosion may pass unnoticed by the average observer, gullies attract immediate attention. They disfigure the landscape and give the impression of land neglect and soil destruction (see Fig. 14-1). Not only do gullies result in soil loss, but the eroded material is usually deposited over more fertile soil at the foot of the slope. Also, fields dissected by gullies offer many problems in farming operations.

Gullying proceeds by three processes: (1) waterfall erosion, (2) channel erosion, and (3) erosion caused by alternate freezing and thawing. Usually, more than one process is active in a gully. Water falling over a soil bank undermines the edges of the bank, which then caves in, and the waterfall moves upstream. This process produces U-shaped gullies, particularly if the

Figure 14-1
Severe gully erosion cuts up fields and makes land unfit for crop production.

underlying soil material is soft and easily cut. Gullies that are V-shaped are produced by *channel erosion* through the cutting away of the soil by water concentrated in a drainageway. This type of gully usually forms when the underlying soil horizons are of a finer texture and more resistant to erosion than surface horizons. Soil loosened from sides of gullies by alternate freezing and thawing sloughs off, and it is then carried away by heavy rains.

A type of erosion that received little attention until recent years is the splashing or scattering of smaller soil particles by the impact of raindrops. At first this action seems trivial, but when consideration is given to the large number of raindrops that strike a square meter of soil surface during a 1-hour rain and the force with which they strike, it is seen that the net effect in loosening and moving soil particles may be considerable.

Geologic Erosion a Natural Process

Geologic erosion is a natural process that tends to bring the earth's surface to a uniform level. Whenever one part of the earth's surface is elevated above its surrounding portions, erosion immediately begins the work of leveling off the high land. The leveling process may result in a very rough topography in the early stages via the cutting of gullies or of canyons in a mountainous region, but the ultimate result is a comparatively level surface. Evidence of this geological process is seen in peneplanes, mesas, valley fills, alluvial plains and deltas, extensive deposits of wind-laid material, and numerous other geologic formations. Some concept of the extent of erosion activity may be gained from the fact that the Appalachian Mountains are about one-half their original height. The loss of surface soil is generally balanced by the gradual incorporation of less weathered material into the lower part of the profile (*see* Fig. 14-2). The net result is the maintenance of soil fertility by the gradual incorporation of more nutrient-rich material within the root zone, and the delay or inhibition in the evolution of highly weathered and infertile soils when viewed on a geologic time scale.

Much eroded material carried by rivers is derived by deep cutting of streams into relatively fresh, unweathered, and nutrient-rich rocks and sediments. Deposition of these materials on alluvial plains has created large areas of fertile soil along the major rivers of the world (*see* Fig. 14-3). The overflowing and deposition of silt along the Nile River is a classic example. Most Chinese live on alluvial soil that is the by-product of erosion.

The pervasiveness of erosion gives rise to the concept of soil as "rock en route to the sea." En route to the sea, however, soil particles may participate in multiple cycles of erosion, deposition, and soil profile evolution. Each cycle results in a progressive depletion of weatherable minerals and increasing soil infertility. This can be observed along a traverse from the Appalachian Mountains to the Piedmont, to the Upper Coastal Plain, and eventually to the Lower Coastal Plain near the Atlantic Ocean. It is obvious that there are many interesting aspects of water erosion. The discussion that follows will focus on erosion, which has accelerated as a result of the removal of vegetation in agriculture, forestry, and urbanization.

Reasons for Employing Water Erosion Control Practices

A relatively short trip through virtually any section of our country reveals evidences of soil mismanagement. In areas of undulating to rolling topography, slopes may be seen that have been denuded of the surface or plow soil, leaving the lighter-colored subsoil exposed. Again, many instances are seen of sandy soils that have been cleared of their forest cover, cropped for a few years until the virgin fertility was exhausted, and then abandoned to become covered with weeds and brush or to be blown about by the wind.

As pointed out in Chapter 15, many na-

Figure 14-2
An example where the rate of erosion nearly balances the formation of soil, resulting in thin soil.

Figure 14-3
Alluvial soils formed by the deposition of eroded material during floods are some of the most productive soils in the world.

tions have inadequate food supplies. A greatly increased food production will be needed to feed the expected world population in the future. This situation affords another reason for giving careful attention to soil conservation.

Damage Done to Agricultural Land

The damage done to farmlands in the United States can only be estimated roughly. Of land used for crop production, it is estimated that around 20,000,000 hectares (50,000,000 acres) have been rendered useless for crop production, and a similar amount is approaching that condition. An additional 40,000,000 hectares (100,000,000 acres), although still being cultivated, have lost one half or more of their surface soil; on a similar area, erosion is carrying on its insidious work of destruction. H. H. Bennett, former chief of the Soil Conservation Service, has expressed the opinion that erosion control "is the first and most essential step in the direction of correct land utilization on about 75 percent of the present and potential cultivated area of the nation."

In considering the damage done by erosion, one should keep in mind that a large share of soil lost by this process is surface or plow soil. It is this soil layer that contains the highest percentage, in an available condition, of many essential plant nutrients. Furthermore, studies of soil eroded from fields in many parts of the country have shown these to be made up largely of finer soil particles (clay, silt, very fine sand, and humus). These particles contain a higher percentage of several plant nutrients than coarser particles. A study in Wisconsin showed that, compared to the original soil, the eroded material contained 2.1 times more organic matter, 2.7 times more nitrogen, 3.4 times more available phosphorus, and 19.3 times more exchangeable potassium.

Erosion Damage Not Confined to Soil Loss

Erosion via water opens the way for at least five types of loss or damage:

1. The loss of water causing the erosion. It might have been useful in crop production had it entered the soil instead of running off over the surface.
2. The soil carried away by erosion frequently ceases to be of value in crop production; furthermore, the remaining soil, denuded of the surface or plow layer, is much decreased in productivity.
3. The soil carried away frequently causes much damage. Especially during gully formation, a layer of infertile subsoil may be deposited over an area of productive soil, thus greatly reducing its crop-producing power.
4. Another damage resulting from gully formation is the cutting up of fields into irregular pieces. As these gullies get too deep to cross with farm implements, a great inconvenience and a loss of efficiency in cultivating the land and planting and harvesting crops result.
5. The soil removed through erosion may be deposited in streams, har-

bors, and reservoirs, thus increasing floods, impeding navigation, and reducing water-storage capacity.
6. Soil particles transport adsorbed nutrients and pesticides into water courses.

Rapid Sedimentation of Reservoirs

The amount of sediment carried by various streams is enormous, as shown by the estimates made by the United States Geological Survey, given in Table 14-1. When water storage reservoirs are built on rivers carrying large amounts of sediment, the reservoirs quickly fill and lose their water-storage capacity. An extreme example of the loss of storage capacity as a result of erosion is furnished by the Washington Mills Reservoir at Fries, Virginia. In the course of 33.5 years, 83 percent of the storage capacity had been lost.

A study of the Lake Decatur municipal water-supply reservoir of Decatur, Illinois, showed that it had lost 1.0 percent of its storage capacity annually between the date of construction (1922) and 1936. Between 1936 and 1946 the annual rate of sedimentation had increased to 1.2 percent of capacity. Also, in a land-use project area near Pierre, South Dakota, the annual rate of silting of stock ponds was found to vary between 1.10 and 5.56 percent.

Roosevelt Dam, which supplies water for electric power and irrigation of the great Salt River Valley in Arizona, lost 7 percent of its storage capacity through sedimentation during the first 24 years of its existence. Similarly, the Elephant Butte Dam on the Rio Grande River in New Mexico decreased in storage capacity about 17 percent during the period between 1915 and 1947. Such losses of storage capacity are serious matters, especially

Table 14-1

Tons of Sediment Carried by Several Rivers

River	Tons Per Year	River	Tons Per Year
Hudson	240,000	Savannah	1,000,000
Susquehanna	240,000	Tennessee	11,000,000
Roanoke	3,000,000	Missouri	176,000,000
Alabama	3,039,000		

when the supply of water is scarcely adequate to meet demands.

Other Environmental Consequences

Sediment or soil is the major pollutant in the country, exceeding by 500 to 700 times the amount of sewage discharged into water. Over 50 percent of the population obtains its municipal water from surface water that is almost universally filtered to remove sediment. Harbors and streams are dredged to maintain navigable waterways. Sediment in streams has a negative effect on fish. Sediment buries fish eggs, reduces the penetration of sunlight, which in turn reduces plant growth (food) and reduces recreational quality and beauty. Sediment particles also serve as carriers of adsorbed pesticide residues and phosphorus from the land to surface waters.

Predicting Water Erosion Losses on Agricultural Land

Since 1930, many controlled studies on field plots and small water sheds have been conducted to study factors affecting erosion. This data forms the basis for the soil-loss prediction equation developed by Wischmeier and Smith. The soil-loss equation is:

$$A = R K L S C P$$

where A is the computed soil loss per unit area (tons per acre).

R, the rainfall factor, is the number of erosion-index units in a normal year's rain. The erosion index is a measure of the erosive force of specific rainfall.

K, the soil-erodibility factor, is the erosion rate per unit of the erosion index for a specific soil in cultivated continuous fallow, on a 9-percent slope 72.6 feet long.

L, the slope-length factor, is the ratio of soil loss from the field slope length to that from a 72.6-foot length on the same soil type and gradient.

S, the slope-gradient factor, is the ratio of soil loss from the field gradient to that from a 9-percent slope.

C, the cropping-management factor, is the ratio of soil loss from a field, with specified cropping and management, to that from the fallow condition on which the factor K is evaluated.

P, the erosion-control practice factor, is the ratio of soil loss with contouring, strip-cropping, or terracing to that with straight-row farming, up-and-down slope.

The six factors in the equation are the significant factors that influence soil loss by rainfall; each will be briefly discussed.

The Rainfall Factor (R)

The rainfall factor is a measure of the erosive force of specific rainfall. The erosive force or available energy is related to both quantity and intensity of rainfall. A 5-centimeter rain falling at 32 kilometers per hour (20 miles per hour) would have 6,000,000 foot-pounds of kinetic energy. The tremendous erosive power of such a rain is apparent, when you consider that it is sufficient to raise a 18-centimeter furrow slice of soil 1 meter (*see* Fig. 14-4). The four most intense storms in a 10-year period at Clarinda, Iowa accounted for 40 percent of the erosion and only 3 percent of the runoff on plots in corn-tilled, up-and-down slope.

The rainfall or R factor is the sum of kinetic energy times the maximum 30-minute intensity for each storm during the year. Rainfall factors have been computed for about 2,000 locations in states east of the Rocky Mountains and used to produce the isoerodent map shown in Fig. 14-5. The values range from 50 in western North Dakota to 600 along the gulf coast.

The rainfall-erosion index measures only the erosivity of rainfall and associated runoff. Therefore, the equation does not predict soil loss that is due solely to thaw, snowmelt, or wind. In areas where such losses are significant, they must be estimated separately and combined with those predicted by the equation for comparison with soil-loss tolerances.

The Soil-Erodibility Factor (K)

Soil factors that influence erodibility by water are: (1) those that affect infiltration rate, permeability, and total water capacity, and (2) those that resist dispersion, splashing, abrasion, and transporting forces of rainfall and runoff. The erodibility factor, K, has been determined experimentally for 23 major soils on which soil erosion studies were conducted since 1930. The soil loss from a plot 72.6 feet long on a 9-percent slope maintained in fallow, with all tillage up and down the slope, is determined and divided by the rainfall factor (for the storms producing the erosion); this is the erodibility factor (K). Values of K determined for 23 major soils are listed in Table 14-2.

Slope Length (L) and Slope Gradient (S) Factors

Slope length is defined as the distance from the point of origin of overland flow to either a point where the slope decreases to the extent that deposition occurs, or the point where runoff enters a well-defined channel. Runoff from the upper part of a slope contributes to the runoff produced on the lower part of the slope. This increases the water running over the lower part of the slope, thus creating more erosion on the lower part of the slope compared to the upper part. Studies have shown that erosion via water increases as the 0.5 power of slope length, and it is used as the basis for calculation of the slope length factor, L. This results in about a 1.3 times greater average soil loss per acre with a doubling of slope length.

Figure 14-4
Columns of soil capped by stones that absorbed the energy of raindrop impact and protected the underlying soil from erosion.

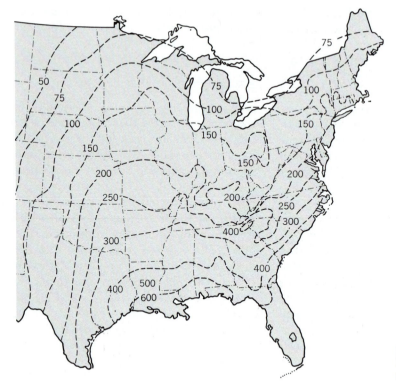

Figure 14-5
Rainfall factors (R) for the eastern United States.

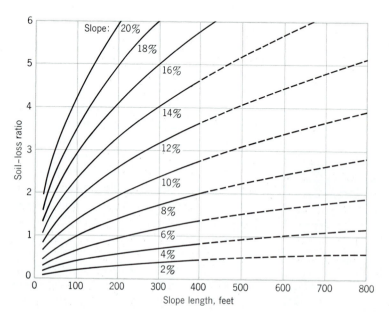

Figure 14-6
Slope-effect chart for the topographic factor LS.

Table 14-2
Computed K Values for Soils on Erosion-Research Stations

Soil	Source of Data	Computed K
Dunkirk silt loam	Geneva, N.Y.	0.69[a]
Keene silt loam	Zanesville, Ohio	0.48
Shelby loam	Bethany, Mo.	0.41
Lodi loam	Blacksburg, Va.	0.39
Fayette silt loam	LaCrosse, Wis.	0.38[a]
Cecil sandy clay loam	Watkinsville, Ga.	0.36
Marshall silt loam	Clarinda, Iowa	0.33
Ida silt loam	Castana, Iowa	0.33
Mansic clay loam	Hays, Kans.	0.32
Hagerstown silty clay loam	State College, Pa.	0.31[a]
Austin clay	Temple, Tex.	0.29
Mexico silt loam	McCredie, Mo.	0.28
Honeoye silt loam	Marcellus, N.Y.	0.28[a]
Cecil sandy loam	Clemson, S.C.	0.28[a]
Ontario loam	Geneva, N.Y.	0.27[a]
Cecil clay loam	Watkinsville, Ga.	0.26
Boswell fine sandy loam	Tyler, Tex.	0.25
Cecil sandy loam	Watkinsville, Ga.	0.23
Zaneis fine sandy loam	Guthrie, Okla.	0.22
Tifton loamy sand	Tifton, Ga.	0.10
Bath flaggy silt loam with surface stones 2 inches removed	Arnot, N.Y.	0.08[a]
Freehold loamy sand	Marlboro, N.J.	0.08
Albia gravelly loam	Beemerville, N.J.	0.03

From Wischmeier and Smith, 1965.
[a]Evaluated from continuous fallow. All others were computed from row crop data.

As the gradient or percent of slope increases, the velocity of runoff water increases, which increases its erosive power. A doubling of velocity of runoff water increases kinetic energy or erosive power four times, and causes a 32-time increase in the amount of material of a given particle size that can be carried. Splash erosion, the splashing into the air of soil particles by raindrop impact, causes a net downslope movement of soil; it also increases with slope gradient. Combined slope length and gradient factors (LS) for use in the soil-loss prediction equation are given in Fig. 14-6.

The Cropping-Management Factor (C)

We have observed the great erosive potential of rainfall on bare land on both long and steep slopes. A vegetative cover, however, can absorb the kinetic energy of falling rain drops and defuse the rain's erosive potential. Furthermore, vegetation by itself retains a significant amount of rain and slows the flow of runoff water.

As a result, the presence or absence of a complete vegetative cover essentially determines whether erosion will be a problem or be zero, for all practical purposes (*see* Fig. 14-7).

The C factor measures the combined effect of all interrelated cover and management variables, including type of tillage, residue management, time of soil protection by vegetation, and so forth. Complicated tables have been devised for calculating the C factor. As an illustration, the C factor for a 4-year rotation of wheat-meadow-corn-corn in central Indiana with conventional tillage, average residue and other management, and average yields is 0.119.

The Erosion Control Practice Factor (P)

In general, whenever sloping soil is to be cultivated and exposed to erosive rains, the protection offered by sod or closely growing crops in the system needs to be supported by practices that will slow runoff water, thus reducing the amount of soil carried. The most important of these practices for croplands are contour tillage, strip-cropping on the contour, and terrace systems.

Limited field studies have shown that contouring alone is effective in controlling erosion during storms of low or moderate intensity, but it provides little protection against the occasional severe storm that causes breakovers of the contoured rows. Contouring alone appears to produce maximum average protection on slopes in the range of 3 to 7 percent. P values for contouring are given in Table 14-3. Strip-cropping, along with contouring, provides more protection. In cases where both strip-cropping and contour tillage are used, the P values listed in Table 14-3 are divided by 2.

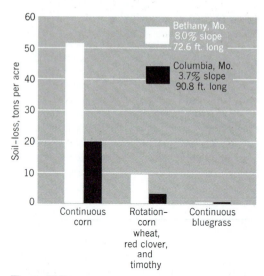

Figure 14-7
A continuous plant cover (e.g., bluegrass) is many times more effective for erosion control than cropping systems that leave the land bare most of the time. Continuous corn production resulted in the most erosion and the longest amount of time with soil unprotected by vegetation.

Terraces are an effective way to reduce slope length. To account for terracing, the slope length used to determine the LS factor should represent the distance between terraces.

Application of the Soil-Loss Equation

A Case Study

The procedure for computing the expected, average annual soil loss, from a given cropping system on a particular field, is illustrated by the following example.

Assume that there is a field in Fountain County, Indiana, on Russell silt loam, having an 8-percent slope about 200 feet long. The cropping system is a 4-year ro-

Table 14-3
Practice Factor Values for Contouring

Land Slope, %	P value
1.1 to 2	0.60
2.1 to 7	0.50
7.1 to 12	0.60
12.1 to 18	0.80
18.1 to 24	0.90

From Wischmeier and Smith, 1965.

tation of wheat-meadow-corn-corn (W-M-C-C) with tillage and rows on the contour, and with corn residues disked for wheat seeding and turned under in the spring for second-year corn. Fertility and residue managment on this farm are such that crop yields are rarely less than 85 bushels of corn, 40 bushels of wheat, or 4 tons of alfalfa-brome hay; the probability of meadow failure is slight.

The first step is to refer to the charts and tables discussed in the preceding section and to select the values of R, K, LS, C, and P that apply to the specific conditions on this particular field.

The value of the rainfall factor, R, is taken from Fig. 14-5. Fountain County, in west-central Indiana, lies between isoerodents 175 and 200. By linear interpolation, R = 185. The value of the soil-erodibility factor, K, is taken from Table 14-2. Soil scientists in the north-central states consider Russell silt loam equal in erodibility to Fayette silt loam, for which Table 14-2 lists K = 0.38.

The slope-effect chart (*see* Fig. 14-6) shows that, for an 8-percent slope, 200-feet long, LS = 1.41. For the productivity level and management practices assumed in this example, factor C for a W-M-C-C rotation in area 16 was shown to be 0.119.

Table 14-3 shows a practice-factor of 0.6 for contouring on an 8-percent slope.

The next step is to substitute the selected numerical values for the symbols in the erosion equation, and solve A. In this example, A = 185 × 0.38 × 1.41 × 0.119 × 0.6 = 7.1 tons of soil loss per acre per year (15.9 metric tons per hectare).

If planting had been up-and-down slope instead of on the contour, the factor P would have equaled 1.0, and the predicted soil loss for this field would have been 185 × 0.38 × 1.41 × 0.119 × 1.0 = 11.8 tons per acre (26.4 metric tons per hectare).

If contour farming had been combined with minimum tillage for all corn in rotation, the value of factor C would have been 0.075. The predicted, average annual soil loss from the field would then have been 185 × 0.38 × 1.41 × 0.075 × 0.6 = 4.5 tons per acre (10.1 metric tons per hectare).

Permissible Soil Loss on Cultivated Land

Soils on level upland sites that have no erosion become infertile in humid regions because of intensive weathering and leaching. Soils on very steep slopes erode so fast that soils remain thin and less fertile than if erosion was slower. Ideally, there is an optimum erosion rate for the maintenance of soil productivity. Unfortunately, this ideal rate is usually much less than the rate of erosion of sloping cultivated lands and the question arises: "What is the maximum rate of erosion permissible that is consistent with long-time maintenance of soil productivity and economic use of land?"

The data in Fig. 14-7 show that a soil loss with continuous blue grass resulted in a loss of about 0.03 tons annually or the loss of 2.5 centimeters of soil in about 5,000 years. By contrast, soil loss for continuous corn resulted in a loss of 2.5 centimeters of soil in about 4 years. We can not maintain all of the land as grass and produce the food we need. On the other hand, we obviously can not tolerate the loss of 2.5 centimeters of soil every 3 to 4 years. Establishment of permissible soil losses for various soils is a matter of collective judgment based on soil properties and economic considerations. Maximum, permissible soil losses for most soils in the United States have been judged to be between 1 and 5 tons per acre annually or 2.2 to 11 metric tons per hectare.

Loess is composed mainly of silt-sized particles that were transported and deposited as a result of wind action. The small size of a silt particle and lack of agents to stick particles together makes loess very erosive. In northern China, there is an extensive loess deposit that is locally over 80 meters thick. This loess is the major source of sediment for the Huang Ho or Yellow River and of sediment deposited on the vast north-China plain, which supports hundreds of millions of people. The landscape is very distinctive, with deeply entrenched streams and gullies. It has been estimated that the loess will be removed by erosion in 40,000 years, which represents a loss of about 15 tons per acre or 33 metric tons per hectare per year. The erosion rate is considered excessive by our standards, but in light of the need for food, the land must be cropped. Luckily, loess has good physical properties and fertility compared to most other common soil-parent materials.

The usefulness of the soil-loss prediction equation rests in its ability to estimate soil losses and to aid in the formulation of managements systems consistent with the long-time maintenance of soil productivity.

Effect of Erosion on Crop Growth and Production Costs

Erosion results in loss of soil with the greatest content of organic matter and nitrogen; thus, erosion is particularly detrimental to nonleguminous grain crops. The reduced nitrogen-supplying power of soil can be restored by the use of a nitrogen fertilizer; however, this increases the cost of production.

Exposure of argillic horizons, because of erosion, reduces infiltration and increases runoff, which results in both more erosion and reduced available water for crops. Seed bed preparation is more difficult and stands are reduced with constant seeding rates. Summarization of data from many studies showed that 2.5 centimeters of soil loss reduced yields of wheat 5.3 percent, of corn 6.3 percent, and of grain sorghum 5.7 percent. Some effects of erosion on crop growth and production costs are given in Table 14-4. The increased cost of production resulting from erosion justifies a certain amount of investment to control erosion within permissible limits.

Rainfall Erosion on Urban Lands

About 50 percent of the sediment in streams and waters throughout the country originates on agricultural land; the other 50 percent originates on urban lands. We have observed that erosion rates can be very high on exposed land on steep slopes. In fact, exposure of land during highway and building construction and subdivision development commonly results in erosion rates many times greater than erosion that typically occurs on agricultural land. Urbanization may result in making 50 percent or more of the land

Table 14-4
Effects of Erosion on Corn Growth and Production Costs

	Degree of Erosion		
	Slight	Moderate	Severe
Depth of topsoil inches	12	7	5
Decrease in height of plant during early stages of growth (from slight), percent	—	13	22
Stand at harvest, percent of planting rate	87	83	76
Corn yield, bushels per acre	112	96	87
Soybean yield, bushels per acre	43	29	16
Reduction in yield for all crops (from slight), percent	—	17	26
Increase in production cost (from slight), percent	—	20	56

From Beasley, 1974.

surface impervious to water because of roads, buildings, and the like. Runoff is greatly increased. Erosion losses as large as 350 metric tons per hectare (100,000 tons per square mile) or about 2.5 centimeters per year have been reported during urbanization. Planning construction that will leave land exposed for the shortest period of time is important in maintaining the quality of local waters (*see* Fig. 14-8).

Wind Erosion

Wind erosion is indirectly related to water conservation in that a lack of water leaves land barren and exposed to wind. Wind erosion reaches its greatest extent in semi-

Figure 14-8
Slurry truck spraying fertilizer and seed on freshly prepared seedbed along a recently completed road in Missouri. A thin layer of straw mulch will be secured with a thin spray of tar to protect the area until vegetation becomes established. (Photo courtesy of USDA.)

arid and arid regions. Nevertheless, much damage is caused to both crops and soils in humid areas by wind-blown soil, although the phenomenon is less spectacular and attracts comparatively little attention in these regions. In the United States, more attention has been given to the control of wind erosion on the Great Plains than in other sections of the country. In North Dakota, wind-erosion damage was reported as early as 1888; in Oklahoma, it was reported 4 years after breaking of the sod. Blowing soil was one of the hazards confronting early settlers. The control of wind erosion was one of the first problems studied by agricultural experiment stations in the Plains states.

Three Types of Wind Erosion

Soil particles move in three ways during wind erosion. Fine soil particles (0.1 to 0.5 millimeters in diameter) are rolled over the surface by direct wind pressure, then suddenly jump up almost vertically from a short distance to 20 to 30 centimeters. Once in the air, particles gain velocity and then descend in an almost straight line, not varying more than 6 to 12 degrees from the horizontal. The horizontal distance traveled by a particle is four to five times the height of its jump. Upon striking the surface, particles may rebound into the air or knock other particles into the air and come to rest themselves. The major part of soil carried by wind moves by this process, which is called *saltation*. It is interesting to note that around 93 percent of the total soil movement via wind takes place below a height of 30 centimeters; probably 50 percent or more occurs between 0 and 5 centimeters.

Very fine dust particles are protected from wind action, because they are too small to protrude above a minute viscous layer of air that clings to the soil surface.

As a result, a soil composed entirely of extremely fine particles is resistant to wind erosion. These dust particles are thrown into the air chiefly by the impact of particles moving in saltation; however, once in the air, their movement is governed by wind action. They may be carried very high and over long distances via *suspension*.

Relatively large particles (between 0.5 and 1.0 millimeter in diameter) are too heavy to be lifted by wind action, but they are rolled or pushed along the surface by the impact of particles in saltation. This process is called *surface creep*. Between 50 and 75 percent of soil is carried in saltation, 3 to 40 percent in suspension, and 5 to 25 percent in surface creep.

From these facts, it is evident that wind erosion is due principally to the effect of wind on particles of a size suitable to move in saltation. Accordingly, wind erosion can be controlled: (1) if soil particles can be built up into clusters or granules of too large a size to move in saltation, (2) if wind velocity near the soil surface can be reduced by ridging the land, by vegetable cover, or even by developing a cloddy surface, and (3) by providing strips of stubble or other vegetative cover sufficient to catch and hold particles moving in saltation. Some management practices designed to provide these conditions are discussed in the following paragraphs.

Factors Affecting Wind Erosion and Its Control

The factors that affect wind erosion are contained in the wind erosion equation:

$$E = f(I, K, C, L, V)$$

where

E = soil loss in tons per acre
I = soil erodibility

K = soil roughness
C = climatic factor
L = length of field
V = quantity of vegetative cover

Soil erodibility is related primarily to texture and structure. As the clay content of soils increases, aggregation of the surface creates clods too large to be transported by wind. A study in western Texas showed that soils with 10 percent clay eroded 30 to 40 times faster than soils with 25 percent clay (*see* Fig. 14-9).

Rough surfaces reduce wind erosion by reducing wind velocity and trapping soil particles. Surface roughness can be increased with the tillage operations shown in Fig. 14-10. Tillage and rows are positioned at right angles to the wind to be most effective.

Climate is a factor in wind erosion via wind frequency and velocity and wetness of soil during high-wind periods. Wind

erosion and blowing dust are persistent features of deserts where plants are widely spaced and dry soil is exposed most of the time. Desert pavements are created as fine particles, blown away, and then gravel accumulates (*see* Fig. 14-11). Consequently, deserts are an important source of loess.

The Dust Bowl of the 1930s was associated with a period of unusually dry years. Both wheat yields and wind erosion were related to rainfall. The demise of the Dust Bowl coincided with years of increased rainfall.

Wind erosion increases from zero, at the edge of a field, to a maximum with increasing distance of soil exposed to prevailing winds. As particles bounce and skip, they create an avalanche effect on soil movement. Windbreaks of trees and alternate strips of crops can be used to reduce field length. Finally, wind erosion is inversely related to the degree of vegetative cover. Growing crops and residues of previous crops, when left on the surface, are effective in controlling wind erosion.

Planting crops in strips at right angles to prevailing winds is valuable in reducing soil erosion. The width of the strips is determined by the nature of the soil, exposure to wind, and similar factors.

Crop residues are an effective soil protection against wind erosion. Small-grain stubble reduces wind velocity; it also catches soil particles moving in saltation. Strips of stubble left at frequent intervals across a field being fallowed or fitted for a spring crop form effective barriers. The height of the stubble, as well as wind velocity, influence the width of the strip needed.

Crops should be cultivated in such a way as to leave residues on the surface; in other words, cultivate beneath them. This result can be accomplished by the use of sweep-type cultivators. The stubble-mulch

Figure 14-9
Amount of wind erosion in relation to the amount of clay in the soil. (Data from Chepil, et al., 1955.)

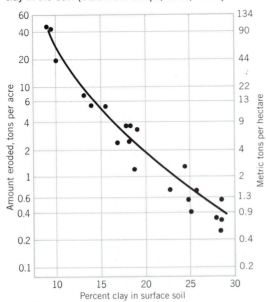

Figure 14-10
Listing on the contour not only protects the listed field from wind action, but it also may collect much soil blowing from an adjoining field, as illustrated in this Oklahoma field. The velocity of the wind is indicated by the posture of the man in the foreground.

system of soil management has many possibilities and should be carefully investigated in areas where wind or water erosion is a serious hazard.

Deep Plowing for Wind Erosion Control

Deep plowing has been used to control wind erosion where sandy surface soils are underlain by Bt horizons that contain 20 to 40 percent clay. Scientists of the United States Department of Agriculture, and the Agricultural Experiment Stations of Kansas and Texas, observed that at least 1 centimeter of subsoil should be plowed up for every 2 centimeters of surface soil thickness to control wind erosion. In many cases, the soil must be plowed to a depth of 50 to 60 centimeters. This kind of plowing increased the clay content of the surface soil by 5 to 12 percent in some

cases. Since it was found that about 27 percent clay in the surface soil was required to halt blowing soil, deep plowing by itself resulted in only partial and temporary control of wind erosion. When

Figure 14-11
Desert pavement created by wind erosion in the desert of the southwestern United States.

deep plowing was not accompanied by other erosion control practices, wind erosion removed clay from the surface soil and the effects of deep plowing were shortlived.

Trees for Windbreaks on Organic Soils

Organic soil areas of appreciable size are frequently protected from wind damage by planting tree windbreaks around them. In addition, rows of trees are often planted across the area at right angles to the prevailing wind. Willows have been found satisfactory for this purpose, especially since they grow very rapidly in organic soil. The use of trees as windbreaks must be limited because of the large amount of soil they take out of crop production due to their extensive root system. A number of shrubs also make good windbreaks. Spirea is frequently used.

Moist muck does not blow appreciably. Accordingly, use of an overhead irrigation system is helpful during windy periods, before the crop cover is sufficient to prevent soil erosion. Rolling muck with a very heavy roller induces moisture to rise more rapidly by capillarity, thus dampening the surface layer. This practice is not effective unless soil layers below the surface are quite moist.

Summary of Wind Erosion Control Principles

Wind erosion is a major problem of sandy soils when used for crop production in regions with ustic and aridic soil moisture regimes. Control is based on one or more of the following:

1. Trap soil particles with rough surface (tillage) and/or the use of crop residues and strip-cropping.

2. Deep-plow to increase clay content of surface soil.

3. Protect surface soil with complete vegetation cover.

Bibliography

Baver, L. D., "How Serious is Soil Erosion?," *Soil Sci. Soc. Am. Proc., 14:* 1–5, 1950.

Beasley, R. P., "How Much Does Erosion Cost?," *Soil Survey Horizons, 15:*8–9, 1974.

Bennett, H. H., *Soil Conservation*, McGraw–Hill, New York, 1939.

Browing, G. M., R. A. Norton, A. G. McCall, and F. G. Bell, "Investigation in Erosion Control and the Reclamation of Eroded Land at the Missouri Valley Loess Conservation Experiment Station, Clarinda, Iowa," *USDA Tech. Bull., 959,* 1948.

Chepil, W. S., N. P. Woodruff, and A. W. Zingg, "Field Study of Wind Erosion in Western Texas," *Kansas and Texas Agr. Expt. Sta., and USDA,* 1955.

Chepil W. S., "Erosion of Soil By Wind," *Soil,* USDA Yearbook, Washington, D.C., 1957, pp. 308–314.

Glymph, L. M., and H. C. Storey, "Sediment—Its Consequences and Control," *Agr. and the Quality of Our Environment,* Am. Assoc. Adv. Sci., Pub. 85, Washington, D.C., 1967.

Lyles, L., "Possible Effects of Wind Erosion on Soil Productivity," *Jour. Soil and Water Con., 30:*279–283, 1975.

Massey, H. F., and M. L. Jackson, "Selective Erosion of Soil Fertility Constituents," *Soil Sci. Soc. Am. Proc., 16:*353–356, 1952.

Robinson, A. R., "Sediment," *Jour. Soil and Water Con., 26:*61–62, 1971.

Tuan, Yi-Fu, *China,* Aldine, Chicago, 1969.

Wischmeier, W. H., and D. D. Smith, "Predicting Rainfall-Erosion Losses from Cropland East of the Rocky Mountains," *USDA Agr. Handbook, 282,* 1965.

Woodruff, N. P., and F. H. Siddoway, "A Wind Erosion Equation," *Soil Sci. Soc. Am. Proc., 29:*602–608, 1965.

CHAPTER 15

POPULATION, FOOD, AND LAND

Two major world problems today are population growth and food supply. These problems are not universal for all regions. Agricultural surpluses in North America have had a depressing effect on the prices of farm products for over 150 years and are continuing to do so. By contrast, during this past decade, there has been a serious decrease in per capita food production in Africa. This chapter is devoted to an analysis of the world's soil or land resources and the major technological problems associated with efforts to increase world food production.

Population Trends

Human population growth and food supply are ancient problems. For several million years, humans migrated onto new lands, slowly improving their hunting and gathering techniques. The population increased slowly; about 10,000 years ago, there were an estimated 5,000,000 people throughout the world. At this time, people had just begun to colonize the last remaining continents—the Americas—by crossing over the land bridge between Asia and Alaska during the last ice age. Settled agriculture is also believed to have developed about 10,000 years ago, which resulted in increased food production and a population explosion. After achieving a population of 5,000,000 over several million years, the human population increased 16 times to 86,000,000 within the next 4,000 years (10,000 – 6,000 BP). Population growth tended to level off, and the world population increased only six to seven times to 545,000,000 over the next 6,000 years. This was followed by the Industrial Revolution, beginning in about 1650, that resulted in improved food production. The Industrial Revolution also added a new dimension to population growth: a reduced death rate. Now the world population increases annually at the rate of 1.7 percent and doubles every 40 years. The most rapid growth is in Africa, where the annual increase is 2.9 percent and the doubling time is 24 years. The slowest population growth is in western Europe, where the annual rate of increase is 0.2 percent and the doubling time is 423 years. Austria and West Germany have negative-population growth rates.

Estimates of world population exceeding 6,000,000,000 by the year 2000 will require about 50 percent more food to just maintain the *status quo*. Considering that 500,000,000 people currently suffer malnutrition, and increased food consumption will occur in areas with rising incomes, it appears that food needs by the year 2000 will be about double of today's needs. Beyond the year 2000, it appears that there will be a world population of 12,000,000,000 or more by the 22nd century.

World Food Production

Historically, increases in world food production were due mainly to expansion onto new lands. Development of settled agriculture and the Industrial Revolution provided the means for greatly increasing food production on existing lands. Today, there is still a significant development of new land; during the period 1950–1975,

Table 15-1

World Grain Production, Total and Per Capita: 1950–1980

Year	Population, Billions	Grain Production, Million Metric Tons	Grain Production Per Capita, kg
1950	2.51	631	251
1960	3.03	863	285
1970	3.68	1,137	309
1971	3.75	1,237	330
1972	3.82	1,197	314
1973	3.88	1,290	332
1974	3.96	1,256	317
1975	4.03	1,275	316
1976	4.11	1,384	337
1977	4.18	1,378	330
1978	4.26	1,494	351
1979	4.34	1,437	331
1980 (prel.)	4.42	1,432	324

From Brown, L.R., "World Food Resources and Population: The Narrowing Margin," *Pop. Ref. Bur. Bull. Vol. 36*, No. 3, 1981.

when world cereal production increased 98 percent, 22 percent of this increase was due to an increased land area.

In recent decades, during a period of rapid population growth, world food production has been able to keep pace with population growth; in fact, it has exceeded the rate of population growth. It appears, however, that the per capita grain production throughout the world is leveling off, and maybe a declining trend is beginning according to the data in Table 15-1.

Increases in food production have been variable from region to region during the decade 1971–1980, as shown in Fig. 15-1. There was a larger annual increase in food production of 3.2 percent in the de-

veloping countries compared to 1.9 percent for the developed countries. This is due, in part, to the low base from which yield increases occurred in the developing countries. There are several interesting observations to be made from the data in Fig. 15-1. The ratio of food increase to population increase is 2:4 for the developed countries and only 1:5 for the developing countries due to differences in population growth. Also, the population in Africa increased much more rapidly than its food production during the 1970s. This caused the per capita food production to decrease 14 percent.

The food situation in the developing regions has worsened compared to the developed regions during the past few de-

Figure 15-1
Annual rates of change of food production and population in developing and developed countries and regions, 1971–1980. (Data from *The State of Food and Agriculture 1980,* FAO, 1981.)

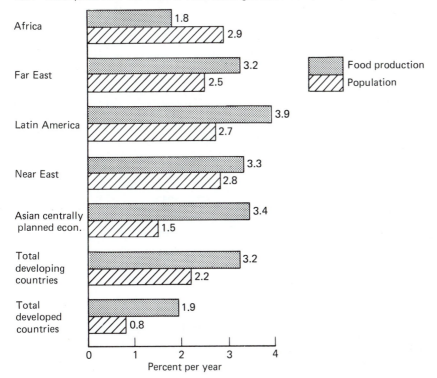

cades. Per capita cereal production in the developing countries increased only 11 percent during the 1950s–1970s, while it increased 38 percent in the developed countries (*see* Fig. 15-2). In addition, per capita production was 318 percent greater in the developed countries during the 1970s. In North America, where over 50 percent of the grain produced is used as livestock feed, the per capita cereal production is over five times greater than in the developing countries.

Before 1940, there was a net flow of grain from Latin America, eastern Europe, the Soviet Union, Africa, and Asia to western Europe. In fact, during the pe-

Figure 15-2

Per capita production of cereals by regions for 1950–1959, 1960–1969, and 1970–1979. (Data from *The State of Food and Agriculture 1980*, FAO, 1981.)

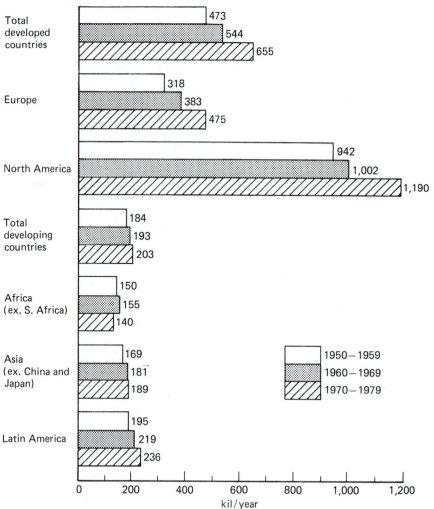

riod 1934–1938, Latin America exported 9,000,000 metric tons of grain, compared to 5,000,000 metric tons for North America. By 1980, the situation had changed drastically, with North America exporting 131,000,000 metric tons and Australia and New Zealand exporting 19,000,000 metric tons. The rest of the world were importers, of millions of metric tons imported as follows: Latin America 10, western Europe 16, eastern Europe and the Soviet Union 46, Africa 15, and Asia 63. By 1980, North America had become the world's "breadbasket." During the past decade, about 50 to 60 nations have developed declining rates of per capita food production. Of the 43 developing nations in Africa, 30 have had a declining per capita food production between 1971–1980 (*see* Fig. 15-3).

Available Land and Soil Resources

About 65 percent of ice-free land has a climate suitable for some cropping. Only 10 or 11 percent of the world's land is cultivated, which suggests a considerable possibility for increasing the world's cultivated land. Many estimates of potential cropland or arable land have been made; several well-documented studies conclude that the cultivated land could at least be doubled. The data from one of the studies which is shown in Table 15-2, lists the potentially arable and cultivated land by continents. The data support the following conclusions:

1. The world's arable land can be increased 100 percent or more.
2. Of the potentially arable land that

Figure 15-3
The geographic distribution of declining annual rates of change in per capita food production, 1971–1980. (Data from *The State of Food and Agriculture, 1980,* FAO, 1981.)

Table 15-2
Present Cultivated Land on Each Continent Compared with Potentially Arable Land

Continent	Area, Billions of Acres[a]				Acres of Cultivated Land per Person
	Total	Potentially Arable	Cultivated	Potentially Arable minus Cultivated	
Africa	7.46	1.81	0.39	1.42	1.3
Asia	6.67	1.55	1.28	0.27	0.7
Australia and New Zealand	2.03	0.38	0.08	0.30	2.9
Europe	1.18	0.43	0.38	0.05	0.9
North America	5.21	1.15	0.59	0.56	2.3
South America	4.33	1.68	0.19	1.49	1.0
U.S.S.R.	5.52	0.88	0.56	0.32	2.4
Total	32.49	7.88	3.47	4.41	1.0

From *The World Food Problem, Vol. 2,* The White House, May 1967.
[a]To convert to hectares multiply by 0.405.

can still be developed, 66 percent is located in Africa and South America.

The 66-percent potentially arable land that can be developed in Africa and South America is located mainly in tropical and subtropical areas. China and India, the two largest and most populated countries in Asia, have latitudes more temperate and subtropical similar to the United States and Canada. India is unique, because as a large country it has, perhaps, the greatest agricultural potential of any country, when soils and climate are taken into consideration. This is reflected by the fact that 50 percent of India's land area is cultivated compared to 10 percent of the total world and 20 percent of the United States.

The developing countries, with 77 percent of the world's population, have 54 percent of the current cultivated land area and 78 percent of the undeveloped, potentially arable land. By contrast, the developed countries have 23 percent of the world's population, 46 percent of its cultivated land, and 28 percent of its undeveloped, potentially arable land. Currently, the developed countries have about two times more cultivated land on a per capita basis.

On a global basis, the main limitations for using the world's soil resources for agricultural production are drought (28%), mineral stress (23%), shallow depth (22%), water excess (10%), and permafrost (6%). Only 11 percent of the world's soils are without serious limitations (*see* Table 15-3). Since the currently cultivated land represents the world's best land, development of unused, potentially arable land would be expected to have greater limitations than those shown in Table 15-3.

Improvements in the world food situation in the future will be mainly a problem, technologically speaking, of increasing production in tropical and subtropical Africa and South America; that is where most undeveloped, potentially arable land

Table 15-3
World Soil Resources and their Major Limitations for Agricultural Use

	Drought	Mineral Stress[a]	Shallow Depth	Water Excess	Permafrost	No Serious Limitations
			% of total land area			
North America	20	22	10	10	16	22
Central America	32	16	17	10	—	25
South America	17	47	11	10	—	15
Europe	8	33	12	8	3	36
Africa	44	18	13	9	—	16
South Asia	43	5	23	11	—	18
North and Central Asia	17	9	38	13	13	10
Southeast Asia	2	59	6	19	—	14
Australia	55	6	8	16	—	15
WORLD	28	23	22	10	6	11

From *The State of Food and Agriculture 1977*, FAO, 1978.
[a]Nutritional deficiencies or toxicities related to chemical composition or mode of origin.

is located and per capita food production is low. Almost 50 percent of the South American continent is centered on the tropical Amazon basin and the central uplands, where soils are mainly Oxisols and Ultisols. Many lands are used by shifting cultivators. This is reflected in the fact that 47 percent of soil-use limitations are for mineral stress and only 17 percent for drought. In Africa, 44 percent of the area has an arid or semiarid climate. Many Oxisols and Ultisols also exist in the Congo basin, but an enormous area of Ustalfs occur along the sub-Saharan region. This reflects the 44-percent drought and only 18-percent mineral stress limitations.

The data presented support the conclusion that the developing countries, compared to the developed countries, presently have about 50 percent as much cultivated land per capita. In addition, there are nearly equal amounts of undeveloped, potentially arable land per cap-

ita, but the development of land will require large imputs, because the soils are largely Oxisols and Utisols with a significant amount of Alfisols. In the developed countries, new development will be largely on high base-status soils, such as Mollisols and Alfisols. The two major soil limitations suppressing production are infertility and drought.

Worldwide Loss of Cropland

Cereals account for 70 percent of the world's cropland. It has been estimated that world cereal area will increase 9.2 percent between 1980 and the year 2000. During this period, the world population is expected to increase 40 percent, so that a considerable increase in crop yields on existing land will be needed to maintain per capita food production. This will be a big task, because the best lands are now being used and the worldwide loss of cropland will make the task more difficult (*see* Fig. 15-4).

The approximately 2,000,000,000 people that will be added to the world population by the year 2000 will use up a significant amount of cropland as cities expand on to an estimated 25,000,000 million hectares (60,000,000 acres). Homes will be built in the countryside of the developing countries, as many farmers currently have small houses completely surrounded by cropland. Other nonagricultural uses that will require expansion and use up some cropland are factories, other urban buildings, transportation facilities, water storage reservoirs, and strip mines.

There will be considerable land abandonment due to soil erosion, salinity and water-logging of irrigated land, and desertification in arid regions. In some irrigated regions, geologic water will be exhausted; in other cases, the high cost of pumping and competition with urban water users will result in the conversion of irrigated cropland to range and pasture land. A recent United Nations report estimated that 10 percent of the world's total irrigated land has had a significant decline in productivity due to water-logging, with another 10-percent productivity decline due to salinity. As agriculture production is pushed onto lands with greater production limitations, land deterioration will accelerate (*see* Fig. 15-5).

At the International Soil Science Congress held in Edmonton, Canada in 1978, concern was expressed about Canada's ability to remain food self-sufficient by the year 2000 and beyond. A population growth of 20 to 45 percent is expected, saline seep is a growing problem on wheatlands, and cities continue to expand onto prime farmland. The amount of land for cereal grains per capita is expected to decline from the present 0.184 hectares to 0.128 hectares by the year 2000. Only three Prairie provinces are surplus food producers; only 10 percent of the food produced is now exported.

Figure 15-4
Farmland in the United States is converted to urban use at the annual rate of about 400,000 hectares or 1,000,000 acres.

Figure 15-5
Farmers are pushed higher and higher on mountain slopes in the Himalayan foothills due to increased population growth. Many such lands are used only a few years before it is no longer profitable to crop the land due to severe soil erosion.

Declining Rate of Yield Increases

The change in North American grain exportation from 5 metric tons in 1931 to 131 metric tons in 1980, despite a large population increase, was largely due to the application of science to the field of agriculture after World War II. In the state of New York, wheat yields increased only 3.1 bushels per acre (209 kilograms per hectare) over the 70 year period from 1865 to 1935. During the next 40 years (1935–1975), wheat yields increased 20.1 bushels per acre (1,352 kilograms per hectare). Fifty-one percent of the increase was due to technology; forty-nine percent was due to improved varieties of wheat (*see* Fig. 15-6).

The development and use of better crop varieties and technological imputs can produce similar yield increases for developing countries with low crop yields. This is, however, a one-time phenomenon because of the difficulty of increasing yields when yields are already high. The situation has been aggravated by the greatly increased cost of fuel and fertilizer since 1973. In the 1950s a kilogram of fertilizer in North America increased grain yield by 15 to 20 kilograms, and a bushel of grain purchased 1 barrel of oil. Now, a kilogram of fertilizer increases yields about 5 kilograms and 10 bushels of grain are needed to buy 1 barrel of oil.

Technological Advances

The future is uncertain with regard to technological development. Some of the most promising research projects for increasing agricultural production are oriented toward: (1) greater photosynthetic efficiency, (2) improved biological nitrogen fixation, (3) new techniques for genetic improvement, (4) more efficient nutrient and water uptake and use by plants, and (5) more resistance of plants and animals to withstand environmental stresses. Large increases in yield levels in developed countries will depend on some technological break-throughs. Yields are cur-

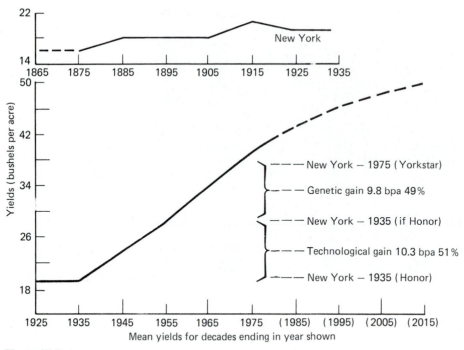

Figure 15-6
Wheat yields in New York state from 1866–1975. Between 1935 and 1975, yield increase was due 51 percent to technology and 49 percent to genetic gain. (Adapted from Jensen, 1978. Used by permission of *Science*.)

Future Outlook

Since World War II, the *Green Revolution* has produced several successful increases in cereal production. During a 6-year period, wheat yields in India doubled. In Mexico, the Green Revolution changed the country from a grain importer, in 1960–1961, to a grain exporter by 1964–1965. At present, however, Mexico has become a net grain importer once again. These successes were mainly due to the development of better grain varieties and the transfer of western technology to high base-status soils. The Green Revolution in Mexico greatly increased yields on the fertile soils of large farms in the alluvial valleys and essentially left the small, hill farmers untouched. Although production was greatly increased, farming methods in Mexico were not significantly altered.

Dr. Norman Borlaug, a Nobel Peace Prize winner for his work with the Green Revolution, has stated many times that his work would buy time to help developing nations stabilize their population problem. From past experience, some highly successful projects have become overwhelmed by exploding populations.

The forecasts for the future are both optimistic and pessimistic. In 1948, the late Dr. Charles Kellogg pointed out that there are sufficient soil resources (and other physical resources) to feed the

rently low for most developing countries, who can greatly benefit from the current stock of technology.

Figure 15-7
The most difficult problem in solving the food-population problem is the creation of an economic, political, and social environment in which subsistence farmers, such as this one, with simple tools can produce crops for the market and produce a profit.

world's people, providing that economic, social, and political problems are solved. That statement is as true today as in 1948. Looking at the food-population problem only from a technological point of view, one can be optimistic. Looking at the problem in its totality, it is more difficult to be optimistic. The food-population problem is, to a large extent, a poverty problem. Technology is available to provide the solution. The difficult task is to provide the economic, social, and political environment in which hundreds of millions of traditional farmers, largely barefoot and with simple tools, can produce enough food for their families and some surplus to sell (*see* Fig. 15-7). The problem is an ancient one and it appears that it will persist for the foreseeable future.

Bibliography

Bentley, C. F., "Canada's Agricultural Land Resources and the World Food Problem," in *Trans. 11th Int. Cong. Soil Sci., 2:* 1–26, Edmonton, 1978.

Brown, H., *The Challenge of Man's Future,* Viking, New York, 1954.

Brown, L. R., "Worldwide Loss of Cropland," *Worldwatch Paper 24,* Worldwatch Institute, Washington, D.C., 1978.

Brown, L. R., "World Food Resources and Population: The Narrowing Margin," *Pop. Ref. Bur. Pop. Rull.,* Vol. 36, No. 3, September 1981.

Buringh, P., "Food Production Potential of the World," in Shina, R., Ed., *The World Food Problem: Concensus and Conflict,* Pergamon, Oxford, 1978, pp. 477–485.

deWit, C. T., "Photosynthesis: Its Relationship to Overpopulation," in *Harvesting the Sun,* Anthony San Pietro, Frances A. Greer, Thomas J. Army, Eds., Academic, New York, 1967, pp. 315–332.

FAO, *The State of Food and Agriculture 1977,* Rome, 1978.

FAO, *The State of Food and Agriculture 1980,* Rome, 1981.

Foth, H. D., "Soil Resources and Food: A Global View," in *Principles and Applications of Soil Geography,* E. M. Bridges, and D. A. Davidson, Eds., Longman, London, 1982.

Jacks, G. V., "Man the Fertility Maker," *Jour. Soil and Water Cons., 17:*147–148, 1962.

Jensen, N. F., "Limits to Growth in World Food Production," *Science, 201:*317–320, July 28, 1978.

Kellogg, C. E., "Modern Soil Science," *Am. Scientist, 38:*517–536, 1948.

Kellogg, C. E., and A. C. Orvedal, "Potentially Arable Soils of the World and Critical Measures for Their Use," in *Advances in Agronomy, 21,* Academic, New York, 1969, pp. 109–170.

Revelle, R., "The Resources Available for Agriculture," *Scien Am, 235:*164–179, September 1976.

Wellhausen, E. J., "The Agriculture of Mexico," *Scien Am, 235:*128–153, September 1976.

The White House, *The World Food Problem,* Vol. 2, U.S. Government Printing Office, Washington, D.C., 1967.

Wittwer, S. H., "The Next Generation of Agricultural Research," *Science, 199:*Editorial, January 27, 1978.

GLOSSARY

A Horizon. Mineral horizon formed at the soil surface or below an O horizon, having accumulations of humified organic matter mixed with mineral matter and not dominated by properties characteristic of E or B horizons.

ABC Soil. A soil with a distinctly developed profile, including an A, a B, and a C horizon.

AC Soil. A soil with an incomplete profile, including an A and a C horizon, but no B horizon. Commonly, such soils are young, such as those developing from alluvium or on steep, rocky slopes.

Acid Soil. A soil with pH < 7.0.

Actinomycetes. A nontaxonomic term applied to a group of organisms with characteristics intermediate between the simple bacteria and the true fungi.

Aerobic. (1) Having molecular oxygen as a part of the environment. (2) Growing only in the presence of molecular oxygen, as aerobic organisms. (3) Occurring only in the presence of molecular oxygen (said of certain chemical or biochemical processes, such as aerobic decomposition).

Agric Horizon. A mineral soil horizon in which clay, silt, and humus derived from an overlying cultivated and ferilized layer have accumulated. The wormholes and illuvial clay, silt, and humus occupy at least 5 percent of the horizon by volume. The illuvial clay and humus occur as horizontal lamellae or fibers, or as coatings on ped surfaces or in wormholes.

Adapted from *Glossary of Soil Science Terms*, published by the Soil Science Society of America, Madison, October 1979.

Albic Horizon. A mineral soil horizon from which clay and free iron oxides have been removed, or in which the oxides have been segregated to the extent that the color of the horizon is determined primarily by the color of the primary sand and silt particles instead of by coatings on these particles. An E horizon.

Albolls. Mollisols that have an albic horizon immediately below the mollic epipedon. These soils have an argillic or natric horizon and mottles, iron-manganese concretions, or both, within the albic, argillic, or natric horizon.

Alfisols. Mineral soils that have umbric or ochric epipedons, argillic horizons, and hold water at less than -15 bars potential for at least 3 months when the soil is warm enough for plants to grow outdoors. Alfisols have a mean annual soil temperature of less than 8°C or a base saturation in the lower part of the argillic horizon of 35 percent or more, when measured at pH 8.2.

Ammonification. The biochemical process whereby ammoniacal nitrogen is released from nitrogen-containing organic compounds.

Ammonium Fixation. The process of converting exchangeable or soluble ammonium ions to those occupying positions similar to K^+ in micas.

Amorphous Material. Noncrystalline constituents that either do not fit the definition of allophane, or it is not certain if the constituent meets allophane criteria.

Anaerobic. (1) The absence of molecular oxygen. (2) Growing in the absence of molecular oxygen (such as anaerobic bacteria). (3) Occurring in the absence of molecular oxygen (as a biochemical process).

Andepts.   Inceptisols that have formed either in vitric pyroclastic materials, or have low bulk density and large amounts of amorphous materials, or both. Andepts are not saturated with water long enough to limit their use for most crops.

Anion Exchange Capacity.   The sum total of exchangeable anions that a soil can adsorb. Expressed as milliequivalents per 100 grams of soil (or of other adsorbing material such as clay).

Anthropic Epipedon.   A surface layer of mineral soil that has the same requirements as the mollic epipedon with respect to color, thickness, organic carbon content, consistence, and base saturation, but has more than 250 parts per million of P_2O_5 soluble in 1 percent citric acid, or is dry more than 10 months (cumulative) during the period when not irrigated. The anthropic epipedon forms under long-continued cultivation and fertilization.

Antibiotic.   A substance produced by one species of organism that, in low concentrations, will kill or inhibit growth of certain other organisms.

Aqualfs.   Alfisols that are saturated with water for periods long enough to limit their use for most crops other than pasture or woodland, unless they are artificially drained. Aqualfs have mottles, iron-manganese concretions, or gray colors immediately below the A or Ap horizons and gray colors in the argillic horizon.

Aquents.   Entisols that are saturated with water for periods long enough to limit their use for most crops other than pasture, unless they are artificially drained. Aquents have low chromas or distinct mottles within 50 centimeters of the surface, or are saturated with water at all times.

Aquepts.   Inceptisols that are saturated with water for periods long enough to limit their use for most crops other than pasture or woodland, unless they are artificially drained. Aquepts have either a histic or umbric epipedon and gray colors within 50 centimeters, or an ochric epipedon underlain by a cambic horizon with gray colors, or have sodium saturation of 15 percent or more.

Aquic.   A mostly reducing soil moisture regime nearly free of dissolved oxygen due to saturation by ground water, or its capillary fringe, and occurring at periods when the soil temperature at 50 centimeters is above 5°C.

Aquods.   Spodosols that are saturated with water for periods long enough to limit their use for most crops other than pasture or woodland, unless they are artificially drained. Aquods may have a histic epipedon, an albic horizon that is mottled or contains a duripan, or mottling or gray colors within or immediately below the spodic horizon.

Aquolls.   Mollisols that are saturated with water for periods long enough to limit their use for most crops other than pasture, unless they are artificially drained. Aquolls may have a histic epipedon, a sodium saturation in the upper part of the mollic epipedon of more than 15 percent, that decreases with depth or mottles or gray colors within or immediately below the mollic epipedon.

Aquox.   Oxisols that have continuous plinthite near the surface, or that are saturated with water sometime during the year if not artificially drained. Aquox have either a histic epipedon, or mottles or colors indicative of poor drainage within the oxic horizon, or both.

Aquults.   Ultisols that are saturated with water for periods long enough to limit their use for most crops other than pasture or woodland unless they are artificially drained. Aquults have mottles, iron-manganese concretions, or gray colors immediately below the A or Ap horizons, and gray colors in the argillic horizon.

Arents.   Entisols that contain recognizable fragments of pedogenic horizons, which have been mixed by mechanical disturbance. Arents are not saturated with water for periods long enough to limit their use for most crops.

Argids.   Aridisols that have an argillic or a natric horizon.

Argillic Horizon.   A mineral soil horizon that is characterized by the illuvial accumulation of layer-lattice silicate clays. The argillic horizon has a certain minimum thickness (depending on the thickness of the solum), a minimum quantity of clay in comparison with an overlying eluvial horizon (depending on the clay content of the eluvial horizon), and usually has coatings of oriented clay on the surface of pores, or peds, or bridging sand grains.

Aridic. A soil moisture regime that has no moisture available for plants for more than one-half the cumulative time that the soil temperature at 50 centimeters is above 5°C, and has no period as long as 90 consecutive days when there is moisture for plants, while the soil temperature at 50 centimeters is continuously above 8°C.

Aridisols. Mineral soils that have an aridic moisture regime, an ochric epipedon, and other pedogenic horizons, but no oxic horizon.

Autotroph. An organism capable of using carbon dioxide and/or carbonates as a sole or major source of carbon, and of obtaining energy for carbon reduction and biosynthetic processes from radiant energy (photoautotroph) or oxidation of inorganic substances (chemoautotroph).

Available Nutrient. Nutrient elements or compounds in the soil that can be readily absorbed and assimilated by growing plants.

Available Water. The portion of water in a soil that can be readily absorbed by plant roots. Considered by most workers as the amount of water released by the soil when the equilibrium soil-water matric potential is decreased from field capacity to −15 bars.

Azonal Soils (Obsolete). Soils without distinct genetic horizons. A soil order in 1949 system.

B Horizon. A horizon that forms below an A, E, or O horizon and dominated by: (1) illuvial concentrations of silicate clay, iron, aluminum, humus, carbonates, gypsum or silica, alone or in combination, (2) residual concentration of sesquioxides, (3) coatings of sesquioxides that make the horizon conspicuously lower in value, higher in chroma, or redder in hue without apparent illuviation of iron, (4) alterations that form silicate clay or liberate ox-

ides, or both, and that form granular, blocky, or prismatic structure if volume changes accompany changes in moisture content, (5) evidence of removal of carbonates, or (6) any combination of these.

Bacteroid. An irregular form of cells of certain bacteria. Refers particularly to the swollen, vaculated cells of *Rhizobium* in nodules of leguminous plants.

Bar. A unit of pressure equal to 1,000,000 dynes per square centimeter.

Base Saturation Percentage. The extent to which the adsorption complex of a soil is saturated with alkali or alkaline earth cations, expressed as a percentage of the cation exchange capacity, which may include acidic cations as H^+ and aluminum.

BC Soil. A soil profile with B and C horizons, but with little or no A horizon.

Biosequence. A sequence of related soils that differ, one from the other, primarily because of differences in kinds and numbers of *soil organisms* as soil-forming factor.

Bleicherde. The light-colored, leached E horizon of Spodosols.

Boralfs. Alfisols that have formed in cool places. Boralfs have frigid or cryic, but no pergelic, temperature regimes, and have udic moisture regimes. Boralfs are not saturated with water for periods long enough to limit their use for most crops.

Borolls. Mollisols with a mean annual temperature of less than 8°C, which are never dry for 60 consecutive days or more within the 3 months following the summer solstice. Borolls do not contain material that has more than 40 percent $CaCO_3$ equivalent, unless they have a calcic horizon, and they are not saturated with water for periods long enough to limit their use for most crops.

Bulk Density, Soil. The mass of dry soil per unit bulk volume. The bulk volume is determined before drying to constant weight at 105°C.

Buried Soil. Soil covered by an alluvial, loessal, or other deposit, usually to a depth greater than the thickness of the solum.

C Horizon.   A mineral horizon or layer, excluding bedrock, that is either like or unlike the material from which solum is presumed to have formed, and is relatively little affected by pedogenic processes.

Calcareous Soil.   Soil containing sufficient free calcium carbonate or calcium-magnesium carbonate to effervesce visibly when treated with cold $0.1N$ hydrochloric acid.

Calcic Horizon.   A mineral soil horizon of secondary carbonate enrichment that is more than 15 centimeters thick, has a calcium carbonate equivalent of more than 15 percent, and has at least 5 percent more calcium carbonate equivalent than the underlying C horizon. A k horizon.

Caliche.   A layer near the surface, more or less cemented by secondary carbonates of calcium or magnesium precipitated from soil solution. It may occur as a soft, thin soil horizon, as a hard, thick bed just beneath the solum, or as a surface layer exposed by erosion (*see* **Croute Calcaire**).

Cambic Horizon.   A mineral soil horizon that has a texture of loamy, very fine sand or finer, has soil structure instead of rock structure, contains some weatherable minerals, and is characterized by the alteration or removal of mineral material, as indicated by mottling or gray colors, stronger chromas or redder hues than in underlying horizons, or the removal of carbonates. The cambic horizon lacks cementation or induration and has too few evidences of illuviation to meet the requirements of the argillic or spodic horizon.

Capillary Fringe.   A zone just above the water table that remains saturated or almost saturated.

Carbon-Nitrogen Ratio.   The ratio of the weight of organic carbon to the weight of total nitrogen (mineral plus organic forms) in soil or organic material.

Catena.   A sequence of soils of about the same age, derived from similar parent material, and occurring under similar climatic conditions, but having different characteristics because of variation in *relief* and in *drainage* (*see* **Clinosequence** and **Toposequence**).

Cation Exchange.   The interchange between a cation in solution and another cation on the surface of any surface-active material, such as clay colloid or organic colloid.

Cation Exchange Capacity (CEC).   The sum total of exchangeable cations that a soil can adsorb. Expressed in milliequivalents per 100 grams or per gram of soil (or of other exchanges such as clay).

Chlorosis.   A condition in plants, resulting from the failure of chlorophyll to develop, caused by a deficiency of an essential nutrient. Leaves of chlorotic plants range from light-green through yellow to almost white.

Chroma.   The relative purity, strength, or saturation of a color. It is directly related to the dominance of the determining wavelength of light and inversely related to grayness; one of the three variables of color (*see* **Munsell Color System, Hue,** and **Value, Color**).

Chronosequence.   A sequence of related soils that differ, one from the other, in certain properties primarily as a result of *time* as a soil-forming factor.

Clay.   (1) A soil separate consisting of particles < 0.002 millimeters in equivalent diameter. (2) A textural class.

Clay Films.   Coatings of clay on the surfaces of soil peds and mineral grains, and in soil pores. (Also called clay skins, clay flows, illuviation cutans, argillans, or tonhäutchen.)

Clay-pan.   A dense, compact layer in subsoil having a much higher clay content than the overlying material from which it is separated by a sharply defined boundary. It is formed by downward movement of clay or by synthesis of clay in place during soil formation. Clay-pans are usually hard when dry, and plastic and sticky when wet. Also, they usually impede the movement of water and air and the growth of plant roots.

Climosequence.   A sequence of related soils that differ, one from the other, in certain properties primarily as a result of the effect of *climate* as a soil-forming factor.

Clinosequence. A group of related soils that differ, one from the other, in certain properties primarily as a result of the effect of the *degree of slope* on which they were formed (*see* **Toposequence**).

Colloid. The term colloid is used in reference to matter, both organic and inorganic, having a very small particle size and a high specific surface.

Colluvium. A deposit of rock fragments and soil material accumulated at the base of steep slopes as a result of gravitational action.

Concretion. A local concentration of a chemical compound, such as calcium carbonate or iron oxide, in the form of a grain or nodule of varying size, shape, hardness, and color.

Consistency. (1) The resistance of a material to deformation or rupture. (2) The degree of cohesion or adhesion of the soil mass.

Consumptive Use. The water used by plants in transpiration and growth, plus water vapor loss from adjacent soil or snow, or from intercepted precipitation in any specified time. Usually expressed as equivalent depth of free water per unit of time.

Contour Tillage. Performing the tillage operations and planting on the contour within a given tolerance.

Cover Crop. A crop used to cover the soil surface, to decrease erosion and leaching, shade the ground, and to offer protection to the ground from excessive freezing and heaving.

Cradle Knoll. A small knoll formed by earth that is raised and left by an uprooted tree. (A microrelief term.)

Creep. Slow mass movement of soil and soil material down relatively steep slopes primarily under the influence of gravity, but facilitated by saturation with water and by alternate freezing and thawing.

Crop Rotation. A planned sequence of crops growing in a regularly recurring succession on the same area of land, as contrasted to continuous culture of one crop or growing different crops in haphazard order.

Crotovina. A former animal burrow in one soil horizon that has been filled with organic matter or material from another horizon (also spelled "krotovina").

Croute Calcaire. Hardened caliche, often found in thick masses or beds overlain by only a few inches of earth (*see* **Caliche**).

Cryic. A soil temperature regime that has mean annual soil temperatures of more than 0°C, but less than 8°C, more than 5°C difference between mean summer and mean winter soil temperatures at 50 centimeters, and cold summer temperatures.

Crystal Structure. The orderly arrangement of atoms in a crystalline material.

Darcy's Law. A law describing the rate of flow of water through porous media.

Deflation. The removal of fine soil particles from soil by wind.

Deflocculate. (1) To separate the individual components of compound particles by chemical and/or physical means. (2) To cause the particles of the *disperse phase* of a colloidal system to become suspended in the *dispersion medium*.

Denitrification. The biochemical reduction of nitrate or nitrite to gaseous nitrogen, either as molecular nitrogen or as an oxide of nitrogen.

Desert Pavement. The layer of gravel or stones left on the land surface in desert regions after the removal of the fine material by wind erosion.

Diatomaceous Earth. A geologic deposit of fine, grayish siliceous material composed chiefly or wholly of the remains of diatoms.

Diatoms. Algae having siliceous cell walls that persist as a skeleton after death. Any of the microscopic unicellular or colonial algae constituting the class *Bacillariaceae*. They occur abundantly in fresh and salt waters, and their remains are widely distributed in soils.

Diffuse Double Layer.   A heterogeneous system that consists of a solid surface having a net electrical charge, together with an ionic swarm under the influence of the solid surface, and in a solution phase that is in direct contact with the surface.

Disperse.   (1) To break up compound particles, such as aggregates, into the individual component particles. (2) To distribute or suspend fine particles, such as clay, in or throughout a dispersion medium, such as water.

Dryland Farming.   The practice of crop production in low-rainfall areas without irrigation.

Durinodes.   Weakly cemented to indurated soil nodules cemented with SiO_2. Durinodes break down in concentrated KOH after treatment with HCl to remove carbonates, but do not break down on treatment with concentrated HCl alone.

Duripan.   A mineral soil horizon that is cemented by silica (usually opal or microcrystalline forms of silica) to the point that air-dry fragments will not slake in water or HCl. A duripan may also have accessory cement such as iron oxide or calcium carbonate.

Dust Mulch.   A loose, finely granular or powdery condition on the soil surface, usually produced by shallow cultivation.

E Horizon.   Mineral horizon in which the main feature is a loss of silicate clay, iron, aluminum, or some combination of these, leaving a concentration of sand and silt particles of quartz or other resistant minerals.

Ecology.   The science that deals with the interrelations of organisms and their environment.

Ecosystem.   A community of organisms and the surroundings in which they live.

Ectotrophic Mycorrhiza.   A mycorrhizal association in which fungal hyphae form a compact mantle on the roots sufaces. Mycelial strands extend inward between cortical cells and outward from the mantle to the surrounding soil.

Edaphic.   (1) Of or pertaining to the soil. (2) Resulting from or influenced by factors inherent in the soil or other substrate, instead of climatic factors.

Electrokinetic (Zeta) Potential.   (1) The difference in electrical potential between the immobile liquid layer attached to the surface of a charged particle and the bulk liquid phase. (2) The work done in bringing a unit charge from infinity (bulk solution) to the plane of shear in the diffuse double layer.

Eluvial Horizon.   A soil horizon that has been formed by the process of eluviation.

Eluviation.   The removal of soil material in suspension (or in solution) from a layer or layers of a soil. (Usually, the loss of material in *solution* is described by the term "leaching.")

Endotrophic.   Nourished or receiving nourishment from within, as fungi or their hyphae receive nourishment from plant roots in a mycorrhizal association.

Entisols.   Mineral soils that have no distinct pedogenic horizons within 1 meter of the soil surface.

Erosion.   (1) The wearing away of the land surface by running water, wind, ice, or other geologic agents, including processes such as gravitational creep. (2) Detachment and movement of soil or rock by water, wind, ice, or gravity. The following terms are used to describe different types of water erosion.

Accelerated Erosion.   Erosion much more rapid than normal, natural, geologic erosion, primarily as a result of the influence of the activities of people or, in some cases, of animals.

Geologic Erosion.   The normal or natural erosion caused by geologic processes acting over long geologic periods and resulting in the wearing away of mountains, the building up of flood plains, coastal plains, etc. Synonymous with *Natural Erosion.*

Gully Erosion.   The erosion process whereby water accumulates in narrow channels and, over short periods, removes the soil from this narrow area to considerable depths, ranging from 1 or 2 feet to as much as 75 to 100 feet.

Natural Erosion.   Wearing away of the earth's

surface by water, ice, or other natural agents under natural environmental conditions of climate, vegetation, etc., undisturbed by humans. Synonymous with *Geological Erosion*.

Normal Erosion. The gradual erosion of land used by people that does not greatly exceed natural erosion *(see Natural Erosion)*.

Rill Erosion. An erosion process in which numerous small channels of only several inches in depth are formed; it occurs mainly on recently cultivated soils *(see **Rill**)*.

Sheet Erosion. The removal of a fairly uniform layer of soil from the land surface by runoff water.

Splash Erosion. The spattering of small soil particles caused by the impact of raindrops on very wet soils. The loosened and spattered particles may or may not be subsequently removed by surface runoff.

Eutrophic. Having concentrations of nutrients optimal, or nearly so, for plant or animal growth. (Said of nutrient or soil solutions and bodies of water.)

Evapotranspiration. The combined loss of water from a given area and during a specified period of time by evaporation from the soil surface and by transpiration from plants.

Exchangeable Bases. Exchangeable basic cations (alkali, alkaline earth, and ammonium) adsorbed by a soil, clay, or organic matter.

Exchangeable-Cation Percentage. The extent to which the adsorption complex of a soil is occupied by a particular cation. It is expressed as follows:

$$ECP = \frac{\text{Exchangeable cation (mEq/100 grams soil)}}{\text{Cation-exchange capacity (mEq/100 grams soil)}} \times 100$$

Exchangeable Phosphate. The phosphate anion reversibly attached to the surface of the solid phase of soil in such form that it may go into solution by anionic equilibrium reaction with isotopes of phosphorus or with other anions of the liquid phase without solution of the colloid phase to which it was attached.

Exchangeable Potassium. The potassium that is held by the adsorption complex of soil and is easily exchanged with a cation of neutral nonpotassium salt solutions.

Exchangeable-Sodium Percentage. The percentage of the cation-exchange capacity of a soil occupied by sodium. It is expressed as follows:

$$ESP = \frac{\text{Exchangeable sodium (mEq/100 grams soil)}}{\text{Cation-exchange capacity (mEq/100 grams soil)}} \times 100$$

Fallow. The practice of leaving land uncropped and weed-free for periods of time to accumulate and retain water and mineralized nutrient elements.

Family, Soil. One of the categories in soil classification.

Ferrods. Spodosols that have more than six times as much free iron (elemental) than organic carbon in the spodic horizon. Ferrods are rarely saturated with water or do not have characteristics associated with wetness.

Fertility, Soil. The status of a soil with respect to its ability to supply the nutrients essential to plant growth.

Fertilizer. Any organic or inorganic material of natural or synthetic origin that is added to a soil to supply one or more elements essential to the growth of plants.

Fertilizer Grade. The guaranteed minimum analysis, in percent, of the major plant nutrient elements contained in a fertilizer material or in a mixed fertilizer (Usually refers to the percentage of N-P_2O_5-K_2O, but proposals are pending to change the designation to the percentage of N-P-K).

Fibrists. Histosols that have a high content of undecomposed plant fibers and a bulk density less than about 0.1. Fibrists are saturated with water for periods long enough to limit their use for most crops, unless they are artificially drained.

Field Capacity. Obsolete in technical work. The percentage of water remaining in a soil 2 or 3 days after having been saturated and after free drainage has practically ceased. (The percentage may be expressed on the basis of weight or volume.)

Fifteen-Bar Percentage. The percentage of water contained in a soil that has been saturated, subjected to, and is in equilibrium with an applied pressure of 15 bars.

Film Water. A layer of water surrounding soil particles and varying in thickness from 1 or 2 to perhaps 100 or more molecular layers. Usually considered as that water remaining after drainage has occurred, because it is not distinguishable in saturated soils.

Fixation. The process of conversion of an element in soil that is essential to plants from a readily available to a less available form.

Flood Plain. The land bordering a stream, built up of sediments from overflow of the stream, and subject to inundation when the stream is at flood stage.

Fluvents. Entisols that form in recent loamy or clayey alluvial deposits, are usually stratified, and have an organic carbon content that decreases irregularly with depth. Fluvents are not saturated with water for periods long enough to limit their use for most crops.

Fluvic Acid. A term of varied usage, but usually referring to the mixture of organic substances remaining in solution upon acidification of a dilute alkali extract from the soil.

Foliar Diagnosis. An estimation of mineral nutrient deficiencies (excesses) of plants based on examination of the chemical composition of selected plant parts, and the color and growth characteristics of the foliage of the plants.

Folists. Histosols that have an accumulation of organic soil materials, mainly as forest litter, that is less than 1 meter deep to rock or to fragmental materials, with interstices filled with organic materials. Folists are not saturated with water for periods long enough to limit their use if cropped.

Forest Floor. All dead vegetable or organic matter, including litter and unincorporated humus, on the mineral soil surface under forest vegetation.

Fragipan. A natural subsurface horizon with high bulk density relative to the solum above, seemingly cemented when dry, but when moist showing a moderate to weak brittleness. The layer is low in organic matter, mottled, slowly or very slowly permeable to water, and usually shows occasional or frequent bleached cracks that form polygons. It may be found in profiles of either cultivated or virgin soils but not in calcareous material.

Friable. A consistency term pertaining to the ease of crumbling of soils (*see* **Consistency**).

Frigid. A soil temperature regime that has mean annual soil temperatures of more than 0°C, but less than 8°C, more than 5°C difference between mean summer and mean winter soil temperatures at 50 centimeters, and warm summer temperatures. Isofrigid is the same except the summer and winter temperatures differ by less than 5°C.

Gilgai. The microrelief of soils produced by expansion and contraction with changes in moisture. It is found in soils that contain large amounts of clay, which swells and shrinks considerably with wetting and drying. Usually, a succession of microbasins and microknolls in nearly level areas or of microvalleys and microridges parallel to the direction of the slope.

Glacial Drift. Rock debris that has been transported by glaciers and deposited, either directly from the ice or from the melt-water. The debris may or may not be heterogeneous.

Glaciofluvial Deposits. Material moved by glaciers and subsequently sorted and deposited by streams flowing from melting ice. The deposits are stratified and may occur in the form of outwash plains, deltas, kames, eskers, and kame terraces.

Gleyzation. A soil-forming process resulting in the development of gley soils.

Gley Soil. Soil developed under conditions of poor drainage, resulting in reduction of iron and other elements and in gray colors and mottles.

Gravitational Water. Water that moves into, through, or out of the soil under the influence of gravity.

Green-Manure Crop. A crop grown for use as green manure; to be incorporated into soil when green.

Ground Water. The portion of the water below the surface of the ground whose pressure is greater than atmospheric.

Gypsic Horizon. A mineral soil horizon of secondary, calcium sulfate enrichment that is more than 15 centimeters thick, has at least 5 percent more gypsum than the C horizon, and in which the product of the thickness in centimeters and the percent calcium sulfate is equal to or greater than 150 percent centimeters.

Gypsum Requirement. The quantity of gypsum, or its equivalent, required to reduce the exchangeable sodium of a given amount of soil to an acceptable level.

Hardpan. A hardened soil layer, in the lower A or in the B horizon, caused by cementation of soil particles with organic matter or with materials such as silica, sesquioxides, or calcium carbonate. The hardness does not change appreciably with changes in moisture content, and pieces of hard layer do not slake in water.

Hemists. Histosols that have an intermediate degree of plant fiber decomposition and a bulk density between about 0.1 and 0.2. Hemists are saturated with water for periods long enough to limit their use for most crops, unless they are artificially drained.

Heterotroph. An organism capable of deriving energy for life processes from the oxidation of organic compounds.

Histic Epipedon. A thin organic soil horizon that is saturated with water at some period of the year unless artificially drained, and that is at or near the surface of a mineral soil. The histic epipedon has a maximum thickness, depending on the kind of materials in the horizon; the lower limit of organic carbon is the upper limit for the mollic epipedon.

Histosols. Organic soils that have organic soil materials in more than 50 percent of the upper 80 centimeters, or that are of any thickness if overlying rock or fragmental materials that have interstices filled with organic soil materials.

Hue. One of the three variables of color. It is caused by light of certain wavelengths and changes with the wavelength (*see* **Munsell Color System, Chroma,** and **Value, Color.**).

Humic Acid. A mixture of dark-colored substances of indefinite composition extracted from soil with dilute alkali and precipitated by acidification.

Humification. The processes involved in the decomposition of organic matter and leading to the formation of humus.

Humin. The fraction of soil organic matter that is not dissolved upon extraction of the soil with dilute alkali.

Humods. Spodosols that have accumulated organic carbon and aluminum, but not iron, in the upper part of the spodic horizon. Humods are rarely saturated with water or do not have characteristics associated with wetness.

Humox. Oxisols that are moist all or most of the time, and that have a high organic carbon content within the upper meter. Humox have a mean annual soil temperature of less than 22°C and a base saturation within the oxic horizon of less than 35%, measured at pH 7.

Humults. Ultisols that have a high content of organic carbon. Humults are not saturated with water for periods long enough to limit their use for most crops.

Humus. That more or less stable fraction of soil organic matter remaining after the major portion of added plant and animal residues have decomposed. Usually, it is dark-colored.

Hydraulic Conductivity. An expression of the readiness of water to flow through a soil in response to a given water potential gradient.

Hydrous Mica. A silicate clay with a 2:1 lattice structure, but of indefinite chemical composition, since usually part of the silicon in the silica-tetrahedral layer has been replaced by aluminum, and containing a considerable amount of potassium that serves as an additional bonding between the crystal units. Sometimes referred to as illite.

Hygroscopic Water. Water adsorbed by a dry soil from an atmosphere of high relative humidity, water remaining in the soil after "air-drying," or water held by the soil when it is in equilibrium with an atmosphere of a specified relative humidity at a specified temperature, usually 98 percent relative humidity at 25°C.

Hyperthermic. A soil temperature regime that has mean annual soil temperatures of 22°C or more and more than 5°C difference between mean summer and mean winter soil temperatures at 50 centimeters. Isohyperthermic is the same, except that summer and winter temperatures differ by less than 5°C.

Igneous Rock. Rock formed from the cooling and solidification of magma, and that has not been changed appreciably since its formation.

Illite. *See* **Hydrous Mica.**

Illuvial Horizon. A soil layer or horizon in which material carried from an overlying layer has been precipitated from solution or deposited from suspension. The layer of accumulation (*see* **Eluvial Horizon**).

Illuviation. The process of deposition of soil material removed from one horizon to another in the soil; usually from an upper to a lower horizon in the soil profile (*see* **Eluviation**).

Immature Soil. A soil with indistinct or only slightly developed horizons because of the relatively short time it has been subjected to the various soil-forming processes. A soil that has not reached equilibrium with its environment.

Immobilization. The conversion of an element from the inorganic to the organic form in microbial tissues or in plant tissues.

Inceptisols. Mineral soils that have one or more pedogenic horizons in which mineral materials other than carbonates or amorphous silica have been altered or removed, but not accumulated to a significant degree. Under certain conditions, Inceptisols may have an ochric, umbric, histic, plaggen, or mollic epipedon. Water is available to plants more than one half of the year or more than 3 consecutive months during a warm season.

Indicator Plants. Plants characteristic of specific soil or site conditions.

Infiltration. The downward entry of water into the soil.

Intergrade. A soil that possesses moderately well-developed, distinguishing characteristics of two or more genetically related, great soil groups.

Iron-Pan. An indurated soil horizon in which iron oxide is the principal cementing agent.

Irrigation. The artificial application of water to soil for the benefit of growing crops.

Irrigation Efficiency. The ratio of the water actually consumed by crops on an irrigated area to the amount of water diverted from the source onto the area.

Isoelectric Point. The pH value of a solution in equilibrium, with a constant potential surface, whose net electrical charge is zero. It is dependent only on the presence of H^+, $OH-$, or H_2O to form species on the surface.

Isomorphous Substitution. The replacement of one atom by another of similar size in a crystal lattice without disrupting or changing the crystal structure of the mineral.

Kame. An irregular ridge or hill of stratified glacial drift.

Kaolinite. An aluminosilicate mineral of the 1:1 crystal lattice group; that is, consisting of one silicon tetrahedral layer and one aluminum oxide-hydroxide octahedral layer.

Landscape. All the natural features (such as fields, hills, forests, water, etc.) that distinguish one part of the earth's surface from another part. Usually that portion of land or territory that the eye can comprehend in a single view, including all its natural characteristics.

Lattice. A three-dimensional grid of lines connecting the points *representing* the centers of atoms or ions in a crystal.

Leaching. The removal of materials in solution from the soil (*see* **Eluviation**).

Leaching Requirement. The fraction of applied irrigation water that should be passed through the root zone to control soil salinity at a specified level.

Liquid Limit. The minimum percentage (by weight) of moisture at which a small sample of soil will barely flow under a standard treatment.

Lithic Contact. A boundary between soil and continuous, coherent, underlying material. The underlying material must be sufficiently coherent to make hand-digging with a spade impractical. If mineral, it must have a hardness of 3 or more (Mohs scale), and gravel-sized chunks that can be broken out do not disperse with 15 hours shaking in water or sodium hexametaphosphate solution.

Lithosequence. A group of related soils that differ, one from the other, in certain properties primarily as a result of differences in the *parent rock* as a soil-forming factor.

Loam. A soil textural class.

Loess. Material transported and deposited by wind and consisting of predominantly silt-sized particles.

Luxury Uptake. The absorption by plants of nutrients in excess of their need for growth. Luxury concentrations during early growth may be used in later growth.

Macronutrient. A chemical element necessary in relatively large amounts (usually < 500 parts per million in the plant) for the growth of plants. These elements consist of C, H, O, Ca, Mg, K, P, S, and N.

Made Land. Areas filled with earth, or with earth and trash mixed, usually by or under the control of humans. A miscellaneous land type.

Mass Flow (Nutrient). The movement of solutes associated with net movement of water.

Mature Soil. A soil with well-developed soil horizons produced by the natural processes of soil formation and essentially in equilibrium with its present environment.

Mesic. A soil temperature regime that has mean annual soil temperatures of 8°C or more, but less than 15°C, and more than 5°C difference between mean summer and mean winter soil temperatures at 50 centimeters. Isomesic is the same, except the summer and winter temperatures differ by less than 5°C.

Metamorphic Rock. Rock derived from preexisting rocks but that differ from them in physical, chemical, and mineralogical properties as a result of natural geologic processes (principally heat and pressure) originating within the earth. The preexisting rocks may have been igneous, sedimentary, or another form of metamorphic rock.

Microclimate. (1) The climatic condition of a small area resulting from the modification of the general climatic conditions by local differences in elevation or exposure. (2) The sequence of atmospheric changes within a very small region.

Micronutrient. A chemical element necessary only in extremely small amounts (usually < 50 parts per million in the plant) for the growth of plants. These elements consist of B, Cl, Cu, Fe, Mn, Mo, and Zn.

Microrelief. Small-scale, local differences in topography, including mounds, swales, or pits, that are only a few feet in diameter and with elevation differences of up to 6 feet (*see* **Cradle Knoll** and **Gilgai**).

Mineralization. The conversion of an element from an organic form to an inorganic state as a result of microbial decomposition.

Mineral Soil. A soil consisting predominantly of, and having its properties determined predominantly by, mineral matter. Usually contains < 20 percent organic matter, but may contain an organic surface layer up to 30 centimeters thick.

Moisture Volume Percentage. The ratio of the volume of water in a soil to the total bulk volume of the soil.

Moisture Weight Percentage. The moisture content expressed as a percentage of the oven dry weight of soil.

Mollic Epipedon. A surface horizon of mineral soil that is dark colored and relatively thick, contains at least 0.58 percent organic carbon, is not massive and hard or very hard when dry, has a base saturation of more than 50 percent when measured at pH 7, has less than 250 parts per million of P_2O_5 soluble in 1 percent citric acid, and is dominantly saturated with bivalent cations.

Mollisols. Mineral soils that have a mollic epipedon overlying mineral material, with a base saturation of 50 percent or more when measured at pH 7. Mollisols may have an argillic, natric, albic, cambic, gypsic, calcic, or petrocalcic horizon, a histic epipedon, or a duripan, but not an oxic or spodic horizon.

Montmorillonite. An aluminosilicate clay mineral with a 2:1 expanding crystal structure; that is, with two silicon tetrahedral layers enclosing an aluminum octahedral layer. Considerable expansion may be caused along the C axis by water moving between silica layers of contiguous units. A smectite clay.

Mor. A type of forest humus in which there is practically no mixing of surface organic matter with mineral soil.

Mottled Zone. A layer that is marked with spots or blotches of different color or shades of color. The pattern of mottling and the size, abundance, and color contrast of the mottles may vary considerably and should be specified in soil description.

Mottling. Spots or blotches of different color or shades or color interspersed with the dominant color.

Muck. Highly decomposed organic material in which the original plant parts are not recognizable. Contains more mineral matter and is usually darker in color than peat (*see* **Muck Soil, Peat,** and **Peat soil**).

Muck Soil. (1) A soil containing between 20 and 50 percent of organic matter. (2) An organic soil in which the organic matter is well decomposed (US usage).

Mulch. (1) Any material (such as straw, sawdust, leaves, plastic film, loose soil, etc.), that is spread on the surface of soil to protect soil and plant roots from the effects of raindrops, soil crusting, freezing, evaporation, etc. (2) To apply mulch to the soil surface.

Mulch Farming. A system of farming in which organic residues are not plowed into or otherwise mixed with soil, but are left on the surface as a mulch.

Mull. A type of forest humus in the A horizon that consists of an intimate mixture of organic matter and mineral soil with gradual transition between the A and the horizon beneath.

Munsell Color System. A color designation system that specifies the relative degrees of the three simple variables of color: hue, value, and chroma. For example: 10YR 6/4 is a color (of soil) with a hue = 10YR, value = 6, and chroma = 4.

Mycorrhiza. Literally, "fungus root." The association, usually symbiotic, of specific fungi with the roots of higher plants.

Natric Horizon. A mineral soil horizon that satisfied the requirements of an argillic horizon, but that also has prismatic, columnar, or blocky structure and a subhorizon having more than 15 percent saturation with exchangeable sodium.

Nitrification. Biological oxidation of ammonium to nitrate and nitrite, or a biologically induced increase in the oxidation state of nitrogen.

Nitrogen Assimilation. The incorporation of nitrogen into organic cell substances by living organisms.

Nitrogen Cycle. The sequence of biochemical changes undergone by nitrogen wherein it is used by a living organism, liberated upon the death and decomposition of the organism, and converted to its original state of oxidation.

Nitrogen Fixation. Biological conversion of molecular dinitrogen (N_2) to organic combinations or to forms usable in biological processes.

Nutrient, Diffusion. The movement of nutrients in soil that results from a concentration gradient.

O Horizon. Layers dominated by organic matter. Some are saturated with water for long periods or were once saturated, but are now artificially drained; others have never been water saturated.

Ochrepts. Inceptisols formed in cold or temperate climates that commonly have an ochric epipedon and a cambic horizon. They may have an umbric or mollic epipedon less than 25 centimeters thick, or a fragipan or duripan under certain conditions. These soils are not dominated by amorphous materials and are not saturated with water for periods long enough to limit their use for most crops.

Ochric Epipedon. A surface horizon of mineral soil that is too light in color, too high in chroma, too low in organic carbon, or too thin to be a plaggen, mollic, umbric, anthropic or histic epipedon, or that is both hard and massive when dry.

Order. The highest category in soil classification.

Organic Soil. A soil that contains a high percentage (> 15 or 20 percent) of organic matter throughout the solum.

Orthents. Entisols that have either textures of very fine sand or finer in the fine earth fraction, or textures of loamy fine sand or coarser and a coarse fragment content of 35 percent or more with an organic carbon content that decreases regularly with depth. Orthents are not saturated with water for periods long enough to limit their use for most crops.

Orthids. Aridisols that have a cambic, calcic, petrocalcic, gypsic, or salic horizon or a duripan, but that lack an argillic or natric horizon.

Orthods. Spodosols that have less than six times as much free iron (elemental) than organic carbon in the spodic horizon, but the ratio of iron to carbon is 0.2 or more. Orthods are not saturated with water for periods long enough to limit their use for most crops.

Orthox. Oxisols that are moist all or most of the time, and that have a low to moderate content of organic carbon within the upper 1 meter or a mean annual soil temperature of 22°C or more.

Ortstein. An indurated layer in the B horizon of Spodosols in which the cementing material consists of illuviated sesquioxides (mostly iron) and organic matter.

Osmotic. A type of pressure exerted in living bodies as a result of unequal concentration of salts on both sides of a cell wall or membrane.

Water will move from the area having the least salt concentration through the membrane into the area having the highest salt concentration, therefore exerting additional pressure on one side of the membrane.

Oven Dry Soil. Soil that has been dried at 105°C until it reaches constant weight.

Oxic Horizon. A mineral soil horizon that is at least 30 centimeters thick and characterized by the virtual absence of weatherable primary minerals or 2:1 lattice clays, the presence of 1:1 lattice clays and highly insoluble minerals such as quartz sand, the presence of hydrated oxides of iron and aluminum, the absence of water-dispersible clay, and the presence of low cation-exchange capacity and small amounts of exchangeable bases.

Oxisols. Mineral soils that have an oxic horizon within 2 meters of the surface or plinthite as a continuous phase within 30 centimeters of the surface, and that do not have a spodic or argillic horizon above the oxic horizon.

Paleosol. A soil formed on a landscape during the geologic past and subsequently buried by sedimentation.

Pan, Pressure or Induced. A subsurface horizon, or soil layer having a higher bulk density and a lower total porosity than soil directly above or below it, as a result of pressure that has been applied by normal tillage operations or other artificial means. Frequently referred to as plowpan, plowsole, or traffic pan.

Pans. Horizons or layers in soils that are strongly compacted, indurated, or very high in clay content. (*see* **Caliche, Clay-pan, Fragipan,** and **Hardpan**).

Paralithic Contact. Similar to a lithic contact, except that the mineral material below the contact has a hardness of less than 3 (Mohs scale), and gravel-sized chunks that can be broken out will partially disperse within 15 hours shaking in water or sodium hexametaphosphate solution.

Parent Material. The unconsolidated and more or less chemically weathered mineral or organic matter from which the solum of soils is developed by pedogenic processes.

Particle Density. The mass per unit volume of soil particles. In technical work, usually expressed as grams per cubic centimeter.

Particle-size Analysis. Determination of various amounts of the different separates in a soil sample, usually by sedimentation, sieving, micrometry, or combinations of these methods.

Parts per Million (ppm). Weight units of any given substance per 1,000,000 equivalent weight units of oven dry soil; or, in the case of soil solution or the solution, the weight units of solute per million weight units of solution.

Peat. Unconsolidated soil material consisting largely of undecomposed, or only slightly decomposed, organic matter accumulated under conditions of excessive moisture.

Peat Soil. An organic soil containing more than 50 percent organic matter. Used in the United States to refer to the stage of decomposition of the organic matter, "peat" refers to the slightly decomposed or undecomposed deposits and "muck" to the highly decomposed materials.

Ped. A unit of soil structure such as an aggregate, crumb, prism, block, or granule, formed by natural processes (in contrast with a clod, which is formed artificially).

Pedon. A three-dimensional body of soil with lateral dimensions large enough to permit the study of horizon shapes and relations. Its area ranges from 1 to 10 square meters. Where horizons are intermittent or cyclic, and recur at linear intervals of 2 to 7 meters, the pedon includes one half of the cycle. Where the cycle is less than 2 meters, or all horizons are continuous and of uniform thickness, the pedon has an area of approximately 1 square meter. If the horizons are cyclic, but recur at intervals greater than 7 meters, the pedon reverts to the 1 square meter size, and more than one soil will usually be represented in each cycle.

Percolation, Soil Water. The downward movement of water through soil. Especially, the downward flow of water in saturated or nearly saturated soil at hydraulic gradients of the order of 1.0 or less.

Pergelic. A soil temperature regime that has mean annual soil temperatures of less than 0°C. Permafrost is present.

Permafrost. (1) Permanently frozen material underlying the solum. (2) A perennially frozen soil horizon.

Permafrost Table. The upper boundary of the permafrost (*see* **Permafrost,** [1]), coincident with the lower limit of seasonal thaw.

Permeability, Soil. The ease with which gases, liquids, or plant roots penetrate or pass through a bulk mass of soil or a layer of soil. Since different soil horizons vary in permeability, the particular horizon under question should be designated.

Petrocalcic Horizon. A continuous, indurated calcic horizon that is cemented by calcium carbonate and, in some places, with magnesium carbonate. It can not be penetrated with a spade or auger when dry, dry fragments do not slake in water, and it is impenetrable to roots.

Petrogypsic Horizon. A continuous, strongly cemented, massive, gypsic horizon that is cemented by calcium sulfate. It can be chipped with a spade when dry. Dry fragments do not slake in water, and it is impenetrable to roots.

pH-Dependent Charge. The portion of the cation- or anion exchange capacity that varies with pH.

pH, Soil. The negative logarithm of the hydrogen-ion activity of a soil. The degree of acidity (or alkalinity) of a soil as determined by means of a glass, quinhydrone, or other suitable electrode or indicator at a specified moisture content or soil-water ratio; expressed in terms of pH scale.

Placic Horizon. A black-to dark-reddish mineral soil horizon that is usually thin, but that may range from 1 to 25 millimeters in thickness. The placic horizon is commonly cemented with iron and is slowly permeable or impenetrable to water and roots.

Plaggen Epipedon. A synthetic suface horizon more than 50 centimeters thick that is formed by long-continued manuring and mixing.

Plaggepts. Inceptisols that have a plaggen epipedon.

Plastic Soil. A soil capable of being molded or deformed continuously and permanently, by relatively moderate pressure, into various shapes (see **Consistency**).

Plinthite. A nonindurated mixture of iron and aluminum oxides, clay, quartz, and other diluents that commonly occurs as red soil mottles usually arranged in platy, polygonal, or reticulate patterns. Plinthite changes irreversibly to ironstone hardpans or irregular aggregates on exposure to repeated wetting and drying.

Primary Mineral. A mineral that has not been altered chemically since deposition and crystallization from molten lava (see **Secondary Mineral**).

Prismatic Soil Structure. A soil structure type with prism-like aggregates that have a vertical axis much longer than the horizontal axes (see **Soil Structure Types**).

Productivity, Soil. The capacity of a soil, in its normal environment, for producing a specified plant or sequence of plants under a specified system of management. The "specified" limitations are necessary, since no soil can produce all crops with equal success nor can a single system of management produce the same effect on all soils. Productivity emphasizes the capacity of soil to produce crops and should be expressed in terms of yields.

Profile, Soil. A vertical section of soil through all its horizons, extending into the parent material.

Psamments. Entisols that have textures of loamy fine sand or coarser in all parts, have less than 35 percent coarse fragments, and that are not saturated with water for periods long enough to limit their use for most crops.

Puddling. Tillage of water-saturated soil.

R Horizon. Underlying consolidated bedrock, such as granite, sandstone, or limestone.

Regolith. The unconsolidated mantle of weathered rock and soil material on the earth's surface; loose earth materials above solid rock. (Approximately equivalent to the term "soil" as used by many engineers.)

Regur. A group of dark calcareous soils high in clay, which is mainly montmorillonitic, and formed mainly from rocks low in quartz; occurring extensively on the Deccan Plateau of India.

Rendolls. Mollisols that have no argillic or calcic horizon, but that contain material with more than 40 percent $CaCO_3$ equivalent within or immediately below the mollic epipedon. Rendolls are not saturated with water for periods long enough to limit their use for most crops. (A suborder in the USDA soil taxonomy.)

Rendzina. A great soil group of the intrazonal order and calcimorphic suborder consisting of soils with brown or black friable surface horizons underlain by light-gray to pale-yellow calcareous material; developed from soft, highly calcareous parent material under grass vegetation or mixed grasses and forest in humid and semiarid climates.

Reticulate Mottling. A network of streaks of different color; most commonly found in the deeper profiles of Lateritic soils.

Rhizobia. Bacteria capable of living symbiotically in roots of legumes, from which they receive energy and often use molecular nitrogen. Collective common name for the genus *Rhizobium*.

Rhizosphere. The zone of soil where the microbial population is altered both quantitatively and qualitatively by the presence of plant roots.

Runoff. The portion of precipitation on an area that is discharged from the area through stream channels. That which is lost without entering the soil is called *surface runoff,* and that which enters the soil before reaching the stream is called *ground-water runoff* or *seepage flow* from ground-water. (In soil science, "runoff" usually refers to water lost by surface flow; in geology and hydraulics, "runoff" usually includes both surface and substrate flow.)

Salic Horizon. A mineral soil horizon of enrichment with secondary salts more soluble in cold water than gypsum. A salic horizon is 15 centimeters or more in thickness and contains at least 2 percent salt. The product of the thickness in centimeters and percent salt by weight is 60 percent centimeters or more.

Salination. The process whereby soluble salts accumulate in soil.

Saline Soil. A nonsodic soil containing sufficient soluble salt to adversely affect the growth of most crop plants. The lower limit of saturation extract, electrical conductivity of such soils is conventionally set at 0.4 siemens per meter (4 mmhos per cm). Actually, sensitive plants are affected at half this salinity and highly tolerant ones at about twice this salinity.

Saline-Sodic Soil. A soil containing a combination of soluble salts and exchangeable sodium sufficient to interfere with the growth of most crop plants. The electrical conductivity and sodium-adsorption ratio of the saturation extract are at least 4 millimohs per centimeter at 25°C and 13, respectively. The pH is usually 8.5 or less in the saturated soil paste. (Formerly called saline-alkali soil.)

Salt Balance. The relation between the quantity of dissolved salts carried to an area in irrigation water and the quantity of dissolved salts removed by drainage water.

Sand. (1) A soil particle between 0.05 and 2.0 millimeters in diameter. (2) Any one of five soil separates: very coarse sand, coarse sand, medium sand, fine sand, very fine sand (see **Soil Separates**). (3) A soil textural class.

Saprists. Histosols that have: (1) a high content of plant materials so decomposed that original plant structures can not be determined, and (2) a bulk density of about 0.2 or more. Saprists are saturated with water for periods long enough to limit their use for most crops unless they are artificially drained.

Saturation Extract. An increment of solution obtained from a saturated soil paste.

Secondary Mineral. A mineral resulting from the decomposition of a primary mineral or from the reprecipitation of the products of decomposition of a primary mineral (see **Primary Mineral**).

Self-Mulching Soil. A soil in which the surface layer becomes so well aggregated that it does not crust and seal under the impact of rain; instead, it serves as a surface mulch upon drying.

Silica-Alumina Ratio. The molecules of silicon dioxide (SiO_2) per molecule of aluminum oxide (Al_2O_3) in clay minerals or in soils.

Silica-Sesquioxide Ratio. The molecules of silicon dioxide (SiO_2) per molecule of aluminum oxide (Al_2O_3) plus ferric oxide (Fe_2O_3) in clay minerals or in soils.

Silt. (1) A soil separate consisting of particles between 0.05 and 0.002 millimeters in equivalent diameter (see **Soil Separates**). (2) A soil textural class.

Site Index. (1) A quantitative evaluation of the productivity of a soil for forest growth under the existing or specified environment. (2) The height in feet of the dominant forest vegetation taken at, or calculated to an index age, usually 50 or 100 years.

Slickensides. Polished and grooved surfaces produced by one mass sliding past another. Slickensides are common in Vertisols.

Slick Spots. Small areas in a field that are slick when wet because of a high content of alkali or exchangeable sodium.

Smectite. Group name for expanding clays. including montmorillonite, beidellite, nontronite, and hectorite.

Sodic Soil. (1) A nonsaline soil containing sufficiently exchangeable-sodium to interfere with the growth of most crop plants. (2) A soil in which the sodium adsorption ratio of the saturation extract is 13 or more. The lower limit of the saturated extract SAR of such soils is conventionally set at 13.

Sodication. The process whereby the exchangeable sodium content of a soil is increased.

Sodium-Adsorption Ratio (SAR). A relation between soluble sodium and soluble divalent cations that can be used to predict the exchangeable-sodium percentage of soil equilibrated with a given solution. It is defined as follows:

$$SAR = \frac{(sodium)}{\left(\dfrac{(calcium + magnesium)}{2}\right)^{\frac{1}{2}}}$$

where concentrations, denoted by parenthesis, are expressed in milliequivalents per liter.

Soil. (1) The unconsolidated mineral material on the immediate surface of the earth that serves as a natural medium for the growth of land plants. (2) The unconsolidated mineral matter on the surface of the earth that has been subjected to and influenced by genetic and environmental factors of: *parent material, climate* (including moisture and temperature effects), *macro-* and *microorganisms,* and *topography,* all acting over a period of *time* and producing a product—soil—that differs from the material from which it is derived in many physical, chemical, biological, and morphological properties and characteristics.

Soil Association. (1) A group of defined and named taxonomic soil units occurring together in an individual and characteristic pattern over a geographic region, comparable to plant associations in many ways. (Sometimes called "natural land type.") (2) A mapping unit used on general soil maps, in which two or more defined taxonomic units occurring together in a characteristic pattern are combined, because the scale of the map or the purpose for which it is being made does not require delineation of the individual soils.

Soil Genesis. (1) The mode or origin of soil, with special reference to the processes or soil-forming factors responsible for the development of the solum, or true soil, from the unconsolidated parent material. (2) A division of soil science concerned with soil genesis (1).

Soil Geography. A specialization of soil science concerned with the areal distributions of soil types.

Soil Horizon. A layer of soil or soil material approximately parallel to the land surface and differing from adjacent, genetically related layers in physical, chemical, and biological properties or characteristics, such as color, structure, texture, consistency, kinds and numbers of organisms present, degree of acidity or alkalinity, and so on.

Soil Management Groups. Groups of taxonomic soil units with similar adaptations or management requirements for one or more specific purposes, such as adapted crops or crop rotations, drainage practices, fertilization, forestry, and highway engineering.

Soil Map. A map showing the distribution of soil types or other soil-mapping units in relation to the prominent physical and cultural features of the earth's surface.

Soil Moisture Regimes. *See* **Aquic, Aridic, Torric, Udic, Ustic,** and **Xeric.**

Soil Monolith. A vertical section of a soil profile removed from the soil and mounted for display or study.

Soil Morphology. (1) The physical constitution, particularly the structural properties, of a soil profile as exhibited by the kinds, thickness, and arrangement of the horizons in the profile, and by the texture, structure, consistency, and porosity of each horizon. (2) The structural characteristics of soil or any of its parts.

Soil Science. The science dealing with soils as a natural resource on the surface of the earth, including soil formation, classification, and mapping; the physical, chemical, biological, and fertility properties of soils *per se;* and these properties in relation to their management for crop production.

Soil Separates. Mineral particles, < 2.0 millimeters in equivalent diameter, ranging between specified size limits. The names and size limits of separates recognized in the United States are: *very coarse sand,* 2.0 to 1.0 millimeters; *coarse sand,* 1.0 to 0.5 millimeters; *medium sand,* 0.5 to 0.25 millimeters; *fine sand,* 0.25 to 0.10 millimeters; *very fine sand,* 0.10 to 0.05 millimeters; *silt,* 0.05 to 0.002 millimeters; and *clay,* < 0.002 millimeters.

Soil Series. The basic unit of soil classification, being a subdivision of a family and consisting of soils that are essentially alike in all major profile characteristics, except the texture of the A horizon.

Soil Solution. The aqueous liquid phase of the soil and its solutes.

Soil Structure. The combination or arrangement of primary soil particles into secondary particles, units, or peds. These secondary units may be, but usually are not, arranged in the profile in such a manner as to give a distinctive characteristic pattern. The secondary units are characterized and classified on the basis of size, shape, and degree of distinctness into classes, types, and grades, respectively.

Soil Survey. The systematic examination, description, classification, and mapping of soils in an area. Soil surveys are classified according to the kind and intensity of field examination.

Soil Temperature Regimes. *See* **Pergelic, Cryic, Frigid, Mesic, Thermic,** and **Hyperthermic.**

Soil Texture. The relative proportions of the various soil separates in a soil.

Soil Water Potential. The amount of work that must be done per unit quantity of pure water to transport reversibly and isothermally an infinitesimal quantity of water from a pool of pure water, at a specified elevation and at atmospheric pressure, to the soil water (at the point under consideration). The total potential (of soil water) consists of the following:

Osmotic Potential. The amount of work that must be done per unit quantity of pure water to transport reversibly and isothermally an infinitesimal quantity of water from a pool of pure water, at a specified elevation and at atmospheric pressure, to a pool of water identical in composition to the soil water (at the point under consideration), but in all other respects being identical to the reference pool.

Gravitational Potential. The amount of work that must be done per unit quantity of pure water to transport reversibly and isothermally an infinitesimal quantity of water, identical in composition to soil water, from a pool at a specified elevation and at atmospheric pressure, to a similar pool at the elevation of the point under consideration.

Matric Potential. The amount of work that must be done per unit quantity of pure water to transport reversibly and isothermally an infinitesimal quantity of water, identical in composition to the soil water, from a pool at the elevation and the external gas pressure of the point under consideration.

Gas Pressure Potential. This potential component is to be considered only when *external* gas pressure differs from atmospheric pressure as, for example, in a pressure membrane apparatus. A specific term and definition is not given.

Solum (plural: **Sola**). The upper and most weathered part of the soil profile; the A and B horizons.

Sombric Horizon. A subsurface mineral horizon that is darker in color than the overlying horizon, but lacks the properties of a spodic horizon. Common in cool, moist soils of high altitude in tropical regions.

Spodic Horizon. A mineral soil horizon that is characterized by the illuvial accumulation of amorphous materials composed of aluminum and organic carbon with or without iron. The spodic horizon has a certain minimum thickness and a minimum quantity of extractable carbon plus aluminum in relation to its content of clay.

Spodosols. Mineral soils that have a spodic horizon or a placic horizon that overlies a fragipan.

Structural Charge. The negative charge on a clay mineral that is caused by isomorphous substitution within the layer. (Expressed as equivalents per formula weight of clay or milliequivalents per 100 grams or per gram of clay.)

Subsoiling. Breaking of compact subsoils, without inverting them, with a special knife-like instrument (chisel) that is pulled through the subsoil.

Sulfur Cycle. The sequence of transformations whereby sulfur is oxidized or reduced through both organic or inorganic products.

Surface-charge Density. The excess of negative or positive charge per unit of surface area of soil or soil mineral.

Talus. Fragments of rock and other soil material accumulated by gravity at the foot of cliffs or steep slopes

Thermic. A soil temperature regime that has mean annual soil temperatures of 15°C or more, but less than 22°C, and more than 5°C difference between mean summer and mean winter soil temperatures at 50 centimeters. Isothermic is the same, except the summer and winter temperature differ by less than 5°C.

Thermosequence. A sequence of related soils that differ, one from the other, primarily as a result of *temperature* as a soil-formation factor.

Till. (1) Unstratified glacial drift deposited directly by the ice and consisting of clay, sand, gravel, and boulders intermingled in any proportion. (2) To plow and prepare for seeding; to seed or cultivate soil.

Tillage. The mechanical manipulation of soil for any purpose; but in agriculture, it is usually restricted to the modifying of soil conditions for crop production.

Tilth. The physical condition of soil as related to its ease of tillage, fitness as a seedbed, and its impedance to seedling emergence and root penetration.

Toposequence. A sequence of related soils that differ, one from the other, primarily because of *topography* as a soil-formation factor (*see* **Clinosequence**).

Topsoil. (1) The layer of soil moved in cultivation (*see* **Surface Soil**). (2) The A horizon. (3) Presumably fertile soil material used to topdress roadbanks, gardens, and lawns.

Torrerts. Vertisols of arid regions that have wide, deep cracks that remain open throughout the year in most years.

Torric. A soil moisture regime defined as an aridic moisture regime, but used in a different category of soil taxonomy.

Torrox. Oxisols that have a torric soil moisture regime.

Tropepts. Inceptisols that have a mean annual soil temperature of 8°C or more, and less than 5°C difference between mean summer and mean winter temperatures at a depth of 50 centimeters below the surface. Tropepts may have an ochric epipedon and a cambic horizon, or an umbric epipedon, or a mollic epipedon under certain conditions but no plaggen epipedon; they are not saturated with water for periods long enough to limit their use for most crops.

Truncated. Having lost all or part of the upper soil horizon or horizons.

Tuff. Volcanic ash, usually more or less stratified and in various states of consolidation.

Udalfs. Alfisols that have a udic soil moisture regime and mesic or warmer soil temperature regimes. Udalfs generally have brownish colors throughout and are not saturated with water for periods long enough to limit their use for most crops.

Uderts. Vertisols of relatively humid regions that have wide, deep cracks that usually remain open continuously for less than 2 months or intermittently for periods that total less than 3 months.

Udic. A soil moisture regime that is neither dry for as long as 90 cumulative days nor for as long as 60 consecutive days in the 90 days following the summer solstice, as periods when the soil temperature at 50 centimeters depth is above 5°C.

Udolls. Mollisols that have a udic soil moisture regime with mean annual soil temperatures of 8°C or more. Udolls have no calcic or gypsic horizon, and are not saturated with water for periods long enough to limit their use of most crops.

Udults. Utisols that have low or moderate amounts of organic carbon, reddish or yellowish-argillic horizons, and a udic soil moisture regime. Udults are not saturated with water for periods long enough to limit their use for most crops.

Ultisols. Mineral soils that have an argillic horizon with a base saturation of less than 35 percent when measured at pH 8.2. Ultisols have a mean annual soil temperature of 8°C or higher.

Umbrepts. Inceptisols formed in cold or temperate climates that commonly have an umbric epipedon, but they may have a mollic or an anthropic epipedon 25 centimeters or more thick under certain conditions. These soils are not dominated by amorphous materials and are not saturated with water for periods long enough to limit their use for most crops.

Umbric Epipedon. A surface layer of mineral soil that has the same requirements as the mollic epipedon, with respect to color, thickness, organic carbon content, consistence, structure, and P_2O_5 content, but that has a base saturation of less than 50 percent when measured at pH 7.

Ustalfs. Alfisols that have an ustic soil moisture regime and mesic or warmer soil temperature regimes. Ustalfs are brownish or reddish throughout and are not saturated with water for periods long enough to limit their use for most crops.

Usterts. Vertisols of temperate or tropical regions that have wide, deep cracks that usually remain open for periods that total more than 3 months, but do not remain open continuously throughout the year, and have either a mean annual soil temperature of 22°C or more or a mean summer and mean winter soil temperature at 50 centimeters that differ by less than 5°C, or have cracks that open and close more than once during the year.

Ustic. A soil moisture regime that is intermediate between the aridic and udic regimes and common in temperate subhumid or semiarid regions, or in tropical and subtropical regions with a monsoon climate. A limited amount of moisture is available for plants, but occurs at times when the soil temperature is optimum for plant growth.

Ustolls. Mollisols that have an ustic soil moisture regime and mesic or warmer soil temperature regimes. Ustolls may have a calcic, petrocalcic, or gypsic horizon, and are not saturated with water for periods long enough to limit their use for most crops.

Ustox. Oxisols that have an ustic moisture regime and either hyperthermic or isohyperthermic soil temperature regimes, or have less than 20-kilogram organic carbon in the surface cubic meter.

Ustults. Ultisols that have low or moderate amounts of organic carbon, are brownish or reddish throughout, and have an ustic soil moisture regime.

Value, Color. The relative lightness or intensity of color, and approximately a function of the square root of the total amount of light. One of the three variables of color (see **Munsell Color System, Hue,** and **Chroma**).

Varve. A distinct band representing the annual deposit in sedimentary materials, regardless of origin, and usually consisting of two layers, one a thick, light-colored layer of silt and fine sand and the other a thin, dark-colored layer of clay.

Vertisols. Mineral soils that have 30 percent or more clay, deep wide cracks when dry, and either gilgai microrelief, intersecting slickensides, or wedge-shaped structural aggregates tilted at an angle from the horizontal.

Vesicular Structure. Soil structure characterized by round or egg-shaped cavities or vesicles.

Water-stable Aggregate. A soil aggregate or ped that is stable to the action of water such as falling drops, or agitation as in wet-sieving analysis.

Water Table. The upper surface of groundwater of that level below which the soil is saturated with water; locus of points in soil water at which the hydraulic pressure is equal to atmospheric pressure.

Water Table, Perched. The water table of a saturated layer of soil that is separated from an underlying saturated layer by an unsaturated layer.

Weathering. All physical and chemical changes produced in rocks, at or near the earth's surface, by atmospheric agents.

Wilt Point (Permanent). The water content of a soil when indicator plants growing on the soil wilt and fail to recover, when placed in a humid chamber. Often estimated by -15 bars water potential.

Xeralfs. Alfisols that have a xeric soil moisture regime. Xeralfs are brownish or reddish throughout.

Xererts. Vertisols of Mediterranean climates that have wide, deep cracks that open and close once each year and usually remain open continuously for more than 2 months. Xererts have a mean annual soil temperature of less than 22°C.

Xeric. A soil moisture regime common to Mediterranean climates that have moist cool winters and warm dry summers. A limited amount of moisture is present, but it does not occur at optimum periods for plant growth. Irrigation or summerfallow is commonly necessary for crop production.

Xerolls. Mollisols that have a xeric soil moisture regime. Xerolls may have a calcic, petrocalcic, or gypsic horizon, or a duripan.

Xerophytes. Plants that grow in or on extremely dry soils or soil materials.

Xerults. Ultisols that have low or moderate amounts of organic carbon, are brownish or reddish throughout, and have a xeric soil moisture regime.

Zero Point of Charge. The pH value of a solution in equilibrium with a particle whose net charge from all sources is zero.

INDEX